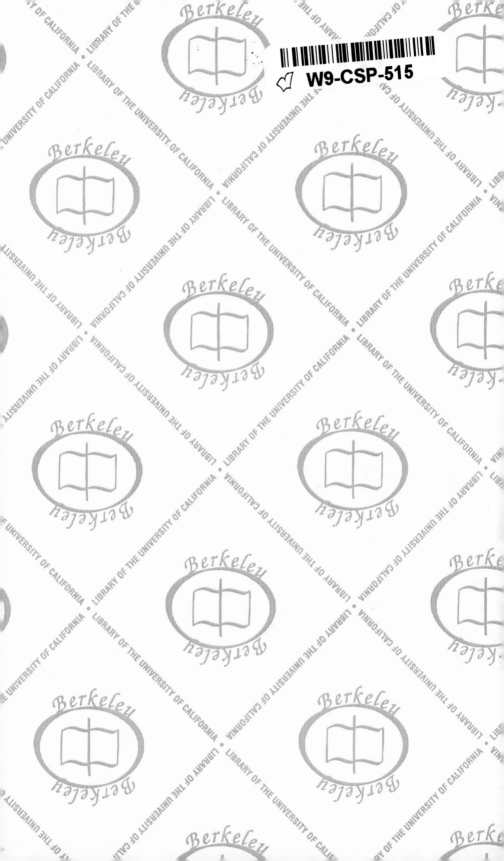

THE PHYSICS AND CHEMISTRY
OF WAVE PACKETS

THE PHYSICS AND CHEMISTRY OF WAVE PACKETS

Edited by

John A. Yeazell
The Pennsylvania State University

Turgay Uzer
Georgia Institute of Technology

A WILEY-INTERSCIENCE PUBLICATION

JOHN WILEY & SONS, INC.

NEW YORK • CHICHESTER • WEINHEIM • BRISBANE • SINGAPORE • TORONTO

Copyright © 2000 by John Wiley & Sons, Inc. All rights reserved.

Published simultaneously in Canada.

For ordering and customer service, call 1-800-CALL-WILEY.

Library of Congress Cataloging-in-Publication Data

The physics and chemistry of wave packets / edited by John Yeazell,
  Turgay Uzer.
      P. cm.
    "A Wiley-Interscience publication."
    Includes bibliographical references and index.
    ISBN 0-471-24684-0 (alk. paper)
    1. Wave packets. I. Yeazell, John, 1959– . II. Uzer, Turgay.
  QC793.3.W3P48   2000
    530.12'4—dc21   99-22019

Printed in the United States of America.

10 9 8 7 6 5 4 3 2 1

◼◼◼◼ CONTENTS

CONTRIBUTORS

Gernot Alber, Universität Ulm, Abteilung für Quantenphysik, Albert Einstein Allee 11, D-89069 Ulm, Germany

Stuart C. Althorpe, University of Houston, Department of Chemistry, Houston, TX 77204-5641

Rosa M. Benito, Ciudad Universitaria, Departamento de Fisica y Mecanica, ETS Ingenieros Agronomos, 28040 Madrid, Spain

Florentino Borondo, Universidad Autonoma de Madrid, Departamento de Quimica, CANTOBLANCO, 28049 Madrid, Spain

Michael Braun, Universität Würzburg, Institut für Physikalische Chemie, Am Hubland, D-97074 Würzburg, Germany

Andrea F. Brunello, Utah State University, Department of Chemistry and Biochemistry, Logan, UT 84322-0300

Philip Bucksbaum, University of Michigan, Physics Department, Randall Lab, Ann Arbor, MI 48109-1120

Charles Cerjan, Lawrence Livermore National Laboratory, Livermore, CA 94550

Holger Dietz, Universität Würzburg, Institut für Physikalische Chemie, Am Hubland, D-97074 Würzburg, Germany

Volker Engel, Universität Würzburg, Institut für Physikalische Chemie, Am Hubland, D-97074 Würzburg, Germany

David Farrelly, Utah State University, Department of Chemistry and Biochemistry, Logan, UT 84322-0300

Eric J. Heller, Harvard University, Department of Physics, Cambridge, MA 02138

David K. Hoffman, Iowa State University, Department of Chemistry and Ames Laboratory, Ames, IA 50011

Martin Koch, Institut für Hochfrequenztechnik, Technische Universität Braunschweig, 38106 Braunschweig, Germany

Donald J. Kouri, University of Houston, Department of Chemistry, Houston, TX 77204

Ernestine Lee, Utah State University, Department of Chemistry and Biochemistry, Logan, UT 84322-0300

Karl Leo, Technische Universität Dresden, Institut für Angewandte Photophysik, D-01062 Dresden, Germany

Stefan Meyer, Universität Würzburg, Institut für Physikalische Chemie, Am Hubland, D-97074 Würzburg, Germany

Michael Nauenberg, University of California, Department of Physics, Santa Cruz, CA 95064

Oliver Rubner, Universität Würzburg, Institut für Physikalische Chemie, Am Hubland, D-97074 Würzburg, Germany

Turgay Uzer, Georgia Institute of Technology, School of Physics, Atlanta, GA 30332-0430

John A. Yeazell, Pennsylvania State University, Physics Department, University Park, PA 16802

■■■■■ PREFACE

When the twentieth century began, classical physicists considered an atom to be made of negatively charged electrons that revolved around a positively charged nucleus, in much the same way that the planets revolve around the Sun. The discovery of quantum mechanics in the mid-1920s quickly led to a more sophisticated picture: The electron was characterized by a wavefunction that contained all of the information needed to compute the probability for it to be found in a certain volume in space. During the ensuing decades research in atomic physics focused on the study of the properties of stationary states wherein the probability density does not evolve in time. That all changed in the mid-1980s; the advent of femtosecond laser pulses opened up a new chapter in the study of atoms and molecules, since it provided the way to prepare, observe, and even manipulate wave packet states that evolve in time. These states are spatially well localized and may behave like classical particles; that is, they follow classical trajectories. Associated with these classical characteristics are other unique properties such as the time and phase dependent absorption of radiation. It is indeed a curious, and compelling, completion of the circle that started just over 100 years ago (with the discovery of the electron) that physicists are again simultaneously fascinated and challenged by the classical mechanics of atoms and molecules. The creation, the detection, and the manipulation of wave packets are the three separate threads that weave this book into a whole.

The goal of the book is to provide a unified overview of a rapidly growing research topic that blurs, to the point of unrecognizably, the border between chemistry and physics. Wave packet states are of as much fundamental interest in laser chemistry as they are in atomic, molecular, optical, and solid state physics. The potential applications of these states seems limitless; some examples of their uses include the exploration of the boundary between classical and quantum mechanics, the investigation of the interpretation of quantum mechanics by creating experimental realizations of such classic *gedanken* experiments as Schrödinger's Cat, the control of chemical reaction dynamics to achieve laser isotope separation, the storage of coherence for quantum computational or communications purposes, and the construction of optical switches and modulators. Undeniably, the manipulation and measurement of wave packet coherence and the effects of decoherence are subjects that lie at the most extreme frontiers of chemistry and physics. New methods and phenomena such as quantum tomography, quantum computing, quantum communication, macroscopic quantum states, and the coherent control of fundamentally quantum processes have all been spawned in the milieu that has grown out of wave packet research. These phenomena are not only of basic interest, but they also offer great technical promise. This book describes the most important advances that have been made in the ability (1) to design and create particular quantum target states, (2) to detect and measure the properties of interest displayed by these states, and (3) to manipulate the properties of these states. Wave packet states have now been observed in a wide variety of systems, and this is reflected in the diversity of content in the contributed chapters. There is no doubt that wave packet technology is a vibrant and an exciting field that is in constant flux. We hope that this book not only captures this sense of excitement but also manages to infect the reader with this spirit.

The opening chapters of the book illustrate the fundamentals using systems drawn from atomic physics. A sometimes astonishing level of complexity and variety of behavior emerges. Having laid the groundwork in the early chapters, extensions and applications to more complex, primarily molecular and mesoscopic, systems are introduced. One of the major challenges of quantum physics in the next century will be the development of a reliable tool box of methods with which to manipulate the behavior of matter at the quantum level. Ergo, the concluding chapters of the book discuss and evaluate progress toward this goal while providing a lively prognosis for further applications. Overall, the advances described herein, whether in the engineering of wave packets in atoms, the coherent control of atomic processes, or the manipulation of the external degrees of freedom of the atom using laser cooling and trapping, all bode well for the future.

The editors would like to thank the contributors for their alacrity in contributing to this volume and, especially, for their clarity of vision which makes this such an exciting field of endeavor. In addition to the contributors, we would like to thank those people with whom discussions over the years have both enlightened and stimulated us: Jonathan Parker, Carlos Stroud Jr., Joseph Eberly, Herbert Walther, David Farrelly, Ernestine Lee, Andrea Brunello, Georg Raithel, and many others. One of us (J.Y.) would also like to thank his wife, Karen Fingerman, for her suggestions and unique perspective on the work.

CHAPTER 1

Wave Packets: Past and Present

M. NAUENBERG

> There was a time when the newspapers reported that only twelve men understood Einstein's theory of relativity. I do not believe that there was ever such a time. . . . On the other hand, it is safe to say that no one understands quantum mechanics.
> —R. Feynman (1967)

> After half a century the relation between quantum mechanics and classical mechanics is still not fully understood. Uncertainty extends even to the foundations.
> —M. Berry (1981)

1.1 INTRODUCTION

Since the time when the quotations above were made, considerable progress has been made in understanding more fully the relation between quantum and classical mechanics, but this correspondence remains an outstanding problem in physics. In classical mechanics the time evolution of a system is determined by Newton's equation of motion, while in quantum mechanics the corresponding evolution is described by a probability distribution given by the modulus square of a wavefunction satisfying Schrödinger's equation. Which of these two mechanics is relevant is usually assumed to depend on whether the system is macroscopic (planets, tennis balls) or microscopic (atoms, molecules, elementary particles), although there are exceptions, such as superconductors which are macroscopic but obey quantum mechanical laws. However, recently physicists and chemists have been investigating an increasingly large number of systems of intermediary size called *mesoscopic systems*, where this distinction of size is no longer valid. Moreover we believe that quantum mechanics is a fundamental theory that should apply to physical systems regardless of size. The key idea to relate these seemingly disparate points of view is the concept of a coherent wave packet in quantum mechanics, which will be elaborated in this book. Such a wave packet gives a probability distribution that can be related in the correspondence limit to an ensemble of orbits in classical mechanics (see Nauenberg, Stroud, and Yeazell 1994). This connection is fundamental, but as we will see, it was not

The Physics and Chemistry of Wave Packets, Edited by John Yeazell and Turgay Uzer
ISBN 0-471-24684-0 © 2000 John Wiley & Sons, Inc.

understood by some of the main founders of quantum mechanics, and subsequent confusion concerning the relation between quantum and classical mechanics persists even to the present.

Already at the birth of quantum mechanics, Schrödinger tried to explain the connection between his wave equation and classical mechanics by discussing the case of a particle moving in a harmonic oscillator potential (Schrödinger 1926). Unfortunately, Schrödinger's example, which is often quoted in the literature, is somewhat misleading because it appears that the evolution of this wave packet describes a single orbit in classical mechanics. A more characteristic example is the case of a free particle which we will discuss in further detail below. In this case the motion in quantum mechanics can be represented by a Gaussian wave packet which is initially confined in space. This wave packet will move along a classical trajectory but spread in position as a function of time. It has been argued that this spreading is an effect that distinguishes quantum from classical behavior, which was the basis for an early criticism by H. A. Lorentz of Schrödinger's equation. However, a more careful examination indicates that the spreading is actually related to classical mechanics. This quantum-classical connection becomes clear if the correspondence is not made with a single classical trajectory, which is deterministic, but with an appropriate ensemble of orbits. Actually it was pointed out already some time ago by M. Born (see Born 1955) that a single orbit is not an appropriate description for the motion of a particle in classical mechanics. Born concluded that determinism in classical mechanics was "a false appearance" because in practice initial conditions are known only approximately. For example, if the initial velocity for a free particle is determined with an uncertainty δv, then the appropriate description of the motion is given by an ensemble of orbits with a distribution of velocity determined with this initial uncertainty. As a consequence at later times there will be a spread in the position x of the particles given by $\delta x = \delta v t$ which increases linearly with time. In fact, we will show that when the initial classical position and velocity uncertainties are described by a Gaussian distribution, the time dependence of the classical and the quantum probability distributions are identically the same, provided there is an appropriate relation between the initial widths of the classical Gaussian distributions. Hence the spreading of a wave packet is an effect which can be understood classically once the appropriate correspondence with quantum mechanics has been made.

Then where does the essential difference between quantum and classical mechanics appear? In quantum mechanics there is a constraint in the minimum value of the product of the root mean square deviation in position, δx, and in momentum, δp, which is determine by Planck's constant $\hbar$. This constraint is Heisenberg's uncertainty relation, $\delta x \delta p \geq \hbar/2$, while in classical mechanics there are no constraints on the magnitude of these deviations. Moreover, if there are any obstacles or slits, or if an external force acts on the particle, which lead to classical trajectories that are curved in space, then it is possible to have interference effects between separate portions of the wave packet. Such interference effects do not have any analog in classical mechanics, and it is at the time when these interference effects first occur that the quantum-classical correspondence fails, and phenomena which are characteristic of quantum mechanics emerge. Finally there are tunneling effects through potential barriers that do not occur in classical mechanics.

A fundamental problem, which already confounded Schrödinger during the early development of his wave mechanics, is to understand how the Bohr orbits in atomic physics or the corresponding Keplerian orbits in planetary physics emerged from wave mechanics. Over the past years there have been many attempts to overcome the technical difficulties that Schrödinger first encountered to obtain a coherent wave packet that corresponds to classical motion in the Coulomb potential (see Brown 1973; Mostowski 1977; Bhaumik, Dutta-Roy, and Ghosh 1986; Gerry 1986). However, a complete solution to this problem was obtained only recently (Gay, Delande, Bommier 1989; Nauenberg 1989). In the following sections coherent wavefunctions that are localized on classical Keplerian ellipses are introduced. It will be shown that the linear superposition of these wavefunctions form coherent wave packets which travel on ellipses with the classical Kepler period t_K, while spreading initially along the orbit, as had been argued originally by H. A. Lorentz. Initially this spreading is in accordance with the behavior of a corresponding ensemble of orbits in classical mechanics (Nauenberg and Keith 1992). However, when the head of the wavepacket catches its own tail interference effects occur, and the quantum-classical correspondence ceases to be valid. Moreover, instead of continuing to spread indefinitely into the future, as had been generally conjectured, the initial wave packet reappears after a definite period t_r (Nauenberg 1989; Alber, Ritsch, and Zoller 1986; Parker and Stroud Jr. 1986; Averbukh and Perelman 1989; Nauenberg 1990, 1992). This recurrence time, which is of purely quantum mechanical origin, depends on the mean principal quantum number n of the wave packet, and $t_r = [n/3]\tau_K \pm 1/2$ where τ_K is the Kepler period. In addition there are more subtle recurrences at rational fractions of this period that do not have any classical analogues. These results have been confirmed experimentally by exciting an electron in alkali atoms to a narrow band of Rydberg states with short-pulsed lasers (Yeazell, Malalieu, and Stroud 1990; Meacher et al. 1991; Yeazell and Stroud 1991). We will discuss the dynamics of these coherent wave packets in the presence of weak electromagnetic fields. These fields can be shaped to create and manipulate coherent wave packets in Rydberg atoms, and during the past few years this has become a very active theoretical and experimental field (Nauenberg 1994; Bellomo, Farrelly, and Uzer 1997, 1998; Bellomo and Stroud Jr. 1998; Chen and Yeazell 1998). We conclude with a new semiclassical time dependent theory of wave packet propagation that demonstrates how the physical origin of the quantum revivals can be understood from the old Bohr quantization of the classical orbits.

1.2 HISTORICAL REMARKS

To describe the stability of atoms and its discrete spectrum of emitted radiation, Niels Bohr postulated in 1913 a quantization rule for Rutherford's classical model of the atom (Bohr 1913). This model consisted of electrons moving in circular classical orbits about a heavy central charged nucleus, and Bohr's quantization rule, which actually had been proposed a year earlier by J. W. Nicholson (1912), required that the angular momentum of the electrons be quantized in units of the Planck's constant $\hbar$. Bohr introduced also a correspondence principle that for a large number of units of $\hbar$ an atomic system should exhibit classical behavior. However, when Schrödinger developed his wave mechanics of the atom Bohr's quantized classical orbits disappeared. This point was emphasized by both Heisenberg and Pauli who believed that such classical concepts as orbits should be eliminated from the atomic realm. Indeed, there does not appear to be any obvious connection between Keplerian orbits and a typical solution of Schrödinger's equation for the hydrogen atom. Actually Schrödinger (1926) already addressed this problem in an article entitled "The Transition from Micro- to Macro-Mechanics," where he treated the harmonic oscillator, and obtained a solution consisting of a time dependent Gaussian wave packet which travels without *spreading* along a classical trajectory. At the end of his paper he wrote

> One can foresee with certainty that similar wave packets can be constructed which will travel along Keplerian ellipses for high quantum numbers; however technical computational difficulties are greater than in the simple example given here.

Schrödinger sent his paper in manuscript form to Lorentz, with whom he had been corresponding about his new wave mechanics (see Przibram 1967). In a letter written on June 6, 1926, he wrote

> Allow me to send you, in an enclosure, a copy of a short note in which something is carried through for the simple case of an oscillator which is also an urgent requirement for all the more complicated cases . . . You see from the text of the note, which was written before I received your letter, how much I too was concerned about the "staying together" of these wave packets. I am very fortunate that now I can at least point to a simple example where, contrary to all reasonable conjectures, it still proves right. I hope that this is so, in any event for all those cases where ordinary mechanics speaks of *quasi-periodic* motion.

Then a surprising statement followed:

> Let us accept this as secured or conceded for once; there still always remains the difficulty of the completely *free* electron in a completely field-free space. Would you consider it a very weighty objection against the theory if it were to turn out that the electron is incapable of existing in a completely field free space? . . .

Lorentz promptly responded that

> . . . with your note . . . you have given me a great deal of pleasure, and as I read it, a first thought came upon me: with a theory which resolves a doubt in such a surprising and beautiful way, one has to be on the right path. Unfortunately my pleasure was soon diminished; namely I can not see, for example, how in the case of the hydrogen atom you

can construct wave packets (I am thinking now of the *very high* Bohr orbits) which travel like the electron.

Earlier Lorentz had written to Schrödinger (Przibram 1967)

Your conjecture that the transformation which our dynamics will have to undergo will be similar to the transition from ray optics to wave optics sounds very tempting, but I have some doubts about it. If I have understood you correctly, then a "particle," an electron for example, would be comparable to a wave packet which moves with the group velocity. But a wave packet can never stay together and remained confined to a small volume in the long run. The slightest dispersion in the medium will pull it apart in the direction of propagation, and even without that dispersion it will always spread more and more in the transverse direction. Because of this unavoidable blurring a wave packet does not seem to me to be very suitable for representing things to which we want to ascribe a rather permanent individual existence.

Lorentz pointed out correctly that the association of the wave packet with the charge density of an electron as Schrödinger had proposed, was not tenable if the wave packet dispersed. This dilemma was resolved by Born's introduction of the probability interpretation, but Schrödinger did not accept it, and as late as 1946 he wrote to Einstein that (Moore 1989)

God knows I am no friend of the probability theory, I have hated it from the first moment our dear friend Max Born gave it birth. For it could be seen how easy and simple it made everything, in principle, every thing ironed out and the true problems concealed.

In the next section we will consider the wave packet for a free particle which resolves the difficulty Schrödinger and Lorentz had in understanding the quantum-classical correspondence. Their misunderstanding, which persists to the present, is due to the association of a quantum wave packet with a single classical trajectory rather than an appropriate ensemble of such trajectories; a situation which led Einstein to regard quantum mechanics as an *incomplete* description of physical reality. As Born (1955) concluded

It is misleading to compare quantum mechanics with deterministically formulated classical mechanics; instead one should first reformulate the classical theory, even for a single particle, in an indeterministic, statistical manner. Then some of the distinctions between the two theories disappear, others emerge with great clarity . . . The essential quantum effects are of two kinds: the reciprocal relation between the maximum of sharpness for coordinate and velocity in the initial and consequently in any later state (uncertainty relations), and the interference of probabilities whenever two (coherent) branches of the probability function overlap. For macro-bodies both these effects can be made small in the beginning and then remain small for a long time; during this period the individualistic description of traditional classical mechanics is a good approximation. But there is a critical moment t_c where this ceases to be true and the quasi-individual is transforming itself into a genuine statistical ensemble.

1.3 WAVE PACKET FOR A FREE PARTICLE

To illustrate some of the basic ideas about coherent quantum wave packets and to show how these wave packets illuminate the fundamental relation between quantum and classical motion, we consider in some detail the simple problem of the motion of a free particle in one dimension. Suppose that the particle has a mean momentum $\bar{p}$ and that it is initially localized near the origin of coordinates at $x = 0$. In quantum mechanics the initial state can be represented by a Gaussian wavefunction

$$\psi(x, 0) = (2\pi\sigma^2)^{-1/4} \exp\left(\frac{i\bar{p}x}{\hbar} - \frac{x^2}{4\sigma^2}\right), \tag{1.1}$$

where σ is the width representing the uncertainty in the initial position of the particle. At a later time the wavefunction is obtained by solving the time dependent Schrödinger equation

$$i\hbar\frac{\partial\psi(x, t)}{\partial t} = -\frac{\hbar^2}{2m}\frac{\partial^2\psi(x, t)}{\partial x^2} \tag{1.2}$$

subject to the initial condition (1.1). For mathematical convenience we will now set Planck's constant $\hbar$ and the mass m of the particle equal to unity (one can recover these parameters in the subsequent equations by the replacement $p \to p/\hbar$ and $t \to t\hbar/m$). We obtain

$$\psi(x, t) = N(t) \exp\left[\frac{i\bar{p}x - i\bar{p}^2 t}{2} - \frac{(x - \bar{p}t)^2}{4\sigma^2 + 2it}\right] \tag{1.3}$$

where $N(t) = (2\pi\sigma^2)^{-1/4}(1 + it/2\sigma)^{-1/2}$. Hence the probability of finding the particle in the interval $(x, x + dx)$ at time t is given by $|\psi(x, t)|^2 dx$, where

$$|\psi(x, t)|^2 = \frac{1}{\sqrt{2\pi}\,\Delta(t)} \exp\left[\frac{-(x - \bar{p}t)^2}{2\Delta^2(t)}\right], \tag{1.4}$$

where $\Delta(t) = \sqrt{\sigma^2 + t^2/4\sigma^2}$ is the time dependent width of the wave packet.

We now consider a corresponding ensemble of free particles in classical mechanics which is described initially by a Gaussian distribution in both momentum and coordinate space, localized near $p = \bar{p}$ and $x = 0$, with corresponding widths σ_p and σ_x,

$$P_c(x, p, 0) = \frac{1}{2\pi\sigma_x\sigma_p} \exp\left[\frac{-(p - \bar{p})^2}{2\sigma_p^2} - \frac{x^2}{2\sigma_x^2}\right], \tag{1.5}$$

where $P_c(x, p, t)$ is the probability distribution for such an ensemble. At later times it must satisfy the Liouville equation for free particles

$$\frac{\partial P_c(x, p, t)}{\partial t} + p \frac{\partial P_c(x, p, t)}{\partial x} = 0, \tag{1.6}$$

where we have set $m = 1$ (to recover the dependence on m in the classical equations replace $t \to t/m$). It can be easily verified that the solution of this equation is given by

$$P_c(x, p, t) = \frac{1}{2\pi\sigma_x\sigma_p} \exp\left[\frac{-(p - \overline{p})^2}{2\sigma_p^2} - \frac{(x - pt)^2}{2\sigma_x^2}\right]. \tag{1.7}$$

To compare the classical and quantum mechanical probability distributions in coordinate space, we now integrate P_c over momentum space to obtain

$$\int P_c(p, q, t)dp = \frac{1}{\sqrt{2\pi}\Delta_c(t)} \exp\left[\frac{-(x - \overline{p}t)^2}{2\Delta_c^2(t)}\right] \tag{1.8}$$

where $\Delta_c(t) = \sqrt{\sigma_x^2 + \sigma_p^2 t^2}$. This expression has the same form as the quantum mechanical probability, (1.4). Furthermore, if we equate the classical and quantum widths in coordinate space, $\sigma_x = \sigma$, and require that σ_p satisfy the minimum quantum mechanical uncertainty relation

$$\sigma_p = \frac{1}{2} \frac{\hbar}{\sigma}, \tag{1.9}$$

we obtain the remarkable result that the time evolution of the classical and the quantum mechanical probability distributions are *identically* the same in coordinate space. In quantum mechanics this relation between σ_p and σ_x can be obtained directly by evaluating the Fourier transform $\phi(p, t)$ of $\psi(x, t)$ which determines the probability distribution in momentum space. It turns out that the Gaussian wave packet corresponds to the minimal uncertainty relation (1.9) which is allowed by the quantum mechanical commutation relations $[x, p] = i\hbar$. Integrating $P_c(x, p, t)$ over the position coordinate x gives the momentum distribution of the classical ensemble which for free particles is independent of time and is equal to $|\phi(p, 0)|^2$ if we identify the classical momentum width in accordance with the uncertainty relation (1.9). It can also be verified that in this special case the classical distribution $P_c(x, p, t)$ is given by the Wigner distribution

$$P_c(x, p, t) = \int \frac{dq}{2\pi} \psi^*\left(\frac{x + q}{2}, t\right) \psi\left(\frac{x - q}{2}, t\right) \exp(ipq). \tag{1.10}$$

Thus we have shown that the main distinction between quantum and classical mechanics for free particles is that quantum mechanics imposes a constraint on the minimal uncertainty (1.9) with which the initial position and momentum of the particle can be determined. Actually this quantum-classical correspondence is exact only for Gaussian distributions, but for large quantum numbers this is also a very good approximation for other distributions which are sharply peaked near mean values of the initial position and the momentum. However, this correspondence breaks down when there is a force or potential acting on the particle. For example, if the quantum wave packet reaches a potential barrier or well, part of the incident wave can be reflected producing interference effects that have no correspondence in classical mechanics. We will see in the next section that it is precisely such interference phenomena that determine the onset for the breakdown of the quantum-classical correspondence. Other well-known phenomena such as quantum mechanical tunneling through a barrier do not have a correspondence in classical mechanics; this topic will not be discussed further here.

1.4 WAVE PACKET FOR A PARTICLE IN A COULOMB POTENTIAL

The Hamiltonian for the Coulomb potential in atomic units ($m = \hbar = e^2 = 1$) is

$$H = \frac{p^2}{2} - \frac{1}{r}, \tag{1.11}$$

where $\mathbf{p}$ is the momentum, and r is the radial distance. This Hamiltonian is rotationally invariant and therefore it commutes with the angular momentum operator $\mathbf{L} = \mathbf{r} \times \mathbf{p}$. From classical mechanics we expect that there exists an additional operator that also commutes with H. This is an operator associated with the Laplace vector

$$\mathbf{A} = \mathbf{p} \times \mathbf{L} - \frac{\mathbf{r}}{r}, \tag{1.12}$$

which is a conserved quantity in classical mechanics. For an elliptic orbit the magnitude of this vector is equal to its eccentricity ε, and its direction is along the major axis of this ellipse. This can be readily seen by multiplying (1.12) by $\mathbf{r}$, which gives the equation of a conic section in polar coordinates,

$$r = \frac{L^2}{1 + \varepsilon \cos(\theta)}, \tag{1.13}$$

where r is the radial coordinate and θ is the angle between $\mathbf{r}$ and $\mathbf{A}$.

In 1926 Pauli, who had been urged by Heisenberg to solve the hydrogen spectrum with his newly developed matrix mechanics, extended the Laplace vector to a matrix operator in quantum mechanics by antisymmetrizing the cross product of $\mathbf{p}$ and $\mathbf{L}$. Applying the Heisenberg-Born-Jordan commutation relations for the components of momentum $\mathbf{p}$ and position $\mathbf{r}$, he then obtained the commutation relations for the components of $\mathbf{L}$ and $\mathbf{A}$,

$$[L_i, L_j] = i\varepsilon_{ijk}L_k, \tag{1.14}$$

$$[L_i, A_j] = i\varepsilon_{ijk}A_k, \tag{1.15}$$

$$[A_i, A_j] = -i2H\varepsilon_{ijk}L_k. \tag{1.16}$$

Setting $\mathbf{A} = \sqrt{(-2H)}\,\mathbf{M}$ for the bound states of the Hamiltonian H, the components of $\mathbf{L}$ and $\mathbf{M}$ satisfy the commutation relation of the generators of the O(4) symmetry group. The Hamiltonian H can then be expressed in terms of these operators in the form

$$H = -\frac{1}{2(L^2 + M^2 + 1)}, \tag{1.17}$$

and in this way one obtains the spectrum of the hydrogen atom.

To simplify our discussion, we restrict these relations to a two-dimensional space, which reduces the symmetry to the rotation group O(3) with the commutation relations

$$[L_z, M_x] = iM_y, \tag{1.18}$$

$$[L_z, M_y] = -iM_x, \tag{1.19}$$

$$[M_x, M_y] = iL_z, \tag{1.20}$$

and

$$H = -\frac{1}{2(L^2 + M^2 + 1/4)}. \tag{1.21}$$

Since the components of the operators $\mathbf{L}$ and $\mathbf{M}$ do not commute, it is not possible to obtain eigenstates of H that are simultaneous eigenstates of these operators. The conventional eigenstates of the Coulomb Hamiltonian found in most quantum mechanics textbooks are chosen to be eigenstates of L^2 and L_z, and for such states the expectation value of the Laplace vector vanishes. This is the fundamental reason why these states do not manifest the properties of classical elliptic orbits even in the correspondence limit of large quantum numbers. For a bound state an angular momentum eigenstate corresponds in the classical limit to an ensemble of elliptic orbits with a fixed value of the angular momentum, but with a uniform distribution of the direction of the major axis. Alternatively, one may consider eigenstates of components of the Laplace vector, but in this case the mean value of the angular momentum vanishes. Hence, to represent a state in quantum mechanics that is related to an elliptic orbit in classical mechanics with fixed values of both the angular momentum and the Laplace vector, we must regard them as mean values in quantum mechanics, and minimize the quantum fluctuations of these operators.

The commutation relations (1.18) to (1.20) suggest that as a starting point we define coherent states Ψ for the Coulomb potential (see Nauenberg 1989) that exhibit the properties of Keplerian orbits as bound eigenstates of H with fixed mean values of $\mathbf{L}$

and $\mathbf{M}$ and we minimize the product $\Delta M_x \Delta M_y$ of the quantum fluctuations of M_x and M_y, where

$$(\Delta M_{x,y})^2 = \langle (M_{x,y} - \langle M_{x,y} \rangle)^2 \rangle. \tag{1.22}$$

The $\langle \ldots \rangle$ refers to the expectation value with respect to the state Ψ. Introducing the auxiliary state

$$\Phi = [M_x - \langle M_x \rangle + i\delta(M_y - \langle M_y \rangle)]\Psi, \tag{1.23}$$

where Ψ is an arbitrary state and δ is a real parameter, we have

$$0 \le (\Phi, \Phi) = (\Delta M_x)^2 + \delta^2 (\Delta M_y)^2 - \delta \langle L_z \rangle. \tag{1.24}$$

Minimizing the right-hand side of this equation with respect to δ, we obtain

$$\delta = \frac{\langle L_z \rangle}{2(\Delta M_y)^2}, \tag{1.25}$$

and substituting this expression for δ in (1.24), we find that

$$\Delta M_x \Delta M_y \ge \tfrac{1}{2} \langle L_z \rangle. \tag{1.26}$$

The minimum value of the product of these fluctuations is obtained when $\Phi = 0$, which implies that the state Ψ is a solution of the eigenvalue equation

$$(M_x + i\delta M_y)\Psi = \eta\Psi, \tag{1.27}$$

where $\eta = \langle M_x \rangle + i \langle M_y \rangle$. Hence the required states Ψ are simultaneous eigenstates of H and the nonhermitian operator $M_x + i\delta M_y$ with eigenvalue η. These eigenstates satisfy the relation

$$\langle (M_x - \langle M_x \rangle)^2 \langle = -\tfrac{1}{2} i\delta \langle [M_x, M_y] \rangle \tag{1.28}$$

and applying the commutation relation (1.20), we obtain

$$(\Delta M_x)^2 = \frac{\delta}{2} \langle L_z \rangle. \tag{1.29}$$

Likewise we find that

$$(\Delta M_y)^2 = \frac{1}{2\delta} \langle L_z \rangle. \tag{1.30}$$

To solve the eigenvalue problem (1.27), we introduce the raising and lowering operator $G_\pm$, where

$$G_{\pm} = \delta M_x + i M_y \mp \sqrt{(1 - \delta^2)}\, L_z, \tag{1.31}$$

which have the desired property that these operators satisfy the commutation relation

$$[M_x + i\Delta M_y, G_{\pm}] = \pm \sqrt{(1 - \delta^2)}\, G_{\pm}. \tag{1.32}$$

Hence, if Ψ is an eigenstate with eigenvalue η, then $G_{\pm}\Psi$ is an eigenstate with eigenvalue $\eta \pm \sqrt{(1 - \delta^2)}$. In particular, there exists eigenstates $\Psi_{\pm}$ such that

$$G_{\pm}\Psi_{\pm} = 0, \tag{1.33}$$

which have real eigenvalues $\eta_{\pm} = \pm l_n \sqrt{(1 - \delta^2)}$, where l_n is an integer. It can be shown that these eigenstates minimize the quantum fluctuation sum

$$(\Delta M_x)^2 + (\Delta M_y)^2 + (\Delta L_z)^2, \tag{1.34}$$

which in this case is equal to l_n; this implies that $\Psi\pm$ are the optimal quantum states to represent classical behavior. Since the eigenvalues η are real, it follows from (1.27) that for these states

$$\langle M_x \rangle = \pm l_n \sqrt{(1 - \delta^2)} \tag{1.35}$$

and

$$\langle M_y \rangle = 0, \tag{1.36}$$

Hence, according to (1.33),

$$\langle L_z \rangle = \pm l_n \delta, \tag{1.37}$$

and applying (1.34), we obtain

$$(\Delta L_z)^2 = \frac{l_n}{2}(1 - \delta^2). \tag{1.38}$$

The mean value of the eccentricity for a bound state of energy E_n is $\langle A_x \rangle = \varepsilon = \sqrt{(-2E_n)}\,\langle M_x \rangle$, where

$$E_n = -\frac{1}{2(l_n + \frac{1}{2})^2}.$$

Hence for large l_n, $\varepsilon \approx \sqrt{(1 - \delta^2)}$, and substituting for δ (1.37), we obtain

$$\varepsilon \approx \sqrt{(1 + 2E_n \langle L_z \rangle^2)}. \tag{1.39}$$

This relation corresponds to the well-known classical relation between eccentricity, angular momentum and energy of a Keplerian orbit, confirming the validity of our criteria for the construction of a coherent wave packet which exhibits classical properties. The coherent Kepler states can be represented as a linear superposition of the conventional eigenstates $\psi_{n,l}(r, \phi)$ of the Hamiltonian H and angular momentum operator L_z,

$$\Psi_{\delta,n} = \sum_{l=-l_n}^{l=l_n} c_{n,l}^{\delta} \psi_{n,l}(r, \phi), \tag{1.40}$$

where the coefficients $c_{n,l}^{\delta}$ are determined from recurrence relations (see Nauenberg 1989) obtained by applying (1.27) and (1.33) to this expansion. We find that

$$c_{n,l}^{\delta} = \frac{1}{2^{l_n}} \sqrt{\frac{(2l_n)!}{(l_n + 1)!(l_n - 1)!}} (1 - \delta^2)^{l_n/2} \left(\frac{1 + \delta}{1 - \delta}\right)^{l/2}. \tag{1.41}$$

For large quantum numbers l_n, these coefficients are well approximated by a Gaussian function

$$c_{n,l}^{\delta} \approx \left[\frac{\pi}{2} l_n (1 - \delta^2)\right]^{1/4} \exp\left[\frac{-(1 - \delta l_n)^2}{l_n(1 - \delta^2)}\right]. \tag{1.42}$$

Notice that this expression is of the form

$$c_{n,l}^{\delta} \propto \exp\left[\frac{-(l - \langle L_z \rangle)^2}{2(\Delta L_z)^2}\right], \tag{1.43}$$

where the mean value $\langle L_z \rangle$ and the root mean square deviation ΔL_z are given by (1.37) and (1.38), respectively. One of the most interesting and nontrivial results of our analysis is that (1.38) determines this width of the Gaussian to be proportional to the mean eccentricity ε and to the square root of the principal quantum number of the state. In particular, for a circular orbit, $\varepsilon = 0$, and in this case the sum for the coherent Coulomb state reduces to a single state of maximum value $l = \pm l_n$.

These coherent states can also be obtained by rotations in the O(3) symmetry group of the Coulomb Hamiltonian. By combining (1.27) and (1.31), we find that

$$(\sqrt{1 - \delta^2} M_x + \delta L_z)\Psi = l_n \Psi. \tag{1.44}$$

Hence, setting $\delta = \cos(\theta)$,

$$L_z' = \exp(-iM_y\theta)L_z \exp(iM_y\theta) = M_x \sin(\theta) + L_z \cos(\theta), \tag{1.45}$$

and

$$M'_x = \exp(-iMy\theta)M_x \exp(iM_y\theta) = M_x \cos(\theta) - L_z \sin(\theta), \tag{1.46}$$

where θ corresponds to a rotation angle about the y-axis, the equation for a coherent state can be written in the equivalent form

$$L'_z\Psi = l_n\Psi. \tag{1.47}$$

It follows that

$$\Psi = \exp(-iM_y\theta)\Psi_c, \tag{1.48}$$

is a general solution of these equations, where Ψ_c is an eigenstate of H and L_z which satisfies the condition

$$\exp(iM_y\theta)G_\pm \exp(-iM_y\theta)\Psi_c = (M_x \pm iM_y)\Psi_c = 0. \tag{1.49}$$

Hence Ψ_c is a circular eigenstate of H and L_z. In three-dimensional space, the coherent Kepler states can also be obtained from circular states by corresponding rotations within a SO(3) subgroup of the SO(4) symmetry group of the Coulomb Hamiltonian, as discussed by Bombier, Delande, and Gay (1988, 1991).
Define

$$\mathbf{J}_\pm = \tfrac{1}{2}(\mathbf{L} \pm \mathbf{M}). \tag{1.50}$$

Then J satisfies the commutation relations

$$[J_{\pm i}, J_{\pm j}] = i\varepsilon_{ijk}J_{\pm k} \tag{1.51}$$

and

$$[J_+, J_-] = 0. \tag{1.52}$$

In particular,

$$\mathbf{J}_+^2 + \mathbf{J}_-^2 = \tfrac{1}{2}(\mathbf{L}^2 + \mathbf{M}^2), \tag{1.53}$$

and the Hamiltonian has the form

$$H = -\frac{1}{(2(\mathbf{J}_+^2 + \mathbf{J}_-^2) + 1)}, \tag{1.54}$$

while

$$\mathbf{J}_+^2 - \mathbf{J}_-^2 = \tfrac{1}{4}(\mathbf{LM} + \mathbf{ML}) = 0. \tag{1.55}$$

Hence the Coulomb eigenstates can be constructed as a linear superposition of product eigenstates of $\mathbf{J}_+^2, J_{+z}$ and $\mathbf{J}_-^2, J_{-z}$, respectively, with the same maximum eigenvalue of $\mathbf{J}_\pm^2 = j(j + 1)$, and the corresponding Coulomb energy

$$E = -\frac{1}{(2j + 1)^2} \tag{1.56}$$

$$\Psi \rangle = \exp(i\theta_+ J_{+x} + \theta_- J_{-x} | j, j + \rangle | j, j - \rangle. \tag{1.57}$$

A numerical example of a coherent Kepler state Ψ_K, equation (1.40), which is localized on a Keplerian ellipse is shown in Figure 1.1 (from Nauenberg 1989); it demonstrates that the probability function $\Psi^\dagger\Psi$ is localized on an elliptical orbit. This distribution is nearly the same as the corresponding classical distribution satisfying a stationary solution of the Liouville equation for an ensemble of classical orbits with energy E and the same distribution in angular momentum and eccentricity. The observed peak in the distribution is due to the slowing down of the particles when reaching the apogee of the orbit, while the minimum occurs, as expected, at perigee where the particles reach maximum velocity.

To obtain a solution that represents an ensemble of classical particles initially localized in angular coordinates on an elliptic orbit, we must consider a linear superposition of these coherent Kepler energy eigenstates. Such a superposition is a time dependent solution of the Schrödinger equation which generally takes the form

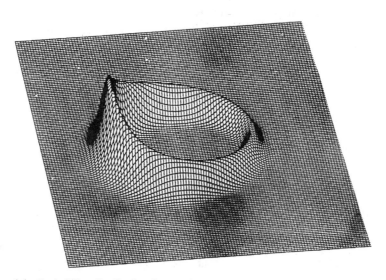

Figure 1.1 Probability distribution for a coherent Coulomb wave packet, equation (1.40), with principal angular momentum $l_n = 40$ and $\delta = 0.8$ which corresponds to eccentricity $\varepsilon = 0.6$ and mean angular momentum $\langle L_z \rangle = 32$.

$$\Psi_K(t) = \sum a_n \psi_n \exp(-iE_n t), \tag{1.58}$$

where the coefficient a_n are constants. We assume that these coefficients have a Gaussian form with a maximum value at $n = \bar{n}$ and a width σ that is determined by experimental conditions,

$$a_n = (2\pi\sigma^2)^{-1/4} \exp\left[\frac{-(n - \bar{n})^2}{4\sigma^2}\right]. \tag{1.59}$$

To show analytically that this superposition gives rise to an angular localized wave packet, we consider the special case of a circular orbit. In this case the coherent state

$$\Psi_{n,l_n}(r, \phi) \propto r^{l_n} \exp\left(\frac{-r}{l_n + 1}\right) \exp(il_n \phi), \tag{1.60}$$

where $n = l_n + \frac{1}{2}$. Expanding the energy eigenvalue E_n up to first order in the difference $n - \bar{n}$, we have

$$E_n \approx E_{\bar{n}} + \frac{2\pi}{\tau}(n - \bar{n}) \tag{1.61}$$

where $\tau = 2\pi\bar{n}^3$ corresponds to the Kepler period for the orbit. Substituting this expression in (1.58), neglecting the dependence on radial coordinates, and approximating the sum over l_n by an integral, we obtain

$$\Psi_K \propto \exp(il_n \phi - iE_{\bar{n}} t) \exp\left[-\left(\phi - \frac{2\pi t}{\tau}\right)^2 \sigma^2\right]. \tag{1.62}$$

Hence the maximum of this wave packet is at $\phi = 2\pi t/\tau$, representing a classical ensemble of particles moving in circular uniform motion with the Kepler period τ. Substituting $t \to t - t_0$ in (1.58) shifts the location of this maximum to $\phi = 2\pi(t - t_0)/\tau$, and corresponds to multiplying the coefficients a_n, equation (1.59), by the complex phase factor $\exp(iE_n t_0)$.

Figure 1.2 shows a numerical evaluation of the probability distribution $\Psi_K^* \Psi_K$ for a corresponding coherent wave packet in the case of a superposition of elliptic states with mean eccentricity $\langle A_x \rangle = \varepsilon$ and angular momentum $\langle L_z \rangle = l_{\bar{n}}\sqrt{1 - \varepsilon^2}$, where $\varepsilon = 0.6$ and $l_{\bar{n}} = 40$ at intervals of one tenth of a Kepler period $T_K = 2\pi n^3$. This wave packet has been launched at perigee at $t = 0$, and it slows down, contracts, and becomes steeper as it reaches apogee at $t = \tau/2$. As the wave packet returns to perigee, it speeds up again, but it also spreads as can be seen from the figure. Then, although it returns to perigee at the end of a Kepler period, $t = \tau$, it has a somewhat more spread out shape. This overall spreading is of classical origin and due to the spread in the position and

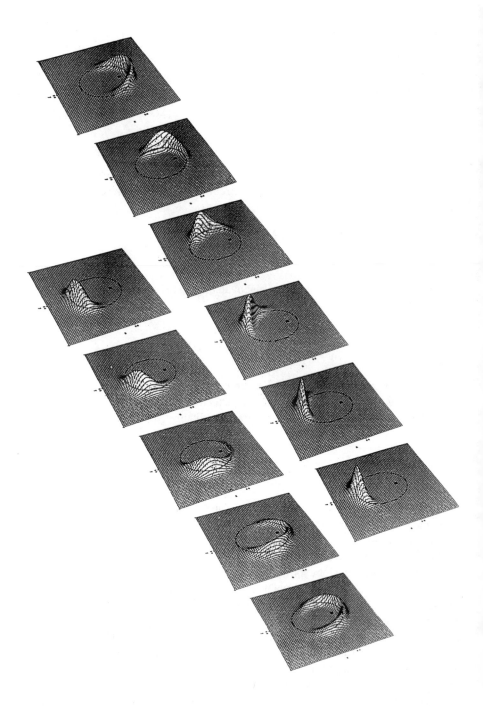

Figure 1.2 Wave packet moving in an elliptic orbit under the action of a Coulomb potential. The probability distribution is shown here each tenth of a Kepler period.

momentum demanded by Heisenberg's uncertainty principle in the corresponding initial ensemble of classical orbits represented by this state. It is this spreading that concerned Lorentz in his correspondence with Schrödinger and led him to question the validity of Schrödinger's theory. (This he assumed would represent the behavior of a single electron in the case of the hydrogen atom rather than that of a classical ensemble of such electrons, as we now properly interpret quantum mechanics.) The spreading of the wave packet can be taken into account by including the second-order terms in the expansion of the energy eigenvalues E_n in powers of $n - \bar{n}$:

$$E_n = E_{\bar{n}} + E'_{\bar{n}}(n - \bar{n}) + \tfrac{1}{2} E''_{\bar{n}}(n - n_{\bar{n}})^2, \tag{1.63}$$

where $E'_{\bar{n}} = 1/\bar{n}^3$ and $E''_{\bar{n}} = -3/\bar{n}^4$. In the approximation the summation over n is replaced by an integral, and we then obtain for the circular wave packet (1.62), with $1/\sigma^2$ replaced by $1/\sigma^2 + 2iE''_{\bar{n}}t$.

1.5 WAVE PACKET REVIVALS

Evidently this spreading must continue in time, and therefore we expect quantum mechanical interference effects to occur when the head of the wave packet catches up with its tail. Indeed, as we pointed out earlier, the onset of this interference marks the breakdown of the quantum-classical correspondence. However, further surprises are in store. After a certain number of Kepler periods, I found that unexpectedly the original wave packet reappeared. This revival had been predicted from numerical calculations of radial wave packets by Alber, Ritsch, and Zoller (1986) and Parker and Stroud Jr. (1986), and it is now well understood analytically (see Averbukh and Perelman 1989; Nauenberg 1990, 1992). The essential point is that the previous approximation of replacing the sum over n by an integral is not valid for times t longer than approximately $1/E''_{\bar{n}}$. According to (1.63), the phase which appears in (1.58) takes the form

$$\theta = 2\pi \frac{t}{\tau} (n - \bar{n}) \left(1 - \frac{3}{2\bar{n}} (n - \bar{n}) \right). \tag{1.64}$$

Setting

$$t = t_{\text{rev}} = \left(\left[\frac{\bar{n}}{3} \right] + \mu \right) \tau, \tag{1.65}$$

where $[\bar{n}/3]$ is the closest integer to $\bar{n}/3$, we obtain

$$\theta \approx 2\pi(n - \bar{n}) \left(\left[\frac{\bar{n}}{3} \right] + \mu + \frac{1}{2} (n - \bar{n}) \right). \tag{1.66}$$

Hence for $\mu = \pm\tfrac{1}{2}$, θ is approximately an integer multiple of 2π, and consequently the initial wave packet reappears at the revival time t_{rev}, equation (1.65).

This behavior is demonstrated numerically in Figure 1.3, which shows the time dependence of the autocorrelation function

$$C(t) = \langle \Psi_K(t), \Psi_K(0) \rangle = \sum |a_n|^2 \exp(iE_n t) \tag{1.67}$$

in units of the Kepler period for the case where $\bar{n} = 80$ at $\sigma = 6$. The autocorrelation function measures the overlap of the initial wave packet with the wave packet at a later time t. In this case $\bar{n} = 80$, and therefore $[\bar{n}/3] = 26$. The results during the first few Kepler periods are those expected from the quantum-classical correspondence principle for an ensemble of classical particles moving on nearby elliptic trajectories. Initially after each Kepler period the increased spreading of the wave packet decreases the overlap with the initial wave packet. Consequently the peak of the autocorrelation decreases, while the width broadens until the head of the wave packet catches up with its tail. The onset of interference is seen in the figure after $4\frac{1}{2}$ Kepler periods. As expected at $[\bar{n}/3] = 26$ the initial overlap reappears displaced by $\frac{1}{2}$ a Kepler period as predicted by (1.65). In addition to this revival, Figure 1.3 show that there are more complex revivals of the wave packet which occur near $t = t_{rev}/2$ and other fractional multiples of t_{rev}. These have been discussed in detail by Averbukh and Perelman (1989). For example, the revival at $t_{rev}/2$ recurs at half the Kepler period and

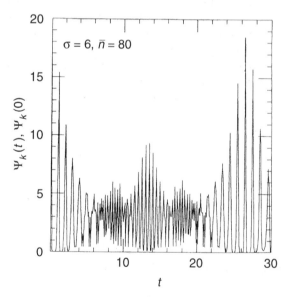

Figure 1.3 Absolute square for the autocorrelation function for a wave packet moving in a Coulomb potential as a function of time in units of the Kepler period. For a mean principal quantum number of $n = 80$, the first full recurrence of the wave packet occurs at $t_r = 26.5$, as predicted by the theory. Also shown are partial recurrences at $t \approx (1/2)t_r$ with half the Kepler period, and at $t \approx (1/3)t_r$ and $t \approx (2/3)t_r$ with a third of the Kepler period.

corresponds to the splitting of the original wave packet into two copies, which appear half a Kepler period apart.

More recently Klauder (1996) has discussed a specific choice of coefficients a_n in (1.58) for time dependent wave packets that form a nonorthogonal but overcomplete set, and provide a representation for the identity operator. However, for a mean principal quantum number n which is typical of current experiments ($n = 50 - 100$), these generalized coherent states spread too rapidly, usually within a Kepler period, and therefore they do not exhibit classical behavior or the interesting revival behavior which we have discussed here.

1.6 DYNAMICS OF COULOMB WAVE PACKETS IN WEAK TIME DEPENDENT ELECTROMAGNETIC FIELDS

Over the past years laser excitation of electronic wave packets in Rydberg atoms, and the manipulation of these states in external electromagnetic fields, has become a very active field. In this section our discussion of the dynamics of coherent quantum wave packets in a Coulomb potential is extended to include the effects of time dependent weak electromagnetic fields, when the energy associated with these fields is small compared to the Coulomb field (Nauenberg 1994; Bellomo, Farrelly, and Uzer 1997, 1998). In particular, we consider periodic fields such as microwaves and radio fields with slowly varying amplitudes relevant to experiments. In this case there are at least five important time scales that must be taken into account associated with the following frequencies: the unperturbed Kepler frequency of the electron in the atom $\omega_K = (-2E)^{3/2}$ where $E = -1/(2n^2)$ is the Kepler energy for a principal quantum number n, the frequency ω of the microwave or radio field, the Stark frequency $\omega_s = (3/2)F/\sqrt{(-2E)}$ where F is the Electric field, the Larmour frequency $\Omega = B$ where B is the magnetic field, and the characteristic time T_s to switch on the fields. I consider the regime where all the applied frequencies are small compared to the unperturbed Kepler frequency, or $\omega << \omega_K$, $\omega_s << \omega_K$ and $\Omega << \omega_K$ in which case a good approximation is to restrict the quantum dynamics to the set of unperturbed degenerate states of a fixed principal quantum number n. It can be readily shown that the reduced Hamiltonian in this approximation is the same as the corresponding classical Hamiltonian obtained by taking a time average over a Kepler period (Kappertz and Nauenberg 1993).

For the case of crossed electric and magnetic fields, let the x-y plane be the polarization plane of the electric field F, and the z-axis the direction of the magnetic field B. Then the reduced Hamiltonian takes the form

$$H = H_K + \omega_x(t)M_x + \omega_y(t)M_y + \omega_z(t)L_z, \qquad (1.68)$$

where H_K is the unperturbed Coulomb Hamiltonian for the electron in the hydrogen atom, $w_x(t)$ and $w_y(t)$ are the x and y components of the Stark frequency, and Ω_z is the Larmor frequency. The time dependent Schrödinger equation for the state $|\Psi(t)\rangle$ is

$$i \frac{\partial}{\partial t} |\Psi(t)\rangle = H |\Psi(t)\rangle, \tag{1.69}$$

which we now solve in the adiabatic approximation, S. We first find an eigenstate of the time dependent Hamiltonian:

$$H |\psi(t)\rangle = E(t) |\psi(t)\rangle, \tag{1.70}$$

where

$$|\psi(t)\rangle = U(t) |\psi(0)\rangle \tag{1.71}$$

and $U(t)$ is a unitary operator in an SO(3) subgroup of SO(4), equation (1.48):

$$U(t) = \exp(iL_z \alpha_t) \exp(iM_x \beta_t). \tag{1.72}$$

Setting

$$w_x = \Delta_t \sin(\theta_t) \sin(\phi_t), \tag{1.73}$$

$$w_y = \Delta_t \sin(\theta_t) \cos(\phi_t), \tag{1.74}$$

and

$$w_z = \Delta_t \cos(\theta_t), \tag{1.75}$$

we require that $U(t)$ have the property

$$U(t)L_z U(t)^\dagger = \sin(\theta_t)(\cos(\phi_t)M_x + \sin(\phi_t)M_y) + \cos(\theta_t)L_z. \tag{1.76}$$

This relation implies that $\alpha_t = \phi_t$ and $\beta_t = \theta_t$. Assuming that the initial state $|\Psi(0)\rangle$ is the maximal circular eigenstate of L_z with eigenvalue $n - 1$, then $|\Psi(t)\rangle$ is an eigenstate of H, equation (1.71) corresponding to an eigenvalue $E = E_K + (n - 1)\Delta(t)$. Then the adiabatic solution of the time dependent Schrödinger equation (1.69) has the form

$$|\Psi(t)\rangle = |\psi(t)\rangle \exp[-iE_K - i(n - 1) \int dt\Delta(t) + i\nu_B], \tag{1.77}$$

where ν_B is the Berry phase (Shapere and Wilczek 1989; Jarzynski 1995). Substituting (1.77) in (1.69), we obtain

$$\dot{\nu}_B = -i \langle \psi(t)| \frac{\partial U(t)}{\partial t} U(t)^\dagger |\psi(t)\rangle \tag{1.78}$$

where

$$i \frac{\partial U(t)}{\partial t} U(t)^\dagger = -\dot{\phi} L_z - \dot{\theta}_t (\cos(\phi_t) M_x - \sin(\phi_t) M_y). \tag{1.79}$$

It is straightforward to evaluate the expectation values of these operators in terms of the initial condition, and we obtain

$$\nu_B = (n - 1) \int dt \dot{\phi}_t \cos(\theta_t). \tag{1.80}$$

The above adiabatic solution is valid only for the special initial condition that the state correspond to a maximum circular state, and the condition that $U(0) = 1$ implies that the external fields must also vanish at $t = 0$. To obtain a more general solution, we introduce a similar unitary transformation

$$U(t) = \exp(i\alpha_t L_z) \exp(i\beta_t M_x) \exp(i\gamma_t L_z), \tag{1.81}$$

which corresponds to a rotation parametrized by the time dependent Euler angles α_t, β_t, and γ_t. We set

$$|\Psi(t)\rangle = U(t)U(0)^\dagger \exp(-iH_K t)|\Psi(0)\rangle. \tag{1.82}$$

Substituting this expression in (1.69), we obtain the relations

$$\omega_z = -[\dot{\alpha}_t + \dot{\gamma}_t \cos(\beta_t)], \tag{1.83}$$

$$\omega_x = -[\dot{\beta}_t \cos(\alpha_t) + \dot{\gamma}_t \sin(\beta_t) \sin(\alpha_t)], \tag{1.84}$$

and

$$\omega_y = [\dot{\beta}_t \sin(\alpha_t) - \dot{\gamma}_t \sin(\beta_t) \cos(\alpha_t)]. \tag{1.85}$$

They lead to first-order differential equations for these parameters,

$$\dot{\alpha}_t = \cot(\beta_t)[\omega_x(t) \sin(\alpha_t) + \omega_y(t) \cos(\alpha_t)] - \omega_z(t), \tag{1.86}$$

$$\dot{\beta}_t = -\omega_x(t) \cos(\alpha_t) + \omega_y(t) \sin(\alpha_t), \tag{1.87}$$

$$\dot{\gamma}_t = -\frac{1}{\sin(\beta_t)} [\omega_x(t) \sin(\alpha_t) + \omega_y(t) \cos(\alpha_t)]. \tag{1.88}$$

The initial conditions α_0, β_0, and γ_0 can now be chosen arbitrarily. In general, these nonlinear equations are solved numerically, but for special cases analytical solutions can also be obtained. For external fields that change adiabatically, we find an approximate solution for these equations. Setting $\omega_x(t) = \omega_s(t) \sin(\phi_t)$ and $\omega_y(t) = \omega_s(t) \cos(\phi_t)$, we simplify equations (1.86) to (1.88) to

$$\dot{\alpha}_t = \omega_s(t) \cot(\beta_t) \cos(\alpha - \phi_t) - \omega_z(t), \tag{1.89}$$

$$\dot{\beta}_t = -\omega_s(t) \sin(\alpha - \phi_t), \tag{1.90}$$

$$\dot{\gamma}_t = -\frac{\omega_s(t)}{\sin(\beta_t)} \cos(\alpha_t - \phi_t). \tag{1.91}$$

In the adiabatic approximation, when the amplitude of the electric and magnetic fields are varying slowly in time, an approximate special solution of these equations is obtained by setting

$$\alpha_t = \phi_t + \delta\alpha. \tag{1.92}$$

Then $\delta\alpha$ must satisfy the equation

$$\delta\alpha \cong \frac{\dot{\beta}}{\omega_s(t)}, \tag{1.93}$$

where to the first order in $\delta\alpha$,

$$\tan(\beta_t) = \frac{\omega_s(t)}{[\dot{\phi}_t + \omega_z(t)]}. \tag{1.94}$$

Hence $\sin(\beta_t) = \omega_s/\Delta_t$, where $\Delta_t = \sqrt{\omega_s(t)^2 + (\dot{\phi}_t + \omega_z(t))^2}$ and

$$\gamma_t = -\int_0^t dt' \Delta_{t'}. \tag{1.95}$$

For $\dot{\phi}_t << \omega_z(t)$ we have

$$\Delta_t \approx \sqrt{\omega_s(t)^2 + \omega_z(t)^2} + \dot{\phi}_t \frac{\omega_z(t)}{\sqrt{\omega_s(t)^2 + \omega_z(t)^2}}, \tag{1.96}$$

where the second term corresponds to the previously calculated Berry phase equation (1.80).

While L_z, M_x, and M_y are the generators of an $SO(3)$ subgroup of $SO(4)$, this is not the case for the remaining operators M_z, L_x, and L_y. However, $L_z^2 + M_x^2 + M_y^2$ and $M_z^2 + L_z^2 + L_y^2$ are separately constants of the motion, whereas $L^2 + M^2 = -1/2E - 1$ is fixed by the Kepler energy $E = -1/(2n^2)$.

The time dependence of the Heisenberg operators L and M can also be obtained from the evolution operator U(t), eg. $L_z(t) = U(0)U(t)^\dagger L_z U(t)U(0)^\dagger$, where $U(t)^\dagger$ is the Hermitian adjoint of U(t). These operators satisfy the equations

$$\dot{L}_z(t) = i[H, L_z(t)] = \omega_x(t)M_y(t) - \omega_y(t)M_x(t), \tag{1.97}$$

$$\dot{M}_x(t) = i[H, M_x(t)] = \omega_y(t)L_z(t) - \omega_z(t)M_y(t), \tag{1.98}$$

$$\dot{M}_y(t) = i[H, M_y(t)] = \omega_z(t)M_x(t) - \omega_x(t)L_z(t), \tag{1.99}$$

which are exactly the same as the corresponding classical equations. The general solution of these equations can be written explicitly in terms of the Euler angles α_t, β_t, γ_t, and the initial operators, $L_z(0)$, $M_x(0)$, and $M_y(0)$. We have

$$L_z(t) = [\cos(\beta_t)\cos(\beta_0) + \sin(\beta_t)\sin(\beta_0)\cos(\gamma_t)]Lz(0)$$

$$+ [\sin(\beta_t)\sin(\gamma_t)]M_x(0)$$

$$+ [\cos(\gamma_t)\sin(\beta_0) - \cos(\beta_0)\sin(\beta_t)\cos(\gamma_t)]M_y(0), \tag{1.100}$$

$$M_x(t) = \cos(\alpha_t)\overline{M}_x(t) + \sin(\alpha_t)\overline{M}_y(t), \tag{1.101}$$

$$M_y(t) = -\sin(\alpha_t)\overline{M}_x(t) + \cos(\alpha_t)\overline{M}_y(t), \tag{1.102}$$

where $\overline{M}_x(t)$ and $\overline{M}_y(t)$ are the operators $M_x(t)$ and $M_y(t)$ in a frame of reference rotated by and angle α_t about the z-axis,

$$\overline{M}_x(t) = -\sin(\gamma_t)\sin(\beta_0)L(0) + \cos(\gamma_t)M_x(0)$$

$$+ \sin(\gamma_t)\cos(\beta_0)M_y(0), \tag{1.103}$$

$$\overline{M}_y(t) = [-\cos(\beta_t)\sin(\beta_0)\cos(\gamma_t) + \sin(\beta_t)\cos(\beta_0)]L_z(0)$$

$$- \cos(\beta_t)\sin(\gamma_t)M_x(0)$$

$$+ [\cos(\beta_t)\cos(\beta_0)\cos(\gamma_t) + \sin(\beta_0)\sin(\beta_t)]M_y(0). \tag{1.104}$$

It can be verified that $\langle L_z \rangle^2 + \langle M_y \rangle^2 + \langle M_x \rangle^2$ and $\langle L_z^2 \rangle + \langle M_y^2 \rangle + \langle M_x^2 \rangle$ are both conserved but have different values. The difference corresponding to quantum fluctuations and the classical states are those for which these quantum fluctuations are a minimum. For example, for an orbit in the plane of polarization, we have $\langle L_z \rangle^2 + \langle M_y \rangle^2 + \langle M_x \rangle^2 = (n-1)^2$ and $\Delta L_z^2 + \Delta M_y^2 + \Delta M_x^2 = (n-1)$.

For the case of a circular polarized electric field, we can simplify the equations by going to a frame of reference rotating with frequency ω. Setting

$$\langle L_z(0) \rangle = (n-1)\cos(\eta), \tag{1.105}$$

$$\langle M_x(0) \rangle = (n-1)\sin(\eta)\cos(\delta), \tag{1.106}$$

and

$$\langle M_y(0) \rangle = (n-1)\sin(\eta)\sin(\delta), \tag{1.107}$$

we obtain from (1.100) to (1.103),

$$\langle L_z(t) \rangle = (n-1)[\cos(\eta)\cos(\beta_t) + \sin(\eta)\sin(\beta_t)\sin(\gamma_t - \delta)], \tag{1.108}$$

$$\langle \overline{M}_x(t) \rangle = (n-1)\sin(\eta))\cos(\gamma_t - \delta), \tag{1.109}$$

and

$$\langle \overline{M}_y(t) \rangle = (n-1)[\cos(\eta)\sin(\beta_t) - \sin(\eta)\cos(\beta_t)\sin(\gamma_t - \delta)]. \tag{1.110}$$

An interesting application of equations (1.108) to (1.109) is the case where the electric field amplitude F is increased slowly to a value such that $\omega_s/\omega \gg 1$. As $\beta_t \to \pi/2$,

$$\langle L_z(t) \rangle \to (n-1)\sin(\eta)\sin(\gamma_t - \delta), \tag{1.111}$$

$$\langle M_x(t) \rangle \to (n-1)\sin(\eta)\cos(\gamma_t - \delta), \tag{1.112}$$

and

$$\langle M_y(t) \rangle \to (n-1)\cos(\eta). \tag{1.113}$$

For $\eta = 0$, corresponding to initial motion in a circular state, the orbit approaches asymptotically maximum eccentricity along the y-axis. In particular, when $\omega_s/\omega \gg 1$,

$$\langle M_y(t) \rangle \to (n-1), \tag{1.114}$$

which implies that asymptotically the component of the Laplace vector along the field in the rotating frame is determined by the mean value of the initial angular momentum (Klauder 1996). Alternatively, if $\eta = \pi/2$, corresponding initially to zero angular momentum, the orbit approaches a circular state when $\gamma_t - \delta = \pi/2$:

$$\langle L_z(t) \rangle \to (n - 1). \tag{1.115}$$

Another interesting solution is obtained for the case where β_t is a constant and $\delta = \pi/2$ (Bellomo and Stroud Jr. 1998) in which case equations (1.92) to (1.95) are exact:

$$\langle L_z(t) \rangle = (n - 1)[\cos(\eta) + (1 - \cos(\gamma_t)) \sin(\beta_0)(\sin(\eta - \beta_0))] \tag{1.116}$$

$$\langle \overline{M}_x(t) \rangle = (n - 1)[\sin(\eta - \beta_0)) \sin(\gamma_t)], \tag{1.117}$$

$$\langle \overline{M}_y(t) \rangle = (n - 1)[\sin(\eta) + (\cos(\gamma_t) - 1) \cos(\beta_0) \sin(\eta - \beta_0)]. \tag{1.118}$$

Then for $\eta = \pi/2$, $\beta_0 = \pi/4$, if γ_t increases from 0 to π, an initial state of zero angular momentum is transformed into a circular state. For the case $\eta = \beta_0$ the evolution is time independent in the rotating frame of reference.

1.7 TIME DEPENDENT SEMICLASSICAL APPROXIMATION FOR COULOMB WAVE PACKETS

We will now discuss a new semiclassical theory that gives some further insight in the correspondence between quantum and classical mechanics and provides a useful approximation method for time dependent calculations in quantum mechanics in the correspondence limit. This method was developed originally by Heller, Tomsovic, and their collaborators (1992) to evaluate correlation functions by the use of multiple periodic classical trajectories, and it is based on a semiclassical approximation to the time dependent Green's function developed by van Vleck. We will describe here a simpler but equivalent form of this technique (see Suarez Barnes et al. 1993; Nauenberg 1995) by solving approximately the time dependent Schrödinger equation for the propagation of a Gaussian wave packet (Heller 1975; Littlejohn 1986) in the Coulomb potential.

The Schrödinger equation in one degree of freedom is

$$i \frac{\partial \psi(q, t)}{\partial t} = -\frac{1}{2} \frac{\partial^2 \psi(q, t)}{\partial q^2} + V(q)\psi(q, t), \tag{1.119}$$

where $V(q) = -1/q$ for the Coulomb potential. A *local* solution of (1.119) near a time dependent trajectory q_t is obtained by expanding the potential $V(q)$ up to quadratic terms in the neigborhood of this trajectory,

$$V(q) \approx V(q_t) + V'(q_t)(q - q_t) + \tfrac{1}{2} V''(q_t)(q - q_t)^2. \tag{1.120}$$

In this approximation the solution of the Schrödinger equation is a Gaussian wave packet

$$\psi_\beta(q, t) = A \exp[i(\xi_t(q - q_t) + \alpha_t(q - q_t)^2 + \gamma_t)], \tag{1.121}$$

where the trajectory q_t is determined by a suitably chosen solution of the classical equations of motion for the Hamiltonian $H = p^2/2 + V(q)$, and the variables ξ_t, α_t, and γ_t are *complex* functions of time that satisfy ordinary nonlinear differential equations obtained by substituting (1.121) in the Schrödinger equation (1.119) with $V(q)$ approximated by (1.120) (see Suarez Barnes et al. 1993; Nauenberg 1995). Their initial conditions are determined by the parameters of the initial Gaussian wave packet and the specific reference trajectory. For example, a normalized initial Gaussian wave packet of variance σ_β^2, centered at q_β and with mean momentum p_β has the form

$$\psi_\beta(q, 0) = (\pi\sigma_\beta^2)^{-1/4} \exp\left(ip_\beta(q - q_\beta) - \frac{(q - q_\beta)^2}{2\sigma_\beta^2} \right). \tag{1.122}$$

In comparing with (1.121) at $t = 0$, we obtain

$$\alpha_0 = \frac{i}{2\sigma_\beta^2},$$

$$\xi_0 = p^\beta + 2\alpha_0(q_0 - q^\beta),$$

$$\gamma_0 = (q_0 - q^\beta)[\alpha_0(q_0 - q^\beta) + p^\beta], \tag{1.123}$$

where q_0 denotes the reference trajectory's initial position and $A = (\pi\sigma_\beta^2)^{-1/4}$. We now evaluate the autocorrelation function $C_\beta(t)$, where

$$C_\beta(t) = \int dq\, \psi_\beta^*(q, 0)\psi_\beta(q, t), \tag{1.124}$$

by constructing the time dependent state as a superposition of wavefunctions (1.121) corresponding to different reference trajectories, with relative phase equal to unity

(Suarez Barnes 1993; Nauenberg 1995). For the special case where $q_t = q_0$, we obtain for $t > \tau_\beta$,

$$C_\beta(t) = \sum_j \left\{ \sqrt{\frac{1}{(1 - i\nu_\beta j^2/t)}} \; \exp\left[3i \frac{\tau_\beta}{q_\beta} j^{2/3} t^{1/3} - \frac{q_\beta^2}{\sigma_\beta} \frac{[(t/j)^{2/3} - 1]^2}{[1 + it/\nu_\beta j^2]} \right] \right\}, \quad (1.125)$$

where the index j refers to a given trajectory, t is the time in units of the mean Kepler period $\tau_\beta = 2\pi q_\beta$, and $\nu_\beta = (3/4)\tau_\beta \sigma_\beta^2/q_\beta^3$. This result is in excellent agreement with exact numerical evaluation of the autocorrelation function (Suarez Barnes et al. 1993; Nauenberg 1995); see Figure 1.4. It can be readily shown that by expanding the phase factors in each term of (1.125) to second order in $(j - N)$ at $t = N + \delta$, where N is an integer, these terms add constructively when $N = [n_\beta/3]$ and $\delta = \pm \frac{1}{2}$, where $q^\beta = 2n_\beta^2$. This implies that the original Gaussian wave packet reappears approximately after a period $t = [n_\beta/3] \pm \frac{1}{2}$. While the reference trajectories depend generally on time, at the recurrence time these trajectories correspond precisely to the quantized Bohr orbits (Nauenberg 1990, 1992).

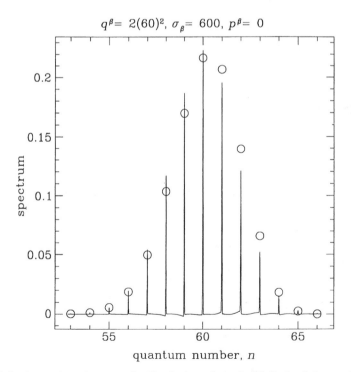

$$q^\beta = 2(60)^2, \; \sigma_\beta = 600, \; p^\beta = 0$$

Figure 1.4 Comparison between the Fourier transform (solid line) of the semiclassical autocorrelation function (1.25) and the corresponding quantum spectra (circles) for the case of $q^\beta = 7200$ and $\sigma_\beta = 600$.

1.8 CONCLUDING REMARKS

We have shown that wave packets elucidate the correspondence between quantum and classical mechanics and give valuable insights in the behavior of Rydberg atoms in weak electromagnetic fields. The formation of such wave packets has been made possible by the development of tunable short-pulsed lasers. This technology has ushered in a fruitful series of experiments in atomic, molecular, and condensed matter physics and chemistry. Unexpected phenomena like wave packet revivals, which had not been anticipated by the founders of quantum mechanics, have now been confirmed experimentally and may turn out to have valuable applications. Although wave packets emerged early in the development of quantum mechanics, they have occupied center stage only recently with the increasing role of the time domain in experiments. In the presence of strong external fields where chaotic dynamics emerges in the classical regime, a wave packet analysis helps clarify the behavior in quantum mechanics, and surprisingly some recent evidence has shown that revivals can occur even in this domain (Tomsovic and Lefebure 1997).

The importance of wave packets extends also to other branches of physics. For example, neutrino oscillations which have been observed more recently (Fukuda et al. 1998) can be understood properly only by a wave packet analysis (Nauenberg 1999). Although these oscillations are due to interference effects between different mass eigenstates of the neutrino, and therefore appear to be a pure wave phenomena, it turns out that the neutrino must also travel along nearly classical trajectories. Wave packets are essential in dealing with such wave-particle duality which has marked the physics of this century.

REFERENCES

Alber, G., Ritsch, H., and Zoller, P. (1986). *Phys. Rev.* A**34**, 1058.

Averbukh, I. Sh., and Perelman, N. F. (1989). *Physics Lett.* A**139**, 449.

Bellomo, P., Farrelly, D., and Uzer, T. (1997). *J. Chem. Phys.* **107**, 2499.

Bellomo, P., Farrelly, D., and Uzer, T. (1998). *J. Chem. Phys.* **108**, 402.

Bellomo, P., Farrelly, D., and Uzer, T. (1998). *J. Chem. Phys.* **108**, 5295.

Bellomo, P., and Stroud, C. R., Jr. (1998). *Phys. Rev.* (to be published).

Bhaumik, D., Dutta-Roy, B., and Ghosh, G. (1986). *J. Phys.* A**19**, 1355.

Bohr, N. (1913). *Philosophical Magazine and Journal of Science* **26**.

Born, M. (1955). *Dan. Mat. Fys. Medd.* **30** (2) p. 12.

Born, M. (1955). *Science* **12**, 675.

Born, M. (1955). *Phys. Blatter* **11**, 49.

Brown, L. S. (1973). *Am. J. Phys.* **41**, 525.

Chen, X., and Yeazell, J. A. (1998). *Phys. Rev. Lett.* **81**, 5772.

Delande, D., and Gay, J. C. (1991). *Europhysics Lett.* **5**, 303.

Fukuda, Y. et al., (1998). *Phys. Rev. Lett.* **81**, 1774.

Gay, J. C., Delande, D., and Bommier, A. (1989). *Phys. Rev. A* A**39**, 6587.

Gerry, C. C. (1986). *Phys. Rev.* A**33**, 6.

Heller, E. J. (1975). *J. Chem. Phys.* **62**, 1544.

Heller, E. J. (1991). *J. Chem. Phys.* **94**, 2723.

Jarzynski, C. (1995). *Phys. Rev. Lett.* **74**, 1264.

Kappertz, P., and Nauenberg, M. (1993). *Phys. Rev.* A**47**, 4749.

Klauder, J. R. (1996). *J. Phys. A: Math. Gen.* **29**, L293.

Lena, C., Delande, D., and Gay, J. C. (1991). *Europhysics Lett.* **15**, 697.

Littlejohn, R. G. (1986). *Physics Reports* **138**, 193.

Meacher, D. R., Meyler, P. E., Hughes, I. G., and Ewart, P. (1991). *J. Phys.* B**24**, L63.

Moore, W. (1989). *Schrödinger, Life and Thought*. Cambridge University Press, Cambridge.

Mostowski, J. (1977). *Lett. Math. Phys.* **2**, 1.

Nauenberg, M. (1989). *Phys. Rev.* A**40**, 1133.

Nauenberg, M. (1990). *J. Phys. B: At. Mol. Opt. Phys.* **23**, L385.

Nauenberg, M. (1992). In *Irregular Atomic Systems and Quantum Chaos*, ed. J. C. Gay. Gordon and Breach.

Nauenberg, M. (1994). In *Coherent States: Past, Present and Future*, eds. D. H. Feng, J. R. Klauder, and M. R. Strayer. World Scientific, Singapore, p. 345.

Nauenberg, M. (1995). *Am. J. Phys.* **63**, 661.

Nauenberg, M. (1999). *Phys. Lett.* B**452**, 434.

Nauenberg, M., and Keith, A. (1992). In *Quantum Chaos—Quantum Measurement*, ed. P. Cvitanovic et al. Kluwer Academic, Dordrecht.

Nauenberg, M., Stroud, C. Jr. and Yeazell, J. (1994). *Scientific Am.* June.

Nicholson, J. W. (1912). *Month. Not. Roy. Astr. Soc.* **72**, p. 49.

O'Connor, P. W., Tomsovic, S., and Heller, E. J. (1992). *Physica* D**55**, 340.

Parker, J., and Stroud, C. R. Jr. (1986). *Phys. Rev. Lett.* **56**, 716.

Przibram, K. (1967). *Letters on Wave Mechanics*. Philosophical Library, New York.

Schrödinger, E. (1926). *Naturwissenchaften* **14**, 664.

Shapere, A., and Wilczek, F. (1989). *Geometric Phases in Physics* World Scientific, Singapore.

Suarez Barnes, I. M., Nauenberg, M., Nockleby, M., and Tomsovic, S. (1993). *Phys. Rev. Lett.* **71**, 1961.

Suarez Barnes, I. M., Nauenberg, M., Nockleby, M., and Tomsovic, S. (1994). *J. Phys.* A**27**, 3299.

Tomsovic, S., and Heller, E. J. (1991). *Phys. Rev. Lett.* **67**, 664.

Tomsovic, S., and Heller, E. J. (1993). *Phys. Rev. E* **47**, 282.

Tomsovic, S., and Lefebvre, J. H. (1997). *Phys. Rev. Lett.* **79**, 3629.

Yeazell, J. A., Mallalieu, M., and Stroud, C. R. (1990). *Phys. Rev. Lett.* **64**, 2007.

Yeazell J. A., and Stroud, C. R. (1991). *Phys. Rev.* A**43**, 5153.

CHAPTER 2

Semiclassical Wave Packets

ERIC J. HELLER

2.1 INTRODUCTION

Experiments on atomic and molecular systems are often conducted in a way that demands theory to be implemented in the time domain. A variety of methods have emerged to handle the time dependent Schrödinger equation, including essentially exact fast fourier transform (FFT) and other basis set methods. These methods are, however, restricted to a few degrees of freedom, and sometimes just give "the answer" without much physical insight. The semiclassical wave packet approaches which we discuss in this chapter are capable of handling many degrees of freedom, and they give much insight into the essential physics.

Even when experiments are done in the energy domain, there is often strong motivation to use the time domain for calculations and understanding. For example, calculating high-resolution Raman scattering for a medium-sized molecule in the energy domain is next to impossible, owing to the huge number of intermediate states that have to be determined. Raman scattering is one of many examples where the essential physics is over in a short time, even though the data are energy-resolved. The time dependent formulation of Raman scattering (Lee and Heller 1979; Meyers et al. 1982) makes clear the essential information in resonance and pre-resonance Raman spectra and makes calculations and simple approximation possible (Heller 1981). Scattering theory is another case in point. Collisions are frequently over quickly, yet the usual formulation of particle scattering is in the energy domain.

Semiclassical wave packet methods are useful tools in the implementation of time dependent formulations of quantum processes. In various ways, Gaussians inevitably creep into any discussion of semiclassical wave packets. The essentials of such wave packets (including exact wave packet dynamics for quadratic potentials and the "thawed" Gaussian semiclassical approximation) are presented again here for completeness (Littlejohn 1986; Heller 1986, 1991; Hase 1991). Phase space pictures are developed next; these are invaluable to understanding approximations and errors.

Following this, we deal with more advanced methods. The full Van Vleck semiclassical Green's function in coordinate space and its generalization to coherent states are discussed. Then the role of chaos and the important concept of multicenter

The Physics and Chemistry of Wave Packets, Edited by John Yeazell and Turgay Uzer
ISBN 0-471-24684-0   © 2000 John Wiley & Sons, Inc.

linearized expansions of the propagator applied to Gaussian wave packets is introduced. Finally, as an application, we discuss the theory of eigenfunctions scarred by classical periodic orbits.

2.2 GAUSSIAN WAVE PACKET DYNAMICS

2.2.1 Anatomy

A general Gaussian wave packet in one dimension can be written as

$$
\psi_1(q) \equiv \left(\frac{\mathrm{Re}\,\alpha_1}{\pi\hbar}\right)^{1/4} \exp\left[-\frac{\alpha_1}{2\hbar}(q-q_1)^2 + \frac{i}{\hbar}p_1(q-q_1) + \frac{i}{\hbar}\phi_1\right]
$$

$$
= \left(\frac{2a_1}{\pi\hbar}\right)^{1/4} \exp\left[-\frac{\alpha_1}{2\hbar}(q-q_1)^2 + \frac{ip_1(q-q_1)}{\hbar} + \frac{i}{\hbar}\phi_1\right], \tag{2.1}
$$

where $\alpha_1/2 \equiv (a_1 + ib_1)$ is a complex number with positive real part a_1 and imaginary part b_1 of either sign. The Gaussian has average position $q_1 = \int \psi_1^*(q)q\psi_1(q)dq = \int |\psi_1(q)|^2 q\,dq$, and average momentum $p_1 = \int \psi_1^*(q)\,(-i\hbar\frac{\partial}{\partial q})\,\psi_1(q)\,dq$. The uncertainties in position and momentum are easily calculated:

$$
\Delta q^2 = \int \psi_1(q)^*(\hat{q}-q_1)^2\psi_1(q)\,dq
$$

$$
= \frac{\hbar}{4a_1} \tag{2.2}
$$

$$
\Delta p^2 = \int \psi_1(q)^*(\hat{p}-p_1)^2\psi_1(q)\,dq
$$

$$
= a_1\hbar + \frac{b_1^2\hbar}{a_1} \tag{2.3}
$$

$$
\Delta q\Delta p = \frac{\hbar}{2}\sqrt{1 + \frac{b_1^2}{a_1^2}}\,. \tag{2.4}
$$

A little "anatomy lesson" is worthwhile. In Figure 2.1 we show a picture of a normalized, general Gaussian in coordinate space and the aspects that each parameter in the Gaussian controls. The parameters a_1, b_1, q_1, p_1, and ϕ_1 are all taken to be real.

The minimum uncertainty of $\hbar/2$ is attained only if the imaginary part b_1 of the "spread" parameter α_1 vanishes. This is evident from (2.4). Note that b_1 controls the "chirp," or position-momentum correlation of the Gaussian. This correlation is seen in Figure 2.1, where the local wavelength decreases to the right. One way of understanding the chirp is to notice that the "local momentum," defined as

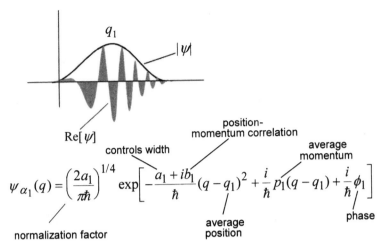

Figure 2.1 Anatomy of a Gaussian. The Gaussian wave packet is displayed in coordinate space along with the roles of the parameters $a_1, b_1, q_1, p_1,$ and ϕ_1.

$$p(q) \sim \hbar \frac{\partial}{\partial q} \text{Im}[\log\psi_1(q)] = p_1 + 2b_1(q - q_1), \tag{2.5}$$

is the mean momentum at the center of the wave packet, and it deviates linearly away from the center according to the sign and magnitude of b_1.

The momentum representation of the Gaussian is obtained by Fourier transform of $\psi_1(q)$:

$$\psi_1(p) \equiv \frac{1}{\sqrt{2\pi\hbar}} \int_{-\infty}^{\infty} e^{-ipq/\hbar} \psi_1(q)\, dq$$

$$= \frac{1}{\sqrt{\alpha_1}} \left(\frac{2a_1}{\pi\hbar}\right)^{1/4} \exp\left[-\frac{(p-p_1)^2}{2\hbar\,\alpha_1} - \frac{iq_1 p}{\hbar} + \frac{i\phi_1}{\hbar}\right]. \tag{2.6}$$

Also

$$\psi_1(q) = \frac{1}{\sqrt{2\pi\hbar}} \int_{-\infty}^{\infty} e^{ipq/\hbar} \psi_1(p)\, dq. \tag{2.7}$$

From this last perspective the localized wavepacket is written as a superposition of plane waves (momentum eigenstates) $e^{ipq/\hbar}$ with a Gaussian amplitude. From

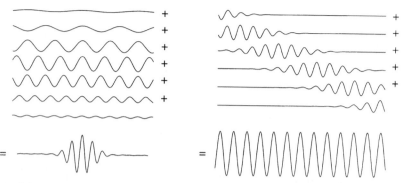

Figure 2.2   Left: Construction of a Gaussian wave packet from plane waves. The plane waves above were added to produce the near Gaussian wave packet seen below them. Right: Construction of plane wave from Gaussian wave packets.

$$\hat{p}\, e^{ipq/\hbar} = -i\hbar\,\frac{\partial e^{ipq/\hbar}}{\partial q} = p\, e^{ipq/\hbar} \tag{2.8}$$

we see that $e^{ipq/\hbar}$ is a momentum eigenstate with eigenvalue p. In Figure 2.2 at the left-hand side we see seven plane waves approximating a Gaussian wave packet, as in equation (2.7), approximated as a discrete sum with seven terms,

$$\psi_1(q) \approx \frac{1}{\sqrt{2\pi\hbar}} \sum_{n=1}^{7} e^{ip_n q/\hbar}\, \psi_1(p_n)\, \delta p_n, \tag{2.9}$$

where $\psi_1(p_n)$ is given in (2.6). Conversely, we can make a plane wave out of Gaussians. This can be done in many ways; the simple relation

$$\sqrt{\frac{\alpha_1}{\pi\hbar}} \int \exp\left[-\frac{\alpha_1(x-x')^2}{2\hbar} + \frac{ip_0 x}{\hbar} \right] dx' = \exp\left(\frac{ip_0 x}{\hbar} \right) \tag{2.10}$$

is shown in Figure 2.2 at the right side. This is our first example of decomposing an extended state in terms of localized Gaussian packets.

2.2.2 Integrals

Often we require the overlap of two wavefunctions or phase space distributions. Every quantum amplitude $\langle a|b\rangle$ is an overlap of two certainties, the certainty of being in state $|a\rangle$ with the certainty of being in state $|b\rangle$. The probability of being in $|a\rangle$ and also being found to be in $|b\rangle$ is just $|\langle a|b\rangle|^2$. An example is the probability (density) of being in the state $|\phi\rangle$ and also in the state $|q\rangle$ which is definitely at the position q is $|\langle q|\phi\rangle|^2$.

A useful Gaussian integral is

$$J = \int_{-\infty}^{\infty} d\mathbf{q} \exp\{-\mathbf{q} \cdot \mathbf{A} \cdot \mathbf{q} + \mathbf{b} \cdot \mathbf{q}\}, \qquad (2.11)$$

where $\mathbf{b}$ is a N-dimensional vector, and $\mathbf{A}$ is a symmetric $N \times N$ matrix. The result is

$$J = \left(\frac{\pi^N}{\det \mathbf{A}}\right)^{1/2} \exp\left\{\frac{1}{4}\mathbf{b} \cdot \mathbf{A}^{-1} \cdot \mathbf{b}\right\}. \qquad (2.12)$$

This may be used to derive the overlap of two completely general Gaussians of the form (2.1) (omitting the phases ϕ_0 and ϕ_1):

$$\int_{-\infty}^{\infty} \psi_{\alpha_1}^*(q)\psi_{\alpha_0}(q)\,dq = \frac{(4a_0a_1)^{1/4}}{\sqrt{\alpha_0 + \alpha_1^*}}$$

$$\exp\left[-\frac{\alpha_0\alpha_1^*(q_0 - q_1)^2}{2(\alpha_0 + \alpha_1^*)\hbar} - \frac{\hbar(p_0 - p_1)^2}{2(\alpha_0 + \alpha_1^*)} - i\frac{(q_0 - q_1)(\alpha_1^*p_0 + \alpha_0p_1)}{(\alpha_0 + \alpha_1^*)}\right]. \qquad (2.13)$$

2.3 PHASE SPACE PICTURES

2.3.1 Classical Phase Space

Phase space representations are powerful tools for understanding errors and formulating approximations in wave packet propagation methods. The Wigner phase space distribution, which we define below, is the "high road" to quantum phase space. However, most of the time a more schematic phase space description is best for developing intuition. The approach we take is to use phase space delta-densities to represent eigenstates of Hermitian operators, such as a position state $|q\rangle$, a propagated position state $|q(t)\rangle = \exp[-iHt/\hbar]|q\rangle$ or a energy eigenstate $|E\rangle$. A phase space Gaussian distribution represents Gaussian wave packets (this is the exact Wigner density for a Gaussian wave packet).

The schematic phase space dynamics we use is just the classical dynamics of a density of classical trajectories chosen appropriate to the initial state. Each trajectory may be viewed as a Dirac delta function phase space distribution,

$$\rho_t(q, p) = h\delta(q - q_t)\,\delta(p - p_t). \qquad (2.14)$$

We introduce Planck's constant as a unit of area in phase space. It is only a convenience, and makes $\rho_t(q, p)$ dimensionless. This choice gives a dimensionless phase space volume element $dp\,dq/h$. The distribution (2.14) is unit normalized, corresponding to unit probability of finding the system somewhere in phase space:

$$\int\limits_{-\infty}^{\infty}\int\limits_{-\infty}^{\infty} \frac{\rho_t(q,p)\,dp\,dq}{h} = 1. \tag{2.15}$$

Classical mechanics supports points in phase space (q_0, p_0) in the sense that they remain localized points (trajectories) (q_t, p_t) as time evolves. For a Hamiltonian $H(p, q)$ the classical trajectory (q_t, p_t) obeys Hamilton's equations of motion

$$\dot{q}_t = \frac{\partial H(q,p)}{\partial p},$$

$$\dot{p}_t = -\frac{\partial H(q,p)}{\partial q}. \tag{2.16}$$

Classical mechanics is equally at home with smooth distributions. The distribution represents a probability density as a function of location in phase space. Such distributions can be visualized as made up of a continuous density $\rho_0(q_0, p_0)$ of point trajectories:

$$\rho_0(q,p) = \int \rho_0(q_0, p_0)\delta(q - q_0)\,\delta(p - p_0)\,dq_0 dp_0. \tag{2.17}$$

This is a uniform density of points in phase space which carry a weight $\rho_0(q_0, p_0)$. Another possibility is to require the density of points to vary according to $\rho_0(q_0, p_0)$, leaving the weight of each point the same. Mathematically the two are equivalent, but numerically they lead to different procedures when the continuous density is approximated by discrete points (each point being the initial condition for a trajectory). The current position and momentum $(q_t(q_0, p_0), p_t(q_0, p_0))$ are functions of the initial conditions (q_0, p_0). The motion of the distribution $\rho_t(q, p)$ is determined by the motion of all the individual point trajectories $\delta(q - q_t(q_0, p_0))\,\delta(p - p_t(q_0, p_0))$:

$$\rho_t(q,p) = \int \rho_0(q_0, p_0)\delta(q - q_t(q_0, p_0))\,\delta(p - p_t(q_0, p_0))\,dq_0 dp_0$$

$$= \rho_0(q_{0,t}(q,p), p_{0,t}(q,p)), \tag{2.18}$$

where $(q_{0,t}(q,p), p_{0,t}(q,p))$ is the unique initial point in phase space corresponding to the final point (q, p). The density $\rho_t(q, p)$ is simply a swarm of classical trajectories weighted by the initial classical density $\rho_0(q_0, p_0)$. It is straightforward to show that $\rho_t(q, p)$ in both (2.14) and (2.18) obeys Liouville's equations

$$\frac{d\rho_t(q,p)}{dt} = \frac{\partial H(p,q)}{\partial q}\frac{\partial \rho_t(q,p)}{\partial p} - \frac{\partial H(p,q)}{\partial p}\frac{\partial \rho_t(q,p)}{\partial q}. \tag{2.19}$$

A distribution $\rho_t(q, p)$ can be thought of as a probability density for finding the system with momentum p and position q. The q (p) space probability is simply the integral over all momenta (positions):

$$P_t(q) = \frac{1}{h} \int \rho_t(q, p) \, dp,$$

$$P_t(p) = \frac{1}{h} \int \rho_t(q, p) \, dq. \tag{2.20}$$

We may consider $P_t(q)$ to be the "overlap" of the two distributions $\rho_t(q, p)$ and $\delta(q - q')$:

$$P_t(q) = \frac{1}{h} \int \rho_t(q', p') \, \delta(q - q') \, dp' dq'$$

$$= \frac{1}{h} \int \rho_t(q, p') \, dp'. \tag{2.21}$$

2.3.2 Wigner Phase Space

There is no substitute for the Wigner phase space distribution, although there are several cheap imitations. For example, the simple phase space density

$$\rho_\psi^{\text{cheap}}(q, p) = \sqrt{2\pi\hbar} \, \psi(q)\psi^*(p) \exp\left[\frac{ipq}{\hbar}\right], \tag{2.22}$$

where $\psi(p) \equiv \langle p|\psi\rangle$ is the momentum representation of the wave function $|\psi\rangle$ and has the desirable property that

$$|\psi(q)|^2 = \int \frac{\rho_\psi(q, p) \, dp}{h},$$

$$|\psi(p)|^2 = \int \frac{\rho_\psi(q, p) \, dq}{h}. \tag{2.23}$$

These imply

$$\int\int \frac{dp, dq}{h} \rho_\psi(q, p) = 1. \tag{2.24}$$

However, ρ_ψ^{cheap} is complex and badly oscillatory. Wigner defined a better phase space density which has many desirable properties. It is

$$\rho^W(q, p) = \left(\frac{h}{\pi\hbar}\right)\int_{-\infty}^{\infty} e^{-2ip\cdot s/\hbar}\,\psi^*(q-s)\psi(q+s)\,ds. \tag{2.25}$$

This obeys the rules (2.23). For example,

$$\frac{1}{h}\int_{-\infty}^{\infty} dp\,\rho^W(q, p) = \left(\frac{1}{\pi\hbar}\right)\int_{-\infty}^{\infty}\int_{-\infty}^{\infty} dp\,ds\,e^{-2ip\cdot s/\hbar}\,\psi^*(q-s)\psi(q+s)$$

$$= 2\pi\left(\frac{1}{\pi\hbar}\right)\int_{-\infty}^{\infty}\delta\left(\frac{2s}{\hbar}\right)e^{-2ip\cdot s/\hbar}\,\psi^*(q-s)\psi(q+s)\,ds$$

$$= |\psi(q)|^2. \tag{2.26}$$

More generally, ρ^W may come from a quantum density ρ_Q:

$$\rho^W(q, p) = 2\int_{-\infty}^{\infty} e^{2ip\cdot s/\hbar}\rho_Q(q+s, q-s)\,ds. \tag{2.27}$$

All of the equations are easily generalized to many dimensions. For example, the Wigner phase space density becomes

$$\rho_t^W(\mathbf{q}, \mathbf{p}) = 2^N\int_{-\infty}^{\infty} e^{2i\mathbf{p}\cdot\mathbf{s}/\hbar}\phi^*(\mathbf{q}+\mathbf{s})\phi(\mathbf{q}-\mathbf{s})\,d\mathbf{s}. \tag{2.28}$$

2.3.3 Gaussians in Phase Space

The Wigner phase space function for the Gaussian (2.1) is obtained by carrying out the integral (2.25), which results in

$$\rho_1^W(q, p) = 2\exp\left[-\frac{a_1}{\hbar}(q-q_1)^2 - \frac{1}{a_1\hbar}(p-p_1)^2 + \frac{2b_1}{\hbar a_1}(q-q_1)(p-p_1)\right]. \tag{2.29}$$

The Wigner phase space distribution of a Gaussian is itself a Gaussian in phase space. The position-momentum correlation appears in this form as a cross term in the exponent, and thus a skewing of the Gaussian in phase space.

Suppose that we start with a Gaussian aligned with the axes, $b_1 = 0$. If $a_1 = 1$, then the ellipse representing the Gaussian degenerates to a circle. If $a_1 > 1$, the major axis of the ellipse stands vertical; if $a_1 < 1$, the major axis is horizontal. If we rotate the ellipse by an angle $\bar{\theta}$, then it is possible to show that

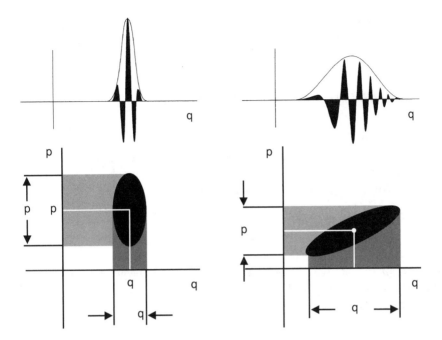

Figure 2.3 Gaussians depicted in coordinate and phase space.

$$a_\theta = \frac{a_1}{d(\theta)}, \quad b_\theta = \frac{(1 - a_1^2)\,\sin\theta\,\cos\theta}{d(\theta)}, \tag{2.30}$$

where

$$d(\theta) = \cos^2\theta + a_1^2\sin^2\theta. \tag{2.31}$$

Note that b_θ remains zero at all angles if $a_1 = 1$.

There is a twofold redundancy in this parameterization, since a Gaussian of the same shape can come for example, from a small $a_1 < 1$ and rotation by $\overline{\theta}$, or a large $a_1 > 1$ and rotation by $\pi/2 - \theta$.

Two of these cases are seen in Figure 2.3, with a schematic diagram of the Wigner phase space density of the Gaussian and the corresponding wave function in coordinate space. On the upper left, we see a narrow Gaussian with no position-momentum correlation, and below its phase space representation. The major axis of the ellipse is parallel to the momentum axis, and the "shadows" on the p and q axes are a minimum, in the sense of that

$$\Delta q \Delta p = \frac{\hbar}{2}. \tag{2.32}$$

Minimum uncertainty is achieved only with no correlation. On the right is an analogous diagram for $b_1 \neq 0$. The correlation of position with momentum is obvious, with the slower component correlating with smaller q in this case. Here $\Delta q \Delta p > \hbar/2$.

2.4 LINEAR AND LINEARIZED DYNAMICS

2.4.1 Linear Dynamics

Now that some of the properties of complex Gaussians have been discussed, we can look at their dynamics. The focus for the moment is on exactly solvable cases. The task is to solve the time dependent Schrödinger equation

$$i\hbar \frac{\partial \psi_1(q, t)}{\partial t} + \frac{\hbar^2}{2m} \frac{\partial^2 \psi_1(q, t)}{\partial q^2} - V(q, t)\, \psi_1(q, t) = 0, \tag{2.33}$$

where $\psi_1(q, 0) = \psi_1(q)$ is a wave packet that starts out as the general Gaussian (2.1). We suppose that the potential is at most quadratic in the coordinate q, but otherwise quite general and even time dependent:

$$V(q, t) = V(q_t, t) + V'(q_t, t)(q - q_t) + \frac{1}{2} V''(q_t, t)(q - q_t)^2. \tag{2.34}$$

Alternatively, we suppose that the wave packet is narrow enough and the potential smooth enough that we can expand the potential quadratically around the (moving) center of the wavepacket q_t. This again results in a potential of the form (2.34).

Since the potential is at most quadratic, it can be expanded about any point; we choose that point to be the parameter q_t. The strategy is to plug the Gaussian form (2.1) into (2.33), assuming that all the parameters $\alpha_{1_t}, q_{1_t}, p_{1_t},$ and ϕ_{1_t} are time dependent. All the terms that result carry the common prefactor of the Gaussian, so it may be ignored. We compare like powers of $(q - q_{1_t})$; this imposes conditions on the parameters $\alpha_{1_t}, q_{1_t}, p_{1_t},$ and ϕ_{1_t} if the time dependent Schrödinger equation is to be satisfied. It is convenient to define $s(t) \equiv \phi_{1_t}$ and $-iA_t/2 \equiv \alpha_{1_t}$. The resulting equations are (Heller 1975, 1976a, b)

Quadratic term:

$$\dot{A}_t = \frac{-2 A_t^2}{m} - \frac{V''(q_t, t)}{2}. \tag{2.35}$$

Linear term:

$$\dot{q}_t = \frac{p_t}{m},$$

$$\dot{p}_t = -V'(q_t, t). \tag{2.36}$$

Constant term:

$$\dot{s}_t = -V(q_t, t) + \frac{i\,\hbar A_t}{m} + p_t\,\dot{q}_t - \frac{p_t^2}{2m}. \tag{2.37}$$

Small adjustments will bring these expressions into more familiar and useful form. Equations (2.36) are just Hamilton's equations of motion for a trajectory which is the expectation value of the position and momentum of the wave packet. This connection was first developed in Heller (1976a, b) and it is given a group theoretical interpretation in (Littlejohn 1986; Littlejohn and Robbins 1987).

We can tame the nonlinear differential equation and reveal an intimate connection between the equation for A_t and the equations for nearby orbits in classical mechanics for A_t by writing

$$A_t \equiv \frac{1}{2}\frac{P_{zt}}{Z_t}, \tag{2.38}$$

which gives, after substitution into (2.35),

$$\dot{P}_{zt} = -V''(q_t, t)Z_t = -\left(\frac{\partial^2 V(q, t)}{\partial q^2}\right)_{q_t} Z_t,$$

$$\dot{Z}_t = \frac{P_{zt}}{m}. \tag{2.39}$$

Since one variable has been replaced by two, there is more than one pair (Z, P_z) corresponding to a given α. It turns out that at the initial time, one is free to choose any (Z, P_z) which correspond to the given initial α_0, with no effect on the physics. (There is a kind of gauge structure tied up in this fact.) These are just the (linear) equations of motion of a harmonic oscillator with (time dependent) force constant $V''(q_t, t)$, if Z_t is the position of the oscillator and P_{zt} is the momentum.

This can be put in matrix form:

$$\frac{d}{dt}\begin{pmatrix} P_Z \\ Z \end{pmatrix} = \begin{pmatrix} 0 & -V''(q_t, t) \\ m^{-1} & 0 \end{pmatrix}\begin{pmatrix} P_Z \\ Z \end{pmatrix} \tag{2.40}$$

or

$$\frac{d}{dt}\begin{pmatrix} P_Z \\ Z \end{pmatrix} = K(t)\begin{pmatrix} P_Z \\ Z \end{pmatrix}. \tag{2.41}$$

This is a *linear* differential equation for P_Z and Z. If K is time independent, namely the potential is constant in time, then it can be integrated to give

$$\begin{pmatrix} P_Z \\ Z \end{pmatrix} = M(t) \begin{pmatrix} P_{Z_0} \\ 1 \end{pmatrix}, \tag{2.42}$$

where

$$M(t) = \exp[Kt]. \tag{2.43}$$

The boundary condition $Z_0 = 1$ has been used. Only the ratio P_Z/Z is important; $Z_0 = 1$ forces the boundary condition $P_Z = 2A_0$. If K is time dependent, the equation (2.42) still holds. However, $M(t)$ cannot be integrated so cleanly as in equation (2.56); it usually must be found by integrating the equations of motion (2.40) numerically.

To proceed with the remaining parameters, we set

$$L_t = -V(q_t, t) + p_t \dot{q}_t - \frac{p_t^2}{2m}$$

$$= p_t \dot{q}_t - E_t, \tag{2.44}$$

where $E_t = p_t^2/2m + V(q_t, t)$ is the energy and $L_t = p_t \dot{q}_t - E_t$ is the Lagrangian for the trajectory (p_t, q_t). Now we can write (2.37) as

$$\dot{s}_t = i\hbar \frac{1}{2m} \frac{P_{zt}}{Z_t} + L_t$$

$$= i\hbar \frac{1}{2} \frac{\dot{Z}_t}{Z_t} + L_t, \tag{2.45}$$

which can be integrated to give

$$s_t = s_0 + i\hbar \frac{1}{2} \mathrm{Tr}\,[\log(Z_t)] + S_t,$$

$$S_t = \int_0^t L_t\, dt. \tag{2.46}$$

The trace in this last equation is superfluous in one dimension, but it shows how the phase-normalization factor s_t is obtained in the N-dimensional case.

In multidimensional form, the Gaussian wave packet is given by

$$\psi_1(\mathbf{q},\, t) = \exp\left\{ \frac{i}{\hbar} [(\mathbf{q} - \mathbf{q_t}) \cdot \mathbf{A}_t \cdot (\mathbf{q} - \mathbf{q_t}) + \mathbf{p_t} \cdot (\mathbf{q} - \mathbf{q_t}) + s_t] \right\}, \tag{2.47}$$

where $\mathbf{A}_t$ is an N-dimensional matrix for N coordinates, and $\mathbf{q}, \mathbf{q_t}, \mathbf{p_t}$ are N-dimensional vectors. The parameters obey

$$\frac{d}{dt}\mathbf{q_t} = \nabla_p H, \tag{2.48}$$

$$\frac{d}{dt}\mathbf{p_t} = -\nabla_q H, \tag{2.49}$$

$$\mathbf{A_t} = \frac{1}{2}\mathbf{P_Z} \cdot \mathbf{Z}^{-1}, \tag{2.50}$$

$$\frac{d}{dt}\begin{pmatrix}\mathbf{P_Z} \\ \mathbf{Z}\end{pmatrix} = \begin{pmatrix}\mathbf{0} & -\mathbf{V}''(t) \\ \mathbf{m}^{-1} & \mathbf{0}\end{pmatrix}\begin{pmatrix}\mathbf{P_Z} \\ \mathbf{Z}\end{pmatrix}, \tag{2.51}$$

$$\dot{s}_t = L_t + \frac{i\hbar}{2}\,\mathrm{Tr}[\dot{\mathbf{Z}} \cdot \mathbf{Z}^{-1}], \tag{2.52}$$

or

$$s_t = s_0 + S_t + \frac{i\hbar}{2}\,\mathrm{Tr}[\ln\mathbf{Z}]. \tag{2.53}$$

$\mathbf{V}''$ and $\mathbf{m}^{-1}$ are N-dimensional matrices of mixed second derivatives of the Hamiltonian with respect to position and momentum coordinates, respectively. That is,

$$[\mathbf{V}'']_{ij} = \frac{\partial^2 H}{\partial q_i \partial q_j}, \tag{2.54}$$

and so forth. Also

$$\begin{pmatrix}\mathbf{P_Z} \\ \mathbf{Z}\end{pmatrix} = \mathbf{M}(t)\begin{pmatrix}\mathbf{P_{Z_0}} \\ 1\end{pmatrix}, \tag{2.55}$$

where from (2.51) and (2.55) we have

$$\frac{d\mathbf{M}(t)}{dt} = \begin{pmatrix}\mathbf{0} & -\mathbf{V}''(t) \\ \mathbf{m}^{-1} & \mathbf{0}\end{pmatrix}\mathbf{M}(t). \tag{2.56}$$

$\mathbf{M}(t)$ is the stability matrix,

$$\mathbf{M}(t) = \begin{pmatrix}m_{11} & m_{12} \\ m_{21} & m_{22}\end{pmatrix}, \tag{2.57}$$

where $m_{11} = \partial p_t / \partial p_0$, and so forth.

2.4.2   Special Cases of Linear Dynamics

We discuss some important special cases of wave packet motion on time independent potentials at most quadratic in q. Free particle dynamics, motion on a linear ramp potential $V(q) = \beta q$, and harmonic motion $V(q) = \pm \frac{1}{2} m\omega^2 q^2$ are examples of linear dynamics where the current positions and momenta are linear function of the initial positions and momenta.

Free Particle   For a free particle with Hamiltonian $H = p^2/2m$ we have

$$q_t = q_0 + \frac{p_0}{m} t, \tag{2.58}$$

$$p_t = p_0. \tag{2.59}$$

This is already tantamount to solving the stability equations, but since we have a time independent Hamiltonian, we can use (2.56),

$$M(t) = \exp[Kt] \tag{2.60}$$

$$M = \exp\left[\begin{bmatrix} 0 & 0 \\ \dfrac{1}{m} & 0 \end{bmatrix} t \right] = \begin{pmatrix} 1 & 0 \\ \dfrac{t}{m} & 1 \end{pmatrix}. \tag{2.61}$$

Thus

$$P_Z = P_{Z_0} Z = 1 + \frac{P_{Z_0} t}{m}, \tag{2.62}$$

so A_t is

$$A_t = P_Z \frac{A_0}{1 + 2A_0 t/m}. \tag{2.63}$$

The classical action integral becomes

$$s_t = s_0 + \frac{p_0^2}{2m} t + \frac{i\hbar}{2} \mathrm{Tr}\left[\ln\left(1 + \frac{2A_0 t}{m}\right) \right]. \tag{2.64}$$

Every parameter in the Gaussian equation (2.1) is now determined.

Figure 2.4 shows the time evolution of the real part of a free Gaussian wave packet. At first, the wavelength is short to the left of the wave packet, corresponding to fast

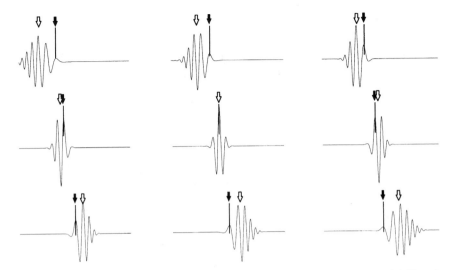

Figure 2.4 Real part of a Gaussian wave packet propagating in a constant potential. Note the position-momentum correlation, which changes during the motion. The wave packet first narrows and then broadens. The phase velocity is less than the group velocity. In a broad wave packet these differ by a factor of two.

motion. This position-momentum correlation causes a narrowing of the wavepacket with time, as the faster part catches up with the middle, and the slower part on the right is caught. The wave packet reaches minimum size and momentarily becomes a minimum uncertainty wave packet with $b_{1t} = 0$. Following this, after leading, the position-momentum correlation reverses and maintains the faster part (shorter wavelength) forever. The wavepacket spreads indefinitely thereafter. Figure 2.4 also shows the distinction between the *phase velocity* and the *group velocity* of the wave packet. The latter is the velocity of the probability density of the wave packet, which we already know is p_t/m. The phase velocity is slower for the time dependent Schrödinger equation. The black arrow points to a local maximum, which is a place of constant phase of the wave from frame to frame in Figure 2.4. The open arrow shows the motion of the center of the wave packet.

Suppose that we take the wavepacket to be very broad, $A_0 \to 0$. This is locally a traveling plane wave. The imaginary part of the exponential becomes, omitting the quadratic term,

$$p_0 \left[q - \left(q_0 + \frac{p_0 t}{m} \right) \right] + \frac{p_0^2}{2mt} = p_0 \left(q - q_0 - \frac{p_0 t}{2m} \right). \tag{2.65}$$

Keeping the phase constant in time means moving q with speed $p_0/2m$. This means that the phase velocity is half the group velocity, p_0/m. In the move of a free particle wave packet, the wavefronts fall behind as the wave packet moves forward.

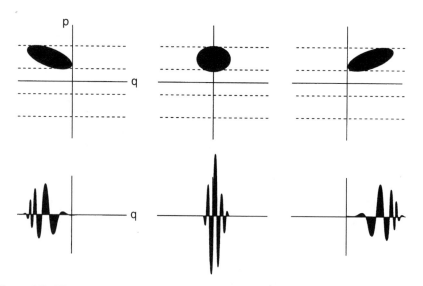

Figure 2.5   Phase space and coordinate space pictures of a Gaussian evolving on a constant potential.

Figure 2.5 shows a Gaussian wave packet with an initial position-momentum correlation for a constant potential. The correlation is revealed by a skewed ellipse in phase space. This ellipse is a filled contour of the corresponding Wigner phase space distribution. Since the center of the distribution has a positive momentum, it moves (on the average) to the right. There is a shear in the phase space distribution corresponding to trajectories with higher momentum traveling farther. This causes an initial narrowing of the wave packet in coordinate space initially, since for this wave packet the faster particles started out in the rear. After the minimum uncertainty wave packet is reached, the faster trajectories continue and give the reverse position-momentum correlation, with the faster trajectories in the lead. The dashed lines show the contours of constant energy. Since the energy of the trajectories is constant, the distribution is confined by the same contours at all times. The phase space distribution has area h at all times.

We now consider the free particle Green's function. As mentioned above, the Green function is closely related to a very narrow initial wave packet. In phase space the "wavefunction" $\delta(q - q')$ is an infinite vertical line. Because the potential is a quadratic form, the phase space distribution evolves classically, with p_t and q_t given by the linear equations, (2.58). This motion is a shear that rotates and stretches the line, which becomes $p(q) = m(q - q')/t$ for $t > 0$.

As soon as the tilting line corresponds to classical trajectories getting to q in time t with velocity $(q - q')/t$ in Figure 2.6. The Green function for $t > 0$ is an undamped oscillating function whose oscillations are more rapid away from the initial position q', in accord with the local deBroglie wavelength given by $\lambda(q, t) h\, t/|(q - q')|m$.

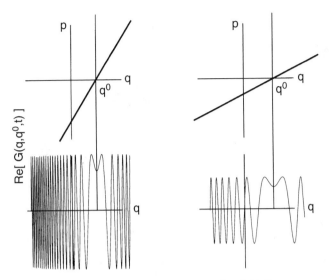

Figure 2.6 Phase space and coordinate space pictures of an initially very narrow Gaussian evolving on an constant potential.

Linear Potential For a particle on a linear ramp with Hamiltonian $H = p^2/2m+\beta q$, we have

$$p_t = p_0 - \beta t$$

$$q_t = q_0 + \frac{p_0}{m} t - \frac{\beta}{2m} t^2,$$

$$s_t = s_0 + \frac{1}{2} i\hbar \ln\left[1 + \frac{P_{Z_0} \cdot t}{m}\right] + \frac{p_0^2}{2m} t + \beta q_0 t \qquad (2.66)$$

$$- \frac{\beta p_0}{m} t^2 + \frac{\beta^2}{3m} t^3.$$

Note that since again $V'' = 0$ the equations for P_Z and Z are unaffected as compared to the constant potential (free particle). Thus A_t is again given simply by (2.63).

The result in phase space is shown in Figure 2.7. The center of the wave packet in position and momentum is now governed by the constant acceleration of the ramp, but since the second derivative vanishes in both, the stability equations of the linear and the constant are identical. Thus the *shape* of the phase space distribution is the same at equal times, but the location in phase space and a plot of the real part of the wavefunction is quite different because q_t and p_t are different. Again the contours of constant energy bound the phase space density.

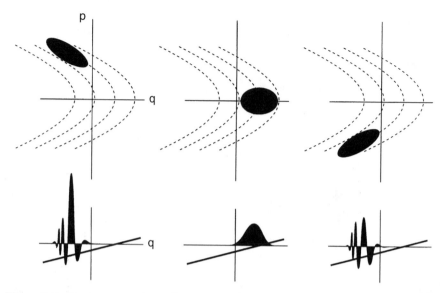

Figure 2.7 Phase space and coordinate space pictures of a Gaussian evolving on a linear potential.

Harmonic Oscillator For a harmonic oscillator with Hamiltonian $H = \frac{1}{2m}p^2 + \frac{1}{2}m\omega^2 q^2$, the standard result, easily verified, is that

$$p_t = p_0 \cos(\omega t) - m\omega q_0 \sin(\omega t),$$

$$q_t = q_0 \cos(\omega t) + \left(\frac{p_0}{m\omega}\right)\sin(\omega t), \tag{2.67}$$

solve the classical equations of motion for the guiding trajectory (q_t, p_t). The 2×2 matrix M is given by

$$M = \exp[Kt] = \exp\left[\begin{pmatrix} 0 & -m\omega^2 \\ \dfrac{1}{m} & 0 \end{pmatrix} t\right]$$

$$= \begin{pmatrix} \cos(\omega t) & -m\omega \sin(\omega t) \\ \dfrac{\sin(\omega t)}{m\omega} & \cos(\omega t) \end{pmatrix}, \tag{2.68}$$

while the phase is

$$s_t = s_0 + \frac{1}{2}\left[p_t q_t - p_0 q_0\right] + \frac{i\hbar}{2}\,\text{Tr}[\log Z]. \tag{2.69}$$

This specifies all the parameters in the Gaussian. From it we calculate A_t as

$$A_t = \frac{1}{2}\,\frac{2A_0\cos(\omega t) - m\omega\sin(\omega t)}{(2A_0/m\omega)\sin(\omega t) + \cos(\omega t)}. \tag{2.70}$$

It is easily seen that A_t oscillates periodically with twice the frequency of the oscillator, unless $A_0 = im\omega/2$, which makes the Gaussian wave packet a "coherent state" of the oscillator. A coherent state is simply the ground state of the oscillator displaced in position and "boosted" in momentum. The displacement by q_0 is accomplished by substituting $q - q_0$ for q in the ground state, and the boost is done by multiplying the ground state by $\exp[ip_0(q - q_0)/\hbar]$. In the special case where $A_0 = im\omega/2$,

$$s_t = s_0 + \frac{1}{2}\left[p_t q_t - p_0 q_0 - \omega t\right]. \tag{2.71}$$

If the displacement and boost are zero, then $q_0 = p_0 = q_t = p_t = 0$, $s_t = s_0 - \frac{1}{2}\omega t$, $A_t = A_0$. The coherent state wave packet *is* the harmonic oscillator ground state eigenfunction. If the Gaussian wave packet starts out as very wide compared to the coherent state, it becomes narrow after one-quarter of a period, wide again after one-half vibrational period, and so forth, as seen in (2.70). In between these mile posts, it develops position-momentum correlation, as evidenced by a complex A_t. The behavior will stand out more clearly when we look at the phase space picture of the dynamics.

Figure 2.8, shows an initial wave packet on the left which is much wider than the coherent state of the oscillator. The coherent state would be circular in this diagram, since the classical energy contours are circular. All the trajectories travel in circles around the origin in phase space with an angular velocity that is independent of energy. This means that the ellipse rigidly rotates around the origin as shown. From the position and orientation of the ellipse at any moment we can sketch the wavefunction as is seen in each panel. Since the dynamics are linear, the classical dynamics of the initial Wigner density are identical to the quantum evolution of the Wigner density. As stated earlier in connection with (2.70), we see the width of the wave packet oscillating at twice the frequency of the classical oscillator. Note the wave packet gets narrow after one-quarter of a period and again at three-quarters of a period, and so forth.

Unstable Harmonic Oscillator   If the potential is a quadratic *barrier* rather than a well, we have a pure imaginary frequency, $\omega \to i\omega$, $V(q) \to -\frac{1}{2}m\omega^2 q^2$. In (2.67) we get $\cos(\omega t) \to \cosh(\omega t)$, $\sin(\omega t) \to i\sinh(\omega t)$, giving

$$p_t = p_0\cosh(\omega t) + m\omega q_0\sinh(\omega t),$$

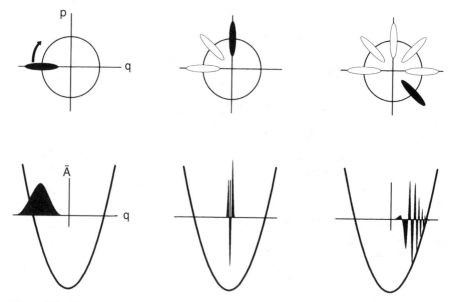

Figure 2.8   Phase space and coordinate space pictures of a Gaussian evolving on an harmonic oscillator potential.

$$q_t = q_0 \cosh(\omega t) + (p_0/m\omega) \sinh(\omega t),$$

$$P_Z = P_{Z_0} \cosh(\omega t) + m\omega \sinh(\omega t),$$

$$Z = \cosh(\omega t) + (P_{Z_0}/m\omega) \sinh(\omega t). \tag{2.72}$$

The phase term carries over directly,

$$s_t = s_0 + \frac{1}{2}[p_t q_t - p_0 q_0] + \frac{i\hbar}{2} \mathrm{Tr}[\log Z], \tag{2.73}$$

with equation (2.72) giving p_t, and so forth. It follows that

$$A_t = \frac{1}{2} \frac{2A_0 \cosh(\omega t) + m\omega \sinh(\omega t)}{(2A_0/m\omega) \sinh(\omega t) + \cosh(\omega t)}. \tag{2.74}$$

This example is a prototype of the important subject of unstable motion near a periodic orbit (the point $q = p = 0$ in this example.) It is only a slight extension to add another, noninteracting stable degree of freedom to the unstable harmonic oscillator, making a two degree of freedom system with an unstable periodic orbit. We will do this in Section (2.7), in a study of a scarring caused by periodic orbits.

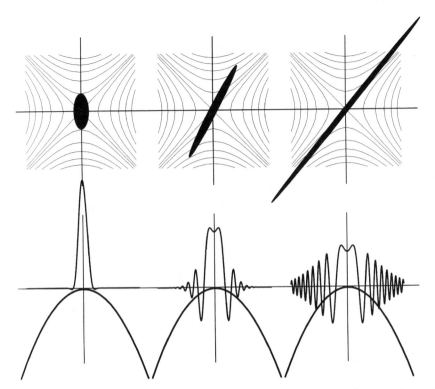

Figure 2.9 Phase space and coordinate space pictures of a Gaussian evolving on an inverted harmonic oscillator.

The wavefunction of the inverted harmonic oscillator was not illustrated before; it is shown in Figure 2.9 at three times along with the corresponding phase space density. The exponential escape of trajectories on the inverted barrier leads to very rapid spreading of the wave packet along the *unstable* axis.

2.5 NONLINEAR DYNAMICS

The examples above are exact solutions to linear dynamics; they are also semiclassical solutions for such linear systems, as we will see momentarily. When the dynamics is nonlinear, as it is for a Morse oscillator or worse a chaotic system, we have many choices as to how to approximate the wave packet dynamics. One simple approximation follows when we suppose that the wave packet is fairly confined in coordinate space, and further that the potential is *locally* nearly expressible as a quadratic function of coordinate. Then *the quadratic potential equation (2.34) is the result of expanding the potential quadratically about the instantaneous center of the wave packet.* If the wave packet is narrow enough, this gives a good representation of what the wave packet "feels." If the wave packet starts out Gaussian, it will remain

Gaussian in this approximation, since it sees an effective time dependent harmonic potential. The equations of motion governing its motion are just equations (2.48). Therefore the center of the Gaussian q_t and its momentum p_t travel completely classically. The time dependent potential arises from a continuous re-expansion of the potential about the moving center of the wave packet. We call this the "thawed" Gaussian wave packet approximation, or TGA (Heller 1975, 1976a, b). It represents the use of a propagator linearized around the instantaneous center of the wave packet. It is very important to recognize that there is a different effective potential for different Gaussian packets, and thus if we have broken up an extended state in terms of many Gaussian pieces, we have a different propagator for each one. This is not as dangerous as it sounds and in fact is an extremely powerful idea that resurfaces often.

The errors clearly will be largest in the wings of the wave packet and will of course be worse if the wave packet is wide, and/or if the anharmonicity is large. Since wave packets eventually tend to spread, the accumulation of errors limits the approximation to relatively short times. Figure 2.10 shows the main idea. The TGA leads to other approximations, such as simple formulas for absorption and Raman spectra (Lee and Heller 1979; Meyers et al. 1982), and relatively accurate description of intrinsically short time photodissociation (Lee and Heller 1982), and scattering (Heller 1975).

We state without proof that in the limit $\hbar \to 0$, the TGA becomes exact for any finite time. It is therefore a semiclassical method, although not necessarily optimal for a given situation at finite $\hbar$. The exactness as $\hbar \to 0$ is clear if we note that the wave packet width can be taken to scale as $\sqrt{\hbar}$, for example, making the local quadratic approximation arbitrarily accurate as $\hbar \to 0$. If the dynamics are unstable, the wave packet will spread exponentially fast so that we only extend the accurate time domain logarithmically in $\hbar$ as $\hbar \to 0$. Nonetheless, small $\hbar$ eventually wins out for any finite time.

If the initial wavefunction is not a single Gaussian, we can write it as a superposition of such Gaussians and propagate each one as if alone, banking on the linearity of the Schrödinger equation. We have already seen the expansion of a plane wave in terms of Gaussians. For a bound state of a harmonic oscillator, suppose that we begin not in the ground state, or a Gaussian wave packet, but in the 40th eigenstate. We can write this state as a known superposition of Gaussians (Heller 1975, 1976a,

Figure 2.10 Motion of a wave packet on an anharmonic potential energy surface is approximated by locally expanding the potential to second order about the instantaneous center of the wave packet. This gives an effective, time dependent harmonic potential, so the wave packet will remain Gaussian. The equations of motion are just those given in (2.48).

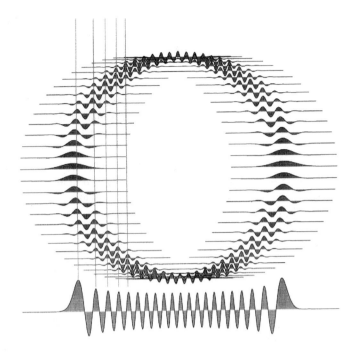

Figure 2.11 Representation of a harmonic oscillator eigenstate for $N = 40$ as a superposition of 60 Gaussian wave packets.

b), as Figure 2.11 illustrates. It may be shown that the *normalized* harmonic oscillator eigenstate is expressible as

$$\psi_n(q) = \left(\frac{m\omega}{\pi\hbar}\right)^{1/4} \frac{\sqrt{n!\,\omega}}{2\pi}\, e^{\mu/2}\, \mu^{(1-n)/2} \int_0^{2\pi} e^{i(n+1/2)\omega t} g(q,\, t)\, dt, \qquad (2.75)$$

where

$$g(q,\, t) = \exp\left[-\frac{m\omega}{2\hbar}\left(q - \frac{p_0}{m\omega}\sin\omega t\right)^2 + i\frac{p_0}{\hbar}\cos\omega t\left(q - \frac{p_0}{m\omega}\sin\omega t\right)\right.$$

$$\left. + \frac{i}{2\hbar}\left(\frac{p_0^2}{m\omega}\cos\omega t \sin\omega t - \hbar\,\omega t\right)\right]. \qquad (2.76)$$

Figure 2.11 is a 60 term numerical quadrature of this expression. Generally, we can take a more numerical approach to express an arbitrary initial state as a superposition of Gaussians (Davis and Heller 1979, 1981).

It is beyond the scope of this chapter to discuss the insights that wave packet approaches have provided to time dependent formulations of spectroscopy. A review is provided in Heller (1981). Much of spectroscopy is governed by relatively short time events, including even high-resolution resonance Raman spectra, and these yield especially useful intuition and formulas if TGA wave packet propagation is used to derive them.

In applications of wave packet approaches to spectroscopy and collision theory, a confusion often arises which we will clear up now. The important point is that time dependent formulation of certain processes (e.g., Raman scattering) is separate from *implementation* in terms of wave packet approaches. Thus, for example, the time dependent formulation of electronic resonance and near-resonance Raman spectroscopy is completely independent of Gaussian wave packet methods, even though the latter are often convenient tools for the implementation of the former.

2.5.1 Extensions

Among the possible ways to improve on the TGA, one could imagine the Gaussian to be one member of a basis set, and apply the time dependent variational principle to propagate the wavefunction (Heller 1976a). The question arises as to what the other members of the basis are. One possibility, considered in (Heller 1976a), is other Gaussian wave packets are launched along with the initial one. These auxiliary packets have no amplitude at first, but the Dirac-McLaughlin variational principle is used to determine the time evolution of the coefficients. This has its drawbacks, including proliferation of basis functions.

Perhaps the most systematic approach is that of Coalson and Karplus (1982, 1990), in which the natural extension to expansions in terms of Hermite polynomials multiplying a single Gaussian is developed. The equations of the Hermites analogous to the FGA are developed in Coalson and Karplus (1982, 1990).

Metiu and coworkers developed ideas for extending the Gaussian wave packet methods to surface crossing situations and have over the years made many innovative applications of wave packet ideas (Heather and Metiu 1986, 1989; Sawada and Metiu 1989).

2.6 VAN VLECK–GUTZWILLER PROPAGATOR AND GENERALIZED GAUSSIAN PROPAGATOR

The Van Vleck propagator (Gutzwiller 1967, 1990) is the stationary phase approximation to the exact Feynman path integral, or equivalently it is the propagator obtained by neglecting terms of order $\hbar^2$ and higher in the expansion of the quantum propagator. The job of the semiclassical Van Vleck propagator is to give the amplitude for starting at q' at time $t = 0$ and arriving at q at time t. Naturally we need trajectories that do this as input. In general, there may be many of them; they differ in their initial momentum for fixed q'. Later Gutzwiller gave the final form (Gutzwiller 1967, 1990), showing how to extend $G_{sc}(\mathbf{q}, \mathbf{q}'; t)$ to situations involving conjugate points (places where trajectories re-focus after having been initially at a given q), and multiple paths, and for that reason we call it the propagator the Van Vleck–Gutzwiller propagator, or VVG. The VVG propagator is

$$G(\mathbf{q}, \mathbf{q}'; t) \approx G_{sc}(\mathbf{q}, \mathbf{q}'; t)$$

$$= \left(\frac{1}{2\pi i\hbar}\right)^{d/2} \sum_j \left| \det\left(\frac{\partial^2 S_j(\mathbf{q}, \mathbf{q}'; t)}{\partial\mathbf{q}\partial\mathbf{q}'}\right) \right|^{1/2}$$

$$\times \exp\left(\frac{iS_j(\mathbf{q}, \mathbf{q}'; t)}{\hbar} - \frac{i\pi\nu_j}{2}\right). \qquad (2.77)$$

The sum over j is over all the above-mentioned trajectories connecting $\mathbf{q}'$ to $\mathbf{q}$ in time t; d is the number of degrees of freedom. The determinant is in fact the square root of the classical probability of starting at $\mathbf{q}'$ at time $t = 0$ and ending at $\mathbf{q}$ at time t. This determinant involves the stability matrix and thus the linearized dynamics in the vicinity of each trajectory connecting q with q' in time t. We can write, in one dimension,

$$\left|\frac{\partial^2 S_j(q, q'; t)}{\partial q\partial q'}\right|^{1/2} = \left|\frac{\partial p}{\partial q'}\right|^{1/2} = |m_{21}|^{-1/2}. \qquad (2.78)$$

The phase is determined by the classical action $S_j(\mathbf{q}, \mathbf{q}'; t)$ and the Maslov index ν_j (see below). The classical action is the time integral of the Lagrangian $\mathcal{L}$:

$$S_j(\mathbf{q}, \mathbf{q}'; t) = \int_0^t dt' \, \mathcal{L} = \int_0^t dt' \{\mathbf{p}(t') \cdot \dot{\mathbf{q}}(t') - H(\mathbf{p}(t'), \mathbf{q}(t'))\} \qquad (2.79)$$

along the jth classical path. H is the classical Hamiltonian which is presumed to be the classical limit of $\hat{H}$. Equation (2.77) was originally written down by Van Vleck in 1928 without the summation or index ν and was thus fundamentally limited to extremely short times at best.

2.6.1 Generalized Semiclassical Gaussian Wave Packet Dynamics

We first derived the TGA directly from quadratic potentials in the time dependent Schrödinger equation. Just as above, we recall the form of the semiclassical VVG propagator. Suppose that we apply the VVG propagator to an initial Gaussian: Does this give FGA? The answer is no, although the TGA and the semiclassical VVG propagator are closely related. Both depend on the same linearized dynamics in the vicinity of reference trajectories, but the VVG expression is more general and more accurate (although more difficult to implement).

There ought to be a full semiclassical analogue to the VVG propagator that does for coherent states what the VVG does for the position representation. Indeed there is such a coherent state semiclassical propagator, which we called generalized Gaussian wave packet dynamics, or GGWPD (Heller 1976c; Huber, Heller, and Littlejohn 1988; Weissman 1982; Klauder 1986). The usual representations, such as position, momentum, or action, are eigenstates of Hermitian operators. Coherent state and Gaussian wave packets are not, and this necessitates some complexification of classical mechanics for a proper semiclassical theory.

For simplicity we work in one dimension and follow Huber, Heller, and Littlejohn (1988), see also Heller (1976c). All the results are easily extended to higher dimensions. We call q_0 and p_0 the "center" position and momentum, and at first we take them to be real (in which case they are the expectation values of the q and p operators). The parameters α_0 and c_0 may be complex (with $\mathcal{R}\alpha > 0$, of course). We note that this initial wave packet can be written in the form $<q|g; q_0, p_0> = e^{iS_0(q;q_0,p_0)/\hbar}$, where the initial action function S_0 is given by

$$S_0(q; q_0, p_0) = i\alpha_0(q - q_0)^2 + p(q - q_0) - ic_0. \tag{2.80}$$

Therefore the initial Gaussian is in the form of an initial wave function in WKB theory, with the unusual feature that the initial action is complex. Note that S_0 is a simple quadratic function of q.

The parameters q_0, p_0, α_0, and c_0 uniquely determine the Gaussian, but the converse is not true, if we allow q_0 and p_0 to be complex. The function S_0, and therefore also the Gaussian wave function itself, is left invariant if q_0, p_0, c_0 are replaced by any values $\bar{q}, \bar{p}, \bar{c}$ that satisfy

$$2\alpha_0 q_0 + ip_0 = \text{const.} = 2\alpha_0\bar{q} + i\bar{p} \tag{2.81}$$

and

$$\bar{c} = c_0 + \alpha_0(\bar{q}^2 - q_0^2) + i(p_0\bar{q} - \bar{p}q_0). \tag{2.82}$$

(S_0 is the same function of q as before, under this replacement.) Here q_0, p_0, and c_0 are considered fixed. We make no replacement of the parameter α_0 because it cannot be changed if S_0 is to remain invariant. We will call the set of points $(\bar{q}, \bar{p})$ that satisfy equation (2.81) the "ket manifold" of points equivalent to (q_0, p_0); note that $(\bar{q}, \bar{p})$ will in general be complex.

Let us now consider the usual initial value problem in WKB theory, in which the initial action $S_0(q)$ is real. The goal of this theory is to find the final action function $S(q, t)$ and amplitude $A(q, t)$ at some final time t such that $\psi(q, t) = A(q, t)e^{iS(q,t)/\hbar}$. (We may take the initial amplitude to be unity, as it is with our initial Gaussian, where the normalization is absorbed in the complex c_0.) The first step in achieving this goal is to consider the initial momentum function

$$p_0(q) = \frac{\partial S_0(q)}{\partial q}, \tag{2.83}$$

which gives an initial momentum $p_i = p_0(q_i)$ for each initial position q_i. The set of initial points $(q_i, p_i) = (q_i, p_0(q_i))$ is a surface in phase space, known as a Lagrangian manifold. (A Lagrangian manifold is defined as a surface on which p as a function of q is a perfect gradient so that the integral of $p\, dq$ on this surface is path independent. In one degree of freedom, a (real) Lagrangian manifold is simply a curve in the two-dimensional phase space. Indeed, since every function in one variable is a perfect gradient, all curves in a two-dimensional phase space are Lagrangian manifolds. In higher dimensions the perfect gradient condition is more restrictive.)

If we apply this idea to our initial Gaussian, ignoring the fact that S_0 is now complex, we find that

$$p_0(q) = 2i\alpha_0(q - q_0) + p, \tag{2.84}$$

which, since it is equivalent to (2.81), shows that our ket manifold is simply the set of points $(\bar{q}, \bar{p}) = (\bar{q}, p_0(\bar{q}))$. Therefore the ket manifold is the same as the initial Lagrangian manifold of WKB theory, but now it is a complex surface in complex phase space. In one degree of freedom this surface has one complex dimension, or two real dimensions. For our initial Gaussian, equation (2.84) shows that the initial Lagrangian manifold $\bar{p} = p_0(\bar{q})$ represents a linear relationship between $\bar{q}$ and $\bar{p}$; it is, in terms of real variables, a two-dimensional plane.

The first step in the GGWPD method, as was illustrated in Heller (1976c), is to search the initial ket manifold for a complex initial condition $(\bar{q}_i, \bar{p}_i)$ which leads *via* Hamilton's equations to a given final, real position q_f in time t. This assumes that we are projecting the wave packet into the coordinate space representation. We could instead use momentum space, in which case we would seek a given final, real momentum p_f. We could also project onto a coherent state or a Gaussian wave packet. In this case we would seek a final complex p_f and q_f satisfying the final "bra manifold," in complete analogy with the ket manifold introduced in (2.81). The integration of trajectories is carried out over real time, but the initial conditions $(\bar{q}_i, \bar{p}_i)$ are in general complex. The trajectory itself, for intermediate times, is also in general complex, as is the final momentum, but the final position q_f is for our purpose real. Usually there will be more than one such initial condition on the ket manifold, or "branch," as we will call it, and there may even be an infinite number. However, as was shown numerically in Heller (1976c) many of these branches can or must be neglected, and often only one or two are needed for short times.

This first step in the GGWPD method will be recognized as the equivalent of the "root search" in WKB theory, in which one searches the initial Lagrangian manifold of points $(q_i, p_0(q_i))$ for initial conditions that lead to the given final q_f in the given amount of time. The only difference between the two methods at this stage is our use of complex variables.

The second step in the GGWPD method is to compute the quantity $\overline{c}_i$ corresponding to the initial conditions $(\overline{q}_i, \overline{p}_i)$ found in step one, according the two equations (2.82). This is

$$\overline{c}_i = c_0 + \alpha_0(\overline{q}_i^2 - q^2) + i(\overline{p}_i\overline{q}_i - pq). \tag{2.85}$$

We have

$$\overline{c}_i = iS_0(q = \overline{q}_i; q, p); \tag{2.86}$$

that is, $-i\overline{c}_i$ is nothing but the initial action S_0 evaluated at the given initial point on the initial Lagrangian manifold. Again, precisely the same step is required in ordinary (real) WKB theory, in which one must compute $S_0(q_i)$, where q_i is the real initial q value, and $p_i = p_0(q_i)$ is the real initial momentum.

The third step in the GGWPD method is to integrate the equations

$$\dot{q} = \frac{p}{m},$$

$$\dot{p} = -V',$$

$$\dot{c} = i\left(L - \frac{\hbar\alpha}{m}\right),$$

$$\dot{\alpha} = \frac{i}{2}\left(V'' - \frac{4\alpha^2}{m}\right), \tag{2.87}$$

subject to the initial conditions $q(0) = \overline{q}_i$, $p(0) = \overline{p}_i$, $c(0) = \overline{c}_i$, and $\alpha(0) = \alpha_0$. (Actually the first two of these equations have already been integrated in step one; they are included here for completeness.) Here we assume a Lagrangian L of the form $\frac{1}{2}m\dot{q}^2 - V(q)$, although more general Lagrangians (e.g., those with magnetic fields) are also easily dealt with by this formalism. Equations (2.87) are precisely the equations of motion for the evolution of a Gaussian wave packet by means of a single (real) trajectory in the quadratic approximation (i.e., the TGA). Here there is a root search and a sum over all the roots, whereas in the TGA a single known trajectory suffices for every final position. This is why the TGA is both less accurate and easier to implement than the GGWPD.

The equations for $\dot{c}$ and $\dot{\alpha}$ also have analogues in ordinary (real) WKB theory for initial value problems. In this theory the final wave function at position q_f at time t is given by

$$\psi(q_f, t) = \sum_{\text{branches}} A(q_f, t)e^{i[S(q_f, t) - \mu\pi/2]/\hbar}, \tag{2.88}$$

where μ is an integer (the Maslov index) and where A, S, and μ all depend on the branch. We will deal with S first, and come back to A and μ in a moment. In standard WKB theory the final action is given by

$$S(q_f, t) = S_0(q_i) + \int_0^t L \, dt, \tag{2.89}$$

where the integration is taken along the orbit. But by combining equations (2.86) and (2.87), we can see that the final value of c in the GGWPD method contains a term that is the complex analogue of the final action, since

$$-ic(t) = S_0(\overline{q}_i; q, p) + \int_0^t L \, dt - \frac{\hbar}{m} \int_0^t \alpha \, dt$$

$$= S(q_f, t) - \frac{\hbar}{m} \int_0^t \alpha \, dt. \tag{2.90}$$

As we will show later, the final term in this equation is essentially the logarithm of the amplitude or Van Vleck determinant.

Finally, the fourth step in the GGWPD method is to compute the final wavefunction by

$$\langle q_f | e^{-iHt/\hbar} | g; q, p \rangle = \sum_{\text{branches}} e^{c(t)/\hbar}. \tag{2.91}$$

Some branches may be excluded from this sum because their contribution is small; this is a numerical matter. Other branches *must* be excluded because, in a sense, they are too large; this is a matter of principle having to do with the phenomenon of Stokes lines.

We again use the auxiliary variables P_z and Z:

$$\alpha = -\frac{i}{2} \frac{P_z}{Z}. \tag{2.92}$$

The values of (Z, P_z) at a later time are determined by equations (2.50) and (2.51). Although this replaces the one equation for α by two for (Z, P_z), it has the salutary effect of replacing a nonlinear equation by linear ones. The q in this equation is actually $q(t)$, the solution of equation (2.87), so the linear system is time dependent. By differentiating P_z/Z with respect to t and using (2.50), it is straightforward to derive (2.87), the given equation for $\dot{\alpha}$, thereby showing that the new equations are equivalent to the old one.

2.6.2 Limiting Cases

If we take $\alpha_0 \to 0$ or $\alpha_0 \to \infty$, we recover the position and momentum bases, respectively. Since the GGWPD expressions are supposed to be generalizations of the VVG, we should recover the VVG in these limits. This recovery should be signified by a return to real valued coordinates. Note that if we set $\alpha_0 \to 0$ in $2\alpha_0 q_0 + i p_0 = \text{const.} = 2\alpha_0 \bar{q} + i \bar{p}$, we have $p_0 = \bar{p}$, and since p_0 is real by definition, we are confined to real momentum. The position becomes undefined in this limit, which is consistent with an infinitely broad wave packet. An intriguing aspect of this limit is that the position can be anywhere in the *complex plane*. One may say that when momentum is perfectly sharp the position is so uncertain that it can be complex! This fact may not have been fully exploited in the semiclassical description of certain classically forbidden processes. If instead we let $\alpha_0 \to \infty$, the position becomes sharp and real, and the comments about completely uncertain complex values now apply to momentum.

Thus we have not quite returned to a requirement that both position and momentum must be real in the "Hermitian" limits $\alpha_0 \to 0$ or $\alpha_0 \to \infty$. In fact we have learned that roots should, in general, be sought for initial complex momentum in the position representation, and vice versa.

2.6.3 TGA Again

The TGA is derivable as an *approximation* to the Van Vleck–Gutzwiller semiclassical propagator acting on an initial Gaussian. A quadratic expansion of equations (2.77) and (2.79) about q_0 is natural if we apply the propagator $G_{sc}(q, q'; t)$ to a narrow Gaussian wave packet, which restricts the effective domain of the integral

$$\psi_0(q') = \left(\frac{\text{Re } \alpha_0}{\pi \hbar} \right)^{1/4} \exp\left[-\frac{\alpha_0}{2\hbar}(q' - q_0)^2 + \frac{i}{\hbar} p_0 (q' - q_0) \right], \tag{2.93}$$

since then the Gaussian cuts off the q' integral. If we also assume that the resulting propagated wave packet is not too extended in its coordinates, then it also makes sense to expand around the final classical position q_t. The quadratically expanded action is

$$S_\gamma(q, q'; t) = S(q_t, q_0; t) + \left(\frac{\partial S}{\partial q} \right)_{q_t} (q - q_t) + \left(\frac{\partial S}{\partial q'} \right)_{q_0} (q' - q_0)$$

$$+ \left(\frac{\partial^2 S}{\partial q^2} \right)_{q_t} \frac{(q - q_t)^2}{2} + \left(\frac{\partial^2 S}{\partial q'^2} \right)_{q_0} \frac{(q' - q_0)^2}{2}$$

$$+ \left(\frac{\partial^2 S}{\partial q \partial q'} \right)_{q_t, q_0} (q - q_t)(q' - q_0), \tag{2.94}$$

where the first derivatives are the final and initial momenta

$$\left(\frac{\partial S}{\partial q}\right)_{q_t} = p_t, \quad \left(\frac{\partial S}{\partial q'}\right)_{q_0} = -p_0. \tag{2.95}$$

The linearized equations of motion are expressed with the stability matrix M. Initial deviations around a reference trajectory γ are expressed as

$$\begin{pmatrix} \partial p_t \\ \partial q_t \end{pmatrix} = M_\gamma \begin{pmatrix} \partial p_0 \\ \partial q_0 \end{pmatrix}, \quad M_\gamma = \begin{pmatrix} m_{11} & m_{12} \\ m_{21} & m_{22} \end{pmatrix}. \tag{2.96}$$

The second derivatives can be expressed as combinations of the stability matrix elements giving

$$\left(\frac{\partial^2 S}{\partial q^2}\right)_{q_t} = \frac{m_{11}}{m_{21}}, \quad \left(\frac{\partial^2 S}{\partial q'^2}\right)_{q_0} = \frac{m_{22}}{m_{21}}, \quad \left(\frac{\partial^2 S}{\partial q \partial q'}\right)_{q_t, q_0} = \frac{-1}{m_{21}}. \tag{2.97}$$

Within the quadratic expansion of the action, one of the terms in the sum (2.77) is

$$G_{sc}(q', q; t) = \left(\frac{1}{2\pi i \hbar}\right)^{d/2} \sum_j \left| \det\left(\frac{\partial^2 S_j(q_t, q_0; t)}{\partial q_t \partial q_0}\right) \right|^{1/2}$$

$$\times \exp\left(iS_{quad}(q, q', q_t, q_0; t)/\hbar + \ldots - \frac{i\pi \nu_j}{2}\right), \tag{2.98}$$

where $S_{quad}(q, q', q_t, q_0; t)$ is the quadratic expansion of S about q_t and q_0 is the linearized propagator corresponding to this trajectory. If only this "on-center" term is used, and applied to the initial Gaussian (2.93), we recover the TGA exactly.

2.6.4 Maslov and Geometric Phase

The often bothersome, sometimes mysterious, but always essential Maslov phase—the integer ν_j in equation (2.98)—does not appear explicitly above in the wave packet derivation of the TGA but appears explicitly in the second derivation just outlined. Accordingly, as noted by Littlejohn (1986), the TGA has in it the seeds of a very simple method to find Maslov indexes. The trick is to keep track of the phase

$$\frac{i\hbar}{2} \mathrm{Tr}[\ln \mathbf{Z}]$$

in equation (2.53). Sudden jumps of the overall phase are to be avoided, yet $\mathrm{Tr}[\ln \mathbf{Z}]$ will jump suddenly in phase if left to its own devices. The addition of a phase correction to cancel a jump of $\pm \pi/2$ in the phase of the trace keeps the phase smooth and gives the Maslov correction automatically.

Child (Ge and Child 1997) has drawn attention to an interesting geometric phase that arises in Gaussian wave packet dynamics with a nonadiabatic time-periodic

Hamiltonian which happens to give a periodic wave packet solution (except for the phase). The phase is automatically included in the equations already given, but it is important to point out a geometric (Berry) phase when present, as Child has done.

One of the great advantages of working with wave packet representations is the automatic uniformization that they provide. This is not to say that all possible caustic singularities are avoided, but some of the most common kinds are ameliorated, much like oil on troubled waters. These advantages were pointed out in the early wave packet work (Heller 1975, 1976a, b), and they are easy to understand in a phase space picture. Take, for example, the semiclassical coordinate space representation of an eigenstate of an oscillator. In the traditional position space, there is a well-known caustic singularity at the classical turning point, with the wavefunction diverging as the inverse velocity. There is no such singularity in the semiclassical coherent state representation. The tangency of the two delta function manifolds $\delta(q - q_0)$ and $\delta(E - H(p, q))$ is seen in Figure 2.12 when q_0 is at the turning point leads to the singularity; a Gaussian amplitude however has no such singularity.

2.6.5 Taming Caustics

2.6.6 Evaluation of the VVG Propagator

Curiously, despite the considerable effort and tests that went into the evaluation of the semiclassical energy Green's function, which reached its pinnacle with Miller's "classical S-matrix theory" (Miller 1970, 1974), no work explicitly evaluating the time dependent semiclassical propagator (the VVG propagator) in nontrivial situations is known to the author prior to 1991, when the "cellular dynamics" method was introduced (Heller 1991; Sepúlveda and Heller 1994). This latency is all the more surprising, since there is a sense in which the time dependent VVG propagator may be more accurate than its energy domain counterpart: The VVG propagator is obtained directly from the exact Feynman path integral by stationary phase over paths; the energy Green's function adds another stationary phase integral (time to energy Fourier transform by stationary phase). We note in passing that the famous Guzwiller trace formula for eigenvalues is another stationary phase integral on top of that, this time over all diagonal elements of the energy propagator.

Miller beautifully showed how the algebra of stationary phase puts all semiclassical expressions on an equal footing, with all Dirac algebra holding, if by each bracket we mean the semiclassical expression (Miller 1970, 1974). Despite this, not all semiclassical expressions are equally efficacious. In the author's opinion, if possible, one should stay in the time domain and use wave packets!

The VVG propagator is seemingly daunting to evaluate; for each q and q' we have to find all the trajectories connecting them, and then do it all over for each time t. But this is the wrong way to look at it. If we follow essentially all the trajectories (using some interpolation between explicitly sampled trajectories) leading from q_0 at $t = 0$, then we cannot fail to find all the final positions q that are reached, and furthermore we can just increment the time in the usual way using all the previous information (i.e., run trajectories to get the next time's data without starting over). These ideas are

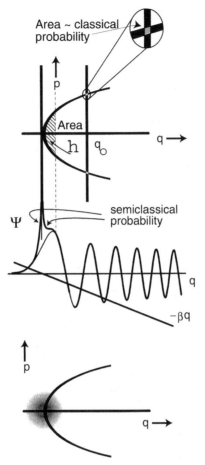

Figure 2.12   Intersection of two delta function manifolds $\delta(q - q_0)$ and $\delta(E - H(p, q))$ leading to a caustic, and the smoother overlap of $\delta(E - H(p, q))$ with a phase space Gaussian. This example corresponds to a linear ramp potential, with an Airy function solution.

contained in the cellular dynamics method, described next, which allowed the first detailed evaluation of the accuracy of the VVG propagator (Heller 1991).

2.6.7   Cellular Dynamics—Wave Packets Again

We can write the time dependent Green's function as an integral over the initial values of momentum, as follows:

$$G^{sc}(q, q_0; t) = \left(\frac{1}{2\pi i \hbar}\right)^{1/2} \int dp_0 \left| \left(\frac{\partial q_t}{\partial p_0}\right)^{1/2}_{q_0} \right| \delta(q - q_t(q_0, p_0))$$

$$\times \exp\left[\frac{iS(q_0, p_0)}{\hbar} - \frac{i\nu\pi}{2}\right]. \tag{2.99}$$

The utility of writing semiclassical amplitudes as integrals over initial values was pointed out long ago by Miller (1970, 1974). We have written the action S as a function of q_0, p_0; $q_t(q_0, p_0)$ is the position of the trajectory that leads from the initial conditions $q = q_0, p = p_0$. The equivalence of this expression to the original VVG formula is verified by performing the integral over p_0. For certain values of p_0 the condition $q = q_t$ is satisfied; this is a trajectory connecting q_0 to q in time t. The Dirac delta function gives a contribution

$$\left|\frac{\partial p_0}{\partial q_t}\right|_{q_0}$$

at each "root" value of p_0.

We can see directly from (2.99) that the Green's function is calculated semiclassically by summing over weighted classical trajectories, all launched from $q = q_0$, with all possible momenta p_0. Each point along the line $q = q_0$ is a separate trajectory that carries a weight $|\partial q_t/\partial p_0|_{q_0}^{1/2}\, dp_0$ and a phase $S(q_0, p_0)/\hbar - \nu\pi/2$. Figure 2.13 shows two roots $q_t = q$ leading from two different initial momenta. The direct implementation of this form of the Green's function is still foreboding, since roots still need to be found for each q and q_0.

Applying (2.99) to an initial wavefunction $\psi(q, 0)$, we have

$$\psi^{sc}(q, t) = \int dq_0\, G^{sc}(q, q_0; t)\psi(q_0, 0)$$

$$= \left(\frac{1}{2\pi i\hbar}\right)^{1/2} \int\int dq_0\, dp_0 \left|\left(\frac{\partial q_t}{\partial p_0}\right)^{1/2}_{q_0} \delta(q - q_t(q_0, p_0))\right|$$

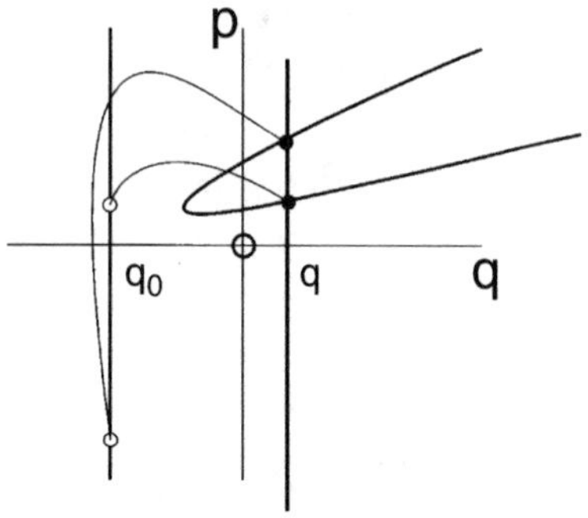

Figure 2.13 Dynamics leading from an initial q = const. manifold via two possible paths toward final positions.

$$\times \exp\left[\frac{iS(q_0, p_0)}{\hbar} - \frac{i\nu\pi}{2}\right]\psi(q_0, 0).\qquad(2.100)$$

We now introduce an idea that will permit the calculation of the semiclassical Green's function without root searches. This idea is related to a method of integration introduced by Filinov (Filinov 1986). We start with the observation that a sum of Gaussians may be made nearly constant, for example,

$$1 \approx \eta \sum_n \exp[-\beta(y - na)^2].$$

By choosing the centers and widths of the Gaussians, it is possible to achieve a nearly constant sum. The factor η normalizes it. The parameter β is taken large enough to allow an expansion near the region around $y = na$ of other functions depending on y (see below). The spacing and width of Gaussians will be determined by physical considerations. In effect, *classical* Gaussian wave packets are being used to calculate the VVG propagator.

We use this device twice in (2.100), obtaining

$$\psi^{sc}(q, t) = \eta\eta'\left(\frac{1}{2\pi i\hbar}\right)^{1/2} \sum_n \sum_m \int\int dq_0\, dp_0 \left|\frac{\partial q_t}{\partial p_0}\right|_{q_0}^{1/2} \delta(q - q_t(q_0, p_0))$$

$$\times \exp[-\alpha(q_0 - q_n)^2]\exp[-\beta(p_0 - p_m)^2]$$

$$\times \exp\left[\frac{iS(q_0, p_0)}{\hbar} - \frac{i\nu\pi}{2}\right]\psi(q_0, 0).\qquad(2.101)$$

The double sum is now a sum over Gaussian cells in phase space. Within each we will linearize the classical dynamics and thus analytically determine the cell's contribution to the propagation of the initial wavefunction. Again, the Maslov phase is $-i\nu\pi/2$. The integer ν increases by one whenever the classical trajectory connecting q_0 to q encounters a focal point or caustic, where $\partial^2 S(q, q_0)/\partial q\partial q_0$ diverges. This happens individually for each cell; the phase is retarded for the cell by $\pi/2$ whenever $(\partial q_t/\partial p_0)|_{q_0}$ changes sign. If the exponential factors α and β are large enough, the (n, m) term will not have much contribution except near $p_0 \approx p_m$ and $q_0 \approx q_n$. This allows us to expand the action $S(q_0, p_0)$ about $p_0 = p_m$, $q_0 = q_n$, which we do to second order. Also we expand q_t about $q_{nmt} \equiv q_t(q_n, p_m)$, to first order, consistent with the second-order expansion of S. The net result of these expansions will be to put $G^{sc}(q, q_0; t)$ into the form of a sum of Gaussians, quadratic q, q_0. If $\psi(q_0, 0)$ is also expanded about q_n, we have only Gaussian integrals to be performed. An alternative ("option B") is to assume that $\psi(q_0, 0)$ has been decomposed as

$$\psi(q_0, 0) = \sum_n a_n \exp\left[-\alpha_n(q_0 - q_n)^2 + \frac{ip_n(q_0 - q_n)}{\hbar}\right],$$

and again only Gaussian integrals need be performed. In either case the same algebra and bookkeeping needs to be done, expanding and collecting terms and doing Gaussian integrals. The action is now expanded about the centers of the cells, defined by the parameters q_n and p_m. Note that the widths of the Gaussians can be as small as we please; namely they are not limited by the uncertainty principle. Of course this is all just bookkeeping on integrals.

In the expansion for S, the exponent of the (n, m) term is (Heller 1991)

$$\frac{1}{\hbar}[S(q_n, p_m) + (m_{22}p_{nmt} - p_m)(q_0 - q_n) + (m_{21}p_{nmt})(p_0 - p_m)$$

$$+ \frac{1}{2}m_{12}m_{22}(q_0 - q_n)^2 + \frac{1}{2}m_{21}m_{11}(p_0 - p_m)^2$$

$$+ m_{12}m_{21}(q_0 - q_n)(p_0 - p_m)] - \alpha(q_0 - q_n)^2 - \beta(p_0 - p_m)^2$$

$$+ \frac{ip_n(q_0 - q_n)}{\hbar} - \frac{i\nu\pi}{2}. \tag{2.102}$$

The term involving p_n is appropriate for option B. This is a pure quadratic form in q and q_0, the arguments of the Green's function. Thus the full semiclassical Green's function is given as a sum of Gaussians quadratic in q and q_0:

$$G^{sc}(q, q_0, t) = \sum_{n,m} \left|\frac{1}{m_{21}}\right|^{1/2} g_{nm}(q, q_0), \tag{2.103}$$

and the wavefunction is given as

$$\psi^{sc}(q, t) = \sum_{n,m} \int \left|\frac{1}{m_{21}}\right|^{1/2} g_{nm}(q, q_0)\psi(q_0, 0)\, dq_0 \tag{2.104}$$

where $g_{nm}(q, q_0)$ is the Gaussian. If option B is used, the wavefunction is given by

$$\psi^{sc}(q, t) = \sum_{n,m} c_n \int \left|\frac{1}{m_{21}}\right|^{1/2} g_{nm}(q, q_0)\, dq_0. \tag{2.105}$$

The Gaussian integral over q_0 is analytic, and the result is a wavefunction that is a sum of Gaussians in q.

As α and β are made larger and more terms in the sum over m and n are taken, the method converges rigorously on the construction of the exact semiclassical time dependent Green's function. For any finite time there is always a linearizable regime around each trajectory so that it is unnecessary to calculate a more dense set of trajectories than the linearizable domains dictate. In a sense we have beaten the uncertainty principle, since we are not tying the linearizable domains to Planck's constant. A grid of initial conditions for the classical trajectories determines all the parameters; there are no trajectories to search for. It may be argued that by running the initial grid of trajectories, we have determined all the relevant dynamics anyway. This is true, but if the entire wavefunction is needed, it is difficult to see how we could propagate it with less than this knowledge. Further, by judicious use of the linearizable domains, we have made global use of pointwise trajectory information.

2.6.8   Other Approaches

An even simpler idea than the TGA is the frozen Gaussian approximation, or FGA (Heller 1981). As the name suggests, in this approach one does not let individual wave packets spread. For a single wave packet, this is sinful. The hope, however, is that if every initial state (even a Gaussian) is expanded in terms of many frozen Gaussians, each with a different classical initial condition, then collectively there might be a proper representation of spreading, and so forth. In addition the notion that the individual Gaussians cannot "get out of hand," as they can in the TGA if they spread too far, is attractive. This approach is unfortunately not exact even as $\hbar \to 0$. Nonetheless, surprisingly, Herman (Herman 1986; Herman and Kluk 1986; see also Kinugawa and Sakurai 1997) made a modification of the method (which affected the prefactors of the frozen Gaussians) that is exact as $\hbar \to 0$. In doing this, he opened up a new avenue to semiclassical approximations, wherein one constructs approximations that are nonstandard in that they do not arise directly from the usual semiclassical procedures but nonetheless have the correct asymptotic limit. More recently Manoloupolis developed a remarkable hybrid of the Herman approach and cellular dynamics (Walton and Manolopoulos 1996; Hulme and Manolopoulos 1997), which has been used to do nontrivial quantum dynamics in 15 degrees of freedom. Kay (Zor and Kay 1996) has given another globally uniform expression inspired by the same notions, in which time independent wavefunctions are written as integrals over Gaussian wave packets. Improvements were noted of Kay's method over the earlier wave packet superposition construction of eigenstates of Davis and Heller (Davis and Heller 1979, 1981).

2.6.9   Off-Center Guiding

A simple idea permits a vast simplification of cellular or VVG approaches to wavefunction propagation when the dynamics is unstable (chaotic). The stretching and folding of phase space, which happens on a short time scale in strongly chaotic systems, dramatically increase the number of linearization cells necessary to accurately describe the motion. However, the stretching can be put to use, in the following way (Tomsovic and Heller 1991, 1992, 1993; Sepúlveda, Tomsovic, and

Heller 1993): In computing an autocorrelation function $\langle\phi|\phi(t)\rangle$, for example, we only require knowledge of the amplitude near the region of phase space spanned by the initial state $|\phi\rangle$. In such a restricted region, the returning amplitude will consist of quasi-linear homoclinic oscillations, which can easily be approximated by a linearized transformation, one for each homoclinic "branch." This is far fewer distinct linearizations than would be required for a direct cellular approach. Each of the linearization corresponds to expansion of the initial wave packet's motion, not about its center but about that homoclinic trajectory which starts on an unstable manifold within the domain of the initial wave packet and returns to a stable manifold therein (see Fig. 2.14).

2.7   WAVE PACKETS AND SCAR THEORY

We begin with a definition and discussion of the scarring phenomenon (Heller 1984, 1991; Kaplan and Heller 1998).

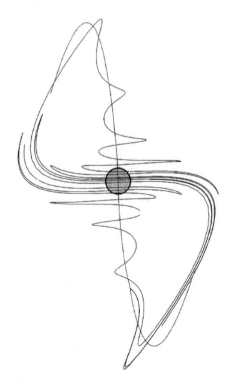

Figure 2.14   Returning (homoclinic) pieces of stretched and folded phase space domains are approximately linear in the region of overlap (circle) with the initial density in phase space.

Definition   A quantum eigenstate of a classically chaotic system has a *scar* of a periodic orbit if its density on the classical invariant manifolds near the periodic orbit is enhanced compared to the statistically expected density.

Scarring is associated with unstable classical orbits embedded in chaos. Scars are not merely associated with a one-dimensional line along a periodic orbit; they are also associated with the stable and unstable manifolds of that orbit. For this reason a phase space study of scarring (which is essentially what a wave packet study is) may often be more illuminating than a coordinate space projection.

Scar strength S, as measured by the projection of scarred eigenstates onto a coherent state centered on the scarring periodic orbit and aligned along the stable and unstable manifolds, is generically a function only of λ, the Lyapunov exponent for one period of the periodic orbit, and is $\hbar$-independent. For small λ, $S \rightarrow C/\lambda$, where C is a constant obtained by considering the linear theory at short times combined with random long-time fluctuations.

Semiclassical Gaussian wave packets were used in the discovery and first theory of the phenomena of scarring of eigenstates by (unstable) periodic orbits embedded in classically chaotic regions. Only short time information is needed to show scarring exists; therefore the semiclassical theory becomes exact as $\hbar \rightarrow 0$.

The spectrum $S(E)$ is defined in terms of the Fourier transform of the correlation function $\langle \phi | \phi(t) \rangle$ involving a stationary wave packet $|\phi\rangle$ and a moving wave packet $|\phi(t)\rangle$:

$$S(E) \equiv \frac{1}{2\pi\hbar} \int_{-\infty}^{\infty} e^{iEt/\hbar} \langle \phi(0) | \phi(t) \rangle \, dt$$

$$= \sum_{n=0}^{\infty} p_n^{\phi} \delta(E_n - E). \tag{2.106}$$

A high-resolution spectrum yields the intensities $p_n^{\phi} = |c_n|^2$. The c_n are the matrix elements

$$c_n = \langle \chi_i | \chi_n \rangle, \tag{2.107}$$

where χ_i is the initial eigenstate. In the case where the state is produced by a photon, as in electronic spectroscopy of molecules, we define the "spectroscopic" state $|\phi\rangle \equiv \mu|\chi_i\rangle$. It is the quantum state of the system before absorption or emission of a photon, modified by the transition moment μ. It is the state that we can produce in the laboratory with a short enough laser pulse, but not so short that other electronic states were populated due to the uncertainty principle breadth of the pulse. Zewail and coworkers have essentially achieved this in the laboratory (Rose, Rosker, and Zewail 1987), literally creating the wave packet $|\phi(t)\rangle$ with femtosecond pulses.

The classical dynamics near the periodic orbit will dictate the shape of the spectrum, up to a resolution determined by the maximum time the linearized analysis of the dynamics is accurate. The fact that such an orbit is embedded in chaos makes no difference to the short-time dynamics and thus to the low-resolution spectrum. If we launch a wave packet $|\phi\rangle$ along the periodic orbit, the wave packet returns at the period of the orbit, in distorted form, but still centered on the orbit. The distortion is controlled by the classical dynamics in the vicinity of the periodic orbit, *via* the stability matrix M_τ. As $\hbar \to 0$, this classical control is exact for any finite time. The magnitude of the recurrence at time τ is a function of the stability matrix and the initial form of the wave packet $|\phi\rangle$. The nth recurrence decays approximately exponentially as $\exp[-n\tau\lambda/2]$, where λ is the classical Lyapunov exponent. The precise form of the decay is a function of the parameters of the wave packet. The recurrences cause structured bands in the spectral envelope, just as we have seen before. The bands are spaced (locally) by $\hbar\omega$, where ω is the classical frequency. The width of the bands is the Lyapunov exponent λ. The bands stand out only if

$$\frac{\omega}{\lambda} > 1.$$

Note that since ω/λ is a purely classical quantity, the argument above will not change as $\hbar$ approaches zero. The spectral oscillations survive the classical limit $\hbar \to 0$ or $E \to \infty$ because the linearized wave packet analysis holds rigorously in this limit. The existence of the local density of states oscillations implies that some or all of the states under the peaks must have a larger projection onto the initial wave packet than would have been expected with a smooth spectrum. The situation is shown in Figure 2.15. It illustrates the spectrum expected from random fluctuations appropriate to a fully chaotic system (top) and the actual spectrum (bottom) given the existence of a periodic orbit on which the wave packet producing the spectrum was launched. The

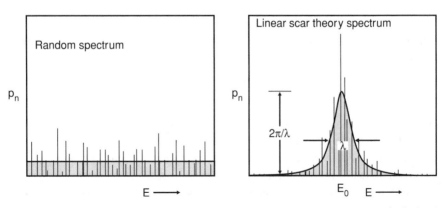

Figure 2.15 Short-time dynamics of a localized wave packet imposes an envelope in the local density of states which the resolved spectrum (coming from long-time dynamics) must obey. The envelope has a peak at quasi-energy E_0, a width $\delta\varepsilon \sim O(\lambda)$, and a height $\sim O(1/\lambda)$.

Figure 2.16 Quantum eigenstate of the stadium billiard scarred by the bowtie-shaped classical periodic orbit.

lumping of the intensity (here just one lump is shown) is what is predicted from the short-time dynamics; however, some individual eigenstates whose energy lies under the lump are therefore necessarily "brighter" than they would be. Since this brightness is a measure of the eigenstates' tendency to be found in a certain region of phase space, we have shown that some states must have a partial localization to the vicinity of the periodic orbit.

No "typical" trajectory spends more time near the periodic orbit than anywhere else in a chaotic system. The time average of any but a set of measure zero trajectories is a uniform, microcanonical distribution, which is itself time invariant under the classical evolution. This is what makes the scarred eigenstates surprising: scarred states *do* exhibit higher density near periodic orbits, yet they too are time invariant distributions.

In Figure 2.16 we see a scarred state along the unstable "bow tie" periodic orbit of the stadium-shaped billiard. The stadium is a well-known example of "hard" classical chaos (no quasi-periodic zones or stable trajectories anywhere). Typical trajectories are wildly chaotic in this system.

The scar enhancement is predicted all along the periodic orbit, since the eigenvalues of the stability matrix M_τ are independent of the starting point on the orbit. We cannot say precisely which states are scarred and by how much, rather that *some* states are scarred by the periodic orbit and by *at least* ω/λ, the local density of states enhancement. This "egalitarian" sharing of the scar intensity is what is expected to happen as $\hbar \to 0$, in general; however, scars can be much more intense due to random fluctuations and classical nonlinear recurrences (Kaplan and Heller 1998).

What is perhaps just as intriguing as scarring is a necessary consequence, namely "antiscarring." Since the average over sufficiently many adjacent eigenstates must be uniform in phase space, the existence of scarring implies antiscarred states. The wings of the spectral envelope have much *less* intensity than statistical theory would predict, and the spectral intensities for states in the wings are correspondingly very low. It was recently realized that this exponentially small amplitude in the wings means some decay processes (e.g., in billiards with holes at the loci of periodic orbits will have many antiscarred states with anomalously long decay times (Kaplan 1998)).

2.8 CONCLUSION

A key principle in all the approaches discussed here is linearization of classical motion about reference trajectories. For a sufficiently smooth Hamiltonian, every trajectory, no matter how nonlinear or chaotic its motion is, has a linearizable region surrounding it; the stability matrix encapsulates this information. Various wave packet propagation methods differ in how the linearized information is used. Even the most sophisticated and powerful methods, such as that of Manoloupolis (Walton and Manolopoulos 1996; Hulme and Manolopoulos 1997), require only this linearized information.

It seems likely that the future will bring have many useful applications and generalizations to time dependent wave packet methods. Major challenges lie ahead: how to best treat curve crossings, finite temperatures, and how best to deal with the proliferation of classical trajectory information when it contains more detail than the quantum mechanics does.

ACKNOWLEDGMENTS

This work was supported by the National Science Foundation and the Institute for Theoretical Atomic and Molecular Physics (ITAMP), Harvard University. Many thanks are owed to S. Tomsovic, Soo-Y. Lee, K. Kulander, R. Littlejohn, E. Stechel, M. Davis, R. Coalson, L. Kaplan, and M. Sepúlveda for inspiring discussions and contributions of important ideas.

REFERENCES

Coalson, R. D., and Karplus, M. (1982). *Chem. Phys. Lett.* **93**, 301.

Coalson, R. D., and Karplus, M. (1990). *J. Chem. Phys.* **93**, 3919.

Davis, M. J., and Heller, E. J. (1979). *J. Chem. Phys.* **71**, 3383.

Davis, M. J., and Heller, E. J. (1981). *J. Chem. Phys.* **75**, 3916.

Filinov, V. S. (1986). *Nuc. Phys.* **B271**, 717.

Ge, Y. C., and Child, M. S. (1997). *Phys. Rev. Lett.* **78**, 2507.

Gutzwiller, M. C. (1967). *J. Math. Phys.* **8**, 1979.

Gutzwiller, M. C. (1990). *Chaos in Classical and Quantum Mechanics.* Springer, New York.

Heather, R., and Metiu, H. (1986). *J. Chem. Phys.* **84**, 3250.

Heather, R., and Metiu, H. (1989). *J. Chem. Phys.* **90**, 6903.

Heller, E. J. (1975). *J. Chem. Phys.* **62**, 1544–1555.

Heller, E. J. (1976a). *J. Chem. Phys.* **64**, 63–73.

Heller, E. J. (1976b). *J. Chem. Phys.* **65**, 4979.

Heller, E. J. (1977). *J. Chem. Phys.* **66**, 5777–5785.

Heller, E. J. (1981). *J. Chem. Phys.* **75**, 2923.

Heller, E. J. (1981). *Accounts of Chem. Res.*, **14**, 368–375.

Heller, E. J. (1984). *Phys. Rev. Lett.* **53**, 1515.

Heller, E. J. (1986). In W. Hase (1991). *Advances in Classical Trajectory Methods* vol. **1**, ed. W. Hase. JAI Press, Greenwich, CT, pp. 165–213.

Heller, E. J. (1991). *J. Chem. Phys.* **94**, 2723.

Heller, E. J. (1991). *Wave Packet Dynamics and Quantum Chaology.* Lectures in the 1989 NATO Les Houches. Summer School on Chaos and Quantum Physics. eds. M-J. Giannoni, A. Voros, and J. Zinn-Justin. Elsevier Science, Amsterdam, p. 547.

Herman, M. F. (1986). *J. Chem. Phys.* **85**, 2069.

Herman, M. F., and Kluk, E. (1986). *Chem. Phys.* **91**, 27.

Huber, D. P., and Heller, E. J. (1987). *J. Chem. Phys.* **87**, 5302.

Huber, D., Heller, E. J., and Littlejohn, R. (1988). *J. Chem. Phys.* **89**, 2003.

Hulme, J. S., and Manolopoulos, D. E. (1997). *J. Chem. Phys.* **106**, 4832.

Kaplan, L. (1998). *Phys. Rev. Lett.* **80**, 2582.

Kaplan, L., and Heller, E. J. (1998). *An. Phys.* **264**, 171–206.

Kinugawa, T., and Sakurai, K. (1997). *J. Phys. Soc. Japan* **66**, 1310.

Klauder, J. R. (1986). *Phys. Rev. Lett.* **56**, 897.

Lee, S.-Y., and Heller, E. J. (1979). *J. Chem. Phys.* **71**, 4777–4788.

Lee, S. Y., and Heller, E. J. (1982). *J. Chem. Phys.* **76**, 3035–3044.

Littlejohn, R. G. (1986). *Phys. Rep.* **138**, 193.

Littlejohn, R. G., and Robbins, J. M. (1987). *Phys. Rev. A.* **36**, 2953.

Meyers, A. B., Mathies, R. A., Tannor, D. J., and Heller, E. J. (1982). *J. Chem. Phys.* **77**, 3857.

Miller, W. H. (1970). *J. Chem. Phys.* **53**, 1949, 3578.

Miller, W. H. (1974). *Adv. Chem. Phys.* **25**, 69.

O'Connor, P. W., Tomsovic, S., and Heller, E. J. (1992). *J. Stat. Phys.* **68**, 131.

Rose, T., Rosker, M., and Zewail, A. H. (1987). *J. Chem. Phys.* **87**, 2395.

Sawada, S., and Metiu, H. (1989). *J. Chem. Phys.* **84**, 6293.

Sepúlveda, M., and Heller, E. (1994). *J. Chem. Phys.* **101**, 8004.

Sepúlveda, M., Tomsovic, S., and Heller, E. J. (1992). *Phys. Rev. Lett.* **69**, 402.

Tomsovic, S., and Heller, E. J. (1991). *Phys. Rev. Lett.* **67**, 664.

Tomsovic, S., and Heller, E. J. (1993). *Phys. Rev. E***47**, 282.

Walton, A. R., and Manolopoulos, D. E. (1996). *Mol. Phys.* **84**, 961.

Weissman, Y. (1982). *J. Chem. Phys.* **66**, 5777.

Zor, D., and Kay, K. G. (1996). *Phys. Rev. Lett.* **76**, 1990.

CHAPTER 3

Excitation of Atomic Wave Packets

JOHN A. YEAZELL

3.1 INTRODUCTION

This chapter focuses on the excitation of wave packets in atomic systems. Much of the original interest in wave packets, and still the vast majority of the work on them, has been in atomic systems. The reason for this emphasis is that atomic systems provide a testbed in which to explore new ideas. Their structure is, in general, well understood, so the focus may be on the dynamics. One example is the ongoing study of how classical mechanics and quantum mechanics fit together. In part, the classical limit of quantum mechanics is approached through the limit of large quantum numbers of these systems. However, the electron's behavior in a highly excited stationary state displays little that can be interpreted as classical dynamics. And yet, by superposing these states, a great variety of electronic states with various classical and quantum mechanical properties may be created. Such superposition states are similar in nature to the coherent state of a harmonic oscillator. That is, the resulting electronic wave packet oscillates like a classical electron in a classical atom.

These wave packets have unique properties. Their interactions with external fields are time dependent, and this property is at the core of many coherent control schemes. For example, the time dependence allows the easy separation of isotopes by arranging for the wave packet of one isotope to have a large ionization cross section at a time when the other has essentially none. In that case a short pulse can selectively photoionize only the single species. The wave packets also offer a way to store coherence for a relatively long time. Such "memory" applications has found use in the field of quantum measurement and may find similar uses in quantum computers and/or quantum communications. In addition fundamental paradoxes have been explored with the excitation of Schrödinger Cat states. Several different varieties of wave packets and the means to excite them will be studied in this chapter. The goal is to provide a general understanding of these processes so that they may be generalized to further experiments in atomic and other systems.

The Physics and Chemistry of Wave Packets, Edited by John Yeazell and Turgay Uzer
ISBN 0-471-24684-0 © 2000 John Wiley & Sons, Inc.

3.2 RYDBERG ATOMIC WAVE PACKET EXCITATION

The defining characteristics of a wave packet are its localization and motion. These characteristics are derived directly from the establishment of a coherence between eigenstates of the system. Some interaction is needed to provide a common time origin or to define the relative phase between the eigenstates of the system. It is these interactions that we explore in this chapter.

3.2.1 Optical Excitation of Radial Wave Packets

Perhaps the simplest Rydberg atomic wave packet to describe (and to excite) is the radially localized wave packet. Its excitation was first proposed by (Parker 1986) and realized experimentally by (Wolde 1988; Yeazell 1990, 1990a). Rydberg atoms are formed by exciting an electron to a weakly bound state of high principal quantum number (n). Such states are nearly macroscopic in size (for $n = 100$, $r_{max} = 1\mu m$). A radial wave packet consists of a coherent superposition of such states with a spread in the principal quantum number but the same angular momentum quantum numbers. For example, consider a Gaussian distribution of the population of the Rydberg states, $|a_n|^2$, with an average n of $\bar{n} = 50$. This coherent superposition may be written as a sum of these eigenstates, u_n, $\Psi(r,t) = \Sigma_n a_n u_n(r)\exp[i\omega_n t]$. The angular quantum numbers have been suppressed since they do not contribute significantly to the dynamics of the radial wave packet. It is the differences in the free precession rate (ω_n) of these states, $\Delta_n = \omega_{\bar{n}} - \omega_n (\omega_n = -1/2n^{-2}$ atomic units are used unless otherwise specified), that govern the evolution of the radial wave packet. This free evolution of the wave packet is depicted in a series of snapshots of its probability density projected onto a plane (see Fig. 3.1). The angular characteristics of the wave packet (in this case, isotropic) are the result of choosing an s-state for the orbital angular momentum quantum number. In one-half of a classical period the wave packet has propagated from near the nucleus to the outer turning point. These oscillations continue relatively unchanged for several periods.

Impulsive excitations, like a rock dropped in a pond, can create these wave packets. For a radial wave packet, a short optical pulse plays the role of the rock. Figure 3.2 shows schematically an optical excitation of a radial wave packet. It depicts the common situation of a single-photon excitation in a single-electron atom. If the pulse is weak (perturbative limit), the coherent bandwidth of the pulse must overlap several states to create the wave packet. For a strong pulse, power broadening can provide the needed bandwidth. In either case, the bandwidth must be greater than the average separation of two neighboring states, $1/\bar{n}^3$. If the bandwidth (or number of states in the superposition) is increased further, the localization at the time of excitation increases. However, the wave packet will disperse more rapidly as it evolves as a consequence of the anharmonicity of the atomic potential. For example, consider the first return of the wave packet to the core which occurs at one classical period, $T_{cl} = 2\pi\bar{n}^3$. For this returning wave packet to be identical to the initial wave packet the phases, $\Delta_n T_{cl}$, must be equal to multiples of 2π. For a harmonic system this is satisfied exactly. However, for the atom, the further the state lies from the center of the distribution, $\bar{n}$, the phase

Figure 3.1 Probability density of a wave packet created from a Gaussian distribution of *s* states. The average principal quantum number is 50 and 7 states are significantly excited. The probability density is projected onto the *x-y* plane.

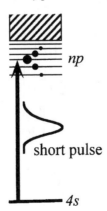

Figure 3.2 Energy level diagram describing excitation of a wave packet via a short pulse. A single electron atom is pictured (in this case potassium). The Gaussian spectrum of the pulse is transferred to the population of the Rydberg states for this perturbative excitation.

error grows more rapidly. The desire to limit the rapid dispersal of the wave packet places an upper limit on the bandwidth of the excitation. If we set a maximum phase error of $\pi/2$ for the state at the $1/e$ point of the distribution, then the standard deviation of the distribution is $\sigma_n = \sqrt{\bar{n}}/6$.

For $\bar{n} = 80$, such a wave packet has 7 states significantly populated. The overlap of the time-evolved wave packet with the initially excited wave packet is shown in Figure 3.3. The evolution displays the characteristic structure of a wave packet propagating in an anharmonic potential. It is initially well localized, and then it slowly disperses as the phase error grows. However, there is no irretrievable loss of coherence, and since only a limited number of states are contained within the superposition, an eventual rephasing of the superposition occurs. These rephasings are known as revivals. Full revivals consist of a relocalization to a single wave packet while fractional revivals consist of localized sub–wave packets, such as the $\frac{1}{2}$ fractional revival (at $\tau = 8T_{cl}$) which consists of two wave packets residing at opposite points of the orbital trajectory (Yeazell 1991). This evolution of the wave packet is well understood and has the been the subject of many experimental and theoretical studies (for a review, see Nauenberg 1994).

This example of Rydberg atomic wave packets, namely the radial wave packet, will serve as the reference for much of the discussion of this chapter. However, there are a wide range of wave packets that have been studied experimentally and theoretically (see Chapter 1). Here a summary of the excitation schemes for a representative selection of these wave packets is given.

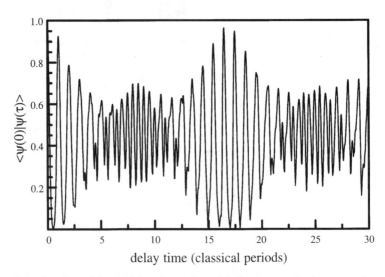

Figure 3.3 Overlap of the initial wave packet with the time-evolved wave packet. On a short-time scale, at every classical orbital period the wave packet nearly regains its initial shape. The wave packet eventually disperses, but the coherence is not lost. The wave packet shows fractional and full revivals as time progresses.

3.2.2   Wave Packets in External Fields—Magnetic, Electric, and Microwave Fields

The huge dipole moments connecting Rydberg states allow relatively weak fields (magnetic, electric, or microwave fields) to be nonperturbative. Eigenstates of both the atom and field serve as intermediate states both for producing zero-field atomic waves and for the producing wave packets directly in these atom-field systems. Examples of the first case are the angularly localized wave packet (Yeazell 1988) and the excitation of momentum-space wave packets with half-cycle pulses (Jones 1996). Chaotic studies of wave packets in strong magnetic fields (Marmet 1994), a variety of studies of the classical evolution of wave packets in strong electric fields (Lankhuijzen 1996), and the nondispersing wave packets found in rotating microwave fields (Kalinski 1996) are examples of the second.

The angularly localized wave packet is excited optically in the presence of an auxiliary circularly polarized radiofrequency (rf) field. Angular localization requires a superposition of eigenstates with different orbital angular momentum and different Zeeman quantum numbers. The rf field acts as the source of angular momentum needed to create such a superposition. Precisely multiphoton rf field processes create atom rf field dressed states that may be excited by an optical pulse. The amplitude and frequency of the rf field are chosen so that the dressed states are a superposition of field-free high angular momentum eigenstates and the particular low angular momentum state that can be excited by the optical pulse from the ground state of the atom. The superposition of dressed states that is created by the optical pulse may be transformed into a field-free state by the adiabatic turn-off of the rf field. The resulting angularly localized wave packet precesses about the nucleus. The ionization of this wave packet by a pulsed electric field is dependent on the angle between the Runge-Lenz vector of the wave packet and the electric field.

Half-cycle pulses (HCP) interacting with a Rydberg atom are an example of the strong field excitation of wave packets. HCP are unipolar pulses of short duration (less than a picosecond). A single Rydberg state is prepared initially by some other means; then this state interacts with the HCP which redistributes the population throughout the eigenstates. For example, when a HCP was applied to the 25d state of sodium (Jones 1996), roughly a thousand states were coherently populated (i.e., states with principal quantum number ranging from 17 to 47 and all the associated orbital angular momentum quantum numbers). These wave packets are observed in momentum space with the aid of an additional HCP that ionizes the wave packet. This probe has shown some promise as a means for retrieving the wavefunction of the wave packet in momentum space. A full description of this type of excitation may be found in Chapter 6 of this book.

The classical evolution of a Rydberg atom in a strong electric and/or magnetic field serves as a simple physical system for the study of nonlinear dynamics. For example, one of the simplest physical systems that classically displays chaotic behavior is the diamagnetic hydrogen atom. The excitation of a wave packet in this system takes place in the presence of a magnetic field whose strength is comparable to the Coulomb field binding the electron to the nucleus. Fortunately, for Rydberg atoms this Coulomb field

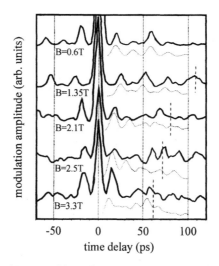

Figure 3.4 Visibility of Ramsey fringes in a strong magnetic field. The wave packets were excited at a fixed detuning from the zero magnetic field ionization threshold (-36.1 cm^{-1}) and at varying values of the magnetic field. The solid line shows the experimental data, and the dotted curve the theoretical model based on a semiclassical trajectory analysis. (This figure is used with permission from Marmet 1994.)

is relatively weak, so for $\bar{n} = 50$ a magnetic field of 3 Tesla approaches the chaotic regime. Therefore the classical trajectories associated with this system are of great variety. They include stable periodic orbits, orbits that approximately recur, and orbits never recur. In contrast, the radial wave packet, discussed above, has only one type of classical trajectory and all have the same recurrence time. This proliferation of classical trajectories, as the magnetic field is increased, is a signature of classical chaos. Experimentally the proliferation can be observed (Marmet 1994) in a phase-dependent Ramsey fringe measurement. A comparison of these data with a theoretical model based on these classical trajectories is shown in Figure 3.4. The general decrease in amplitude of the signal is one indicator of this proliferation of orbits. In addition the growth of a new class of orbits with a distinct recurrence time was observed in that paper.

The work studying the nondispersing wave packet is discussed in a later chapter of this book in detail. There a crossed electric and magnetic field produces a nearly harmonic potential in a certain region in the vicinity of the atom. These regions are similar to the Lagrange points studied in astrophysics. In these harmonic potentials, it is possible to generate coherent states similar to those of a harmonic oscillator, and as such, these coherent states do not disperse.

3.2.3 Multi-electron Wave Packets

Wave packets may of course be excited in more complicated atomic systems. Several recent experimental studies have observed the behavior of the wave packets in two-electron systems. The wave packet may undergo collisions with the remaining valence electron upon returning to the core and exhibit changes in character (Campbell 1998). The presence of a perturber state can also significantly alter its behavior (Schumacher 1997). Recently, radially localized wave packets in two-electron systems have been excited that demonstrate nearly identical temporal behavior (Lyons 1998; Strehle 1998) to that described above for the single-electron atom. A point of significant interest is the use of the electron-electron correlations to control the behavior of such a wave packet.

The general interest in the ability to control processes and to design special wave packets has been seen in this and other chapters of this book. In this section the control of the autoionization process and the shaping of a nondecaying wave packet will be used as an example of the work with wave packets in two-electron atoms. The control of the behavior of a radially localized wave packet in a two-electron system can be accomplished by driving the remaining, correlated valence electron. A short pulse is used to excite one of the valence electrons to a radially localized wave packet state. The remaining core electron is driven so that it oscillates between its ground state and its first excited state. The states of this core electron are much like the states of the singly ionized atom or like that of a single-electron atom so that this type of excitation is described as an isolated core excitation (ICE) (Cooke 1978).

When the core electron is in the ground state the two-electron system is bound. In contrast, if the core electron is excited, the system may autoionize. Thus, when the wave packet approaches the core, it sees one of two possible core configurations. If the core oscillation (Rabi oscillation) between these two configurations is synchronous with the classical oscillation of the wave packet, then the entanglement between the two electrons leads to several novel results.

Two of the more dramatic predictions are the control of autoionization and the shaping of nondispersing or nondecaying wave packets (Hanson 1995; Chen 1998). The shaping effects are striking when the Rabi frequency for the ICE is equivalent to and synchronized to the classical orbital frequency of the radial wave packet (see Fig. 3.5). Consider the situation in which the goal is to suppress decay or autoionization (Hanson 1995). There are two ways to achieve this goal. One is almost trivial. When the core electron is in the ground state, the autoionization process is not energetically possible. On the other hand, if the wave packet is far from the core, there is little overlap between the wave packet and the core. In that case, even if the core is excited, the probability of autoionization is low. Therefore the goal can be achieved by timing the wave packet's motion so that it is localized away from the core when the core is in the excited state. If the Rabi frequency is equal to the classical oscillation frequency and the oscillations are also in phase, then the timing is accomplished. That is, if the core is in the ground state and the wave packet is near the core, no autoionization occurs. Half a period later, the core will be excited, but the wave packet will be at the outer turning point. The resulting lack of overlap between the wave packet and the excited core limits the autoionization.

This view gives a clear picture of the suppression of autoionization. However, the shaping or modification of the wave packet is perhaps less apparent. Figure 3.6*a*

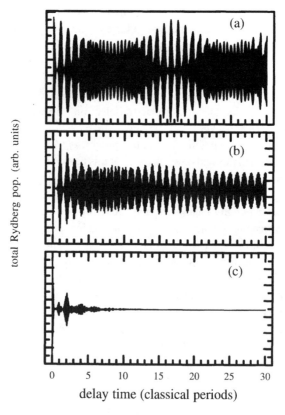

Figure 3.5   Ramsey fringes in the presence of strong, laser-induced core excitation. In (*a*) the strength of the ICE field is zero. In (*b*) the Rabi frequency is chosen equal to the Kepler frequency of the wave packet. This leads to the suppression of the autoionization process and to the formation of a nondecaying wave packet. In (*c*) the strength of the ICE field is half that of (*b*).

provides an energy level diagram of the ICE effect on the superposition of Rydberg states that forms the wave packet. Analysis of this system may be simplified by viewing the Rabi oscillations of the core as a Rabi oscillation between bound and autoionizing quasi-continua (dashed arrow). It is the channel interaction coupling the autoionizing Rydberg series to the continuum that provides the shaping. The channel interaction provides a coherent coupling, via the continuum, between the states making up the wave packet. This allows the amplitudes and the relative phases of these states to be modified. In the situation described above, this modification results in the formation of a nondecaying state. The precise phasing of the states results in a quantum interference between the decay routes of the various states of the wave packet. If the synchronization were to be removed, the dispersion of the wave packet would disrupt this precise phasing, and the autoionization would proceed. However, as long as the synchronizing field is maintained, the channel interaction continually reshapes the wave packet and maintains the nondecaying form (Chen 1998). It is this compensation of the dispersion of the wave packet that has led to the appellation of nondispersing wave packets. Note that these nondecaying, multilevel superposition states are analogous to those known as "dark" states (Arimondo 1996) that have been studied in three-level atomic systems. The evolution of a nondecaying wave packet is compared to the evolution of a wave packet without the ICE in Figure 3.5. The shape of the nondecaying wave packet evolves to a stable form, whereas the ordinary wave packet goes through dispersion, fractional revivals and revivals.

The formation of a nondecaying wave packet has been observed. A Rydberg electronic wave packet was excited in atomic calcium and further manipulated via a strong core excitation into the nondecaying form. Figure 3.6*b* shows the experimental setup for this experiment. A cw laser is tuned to the resonance of the isolated core transition. Note that until the pulsed laser excites the Rydberg wave packet, the cw field is far from any resonance in the Ca atom. A two-photon process driven by the pulsed laser (~ 406 nm) excites a wave packet with predominantly *nd* character. This Rydberg wave packet sees a core that oscillates at the Rabi frequency between a ground state and an excited state configuration.

The preservation of the wave packet was observed (Chen 1998a) by measuring the amplitude of the Ramsey fringes as a function of $\bar{n}$ (see Fig. 3.7). The fringes were examined at a time long compared to the autoionization decay time of a single

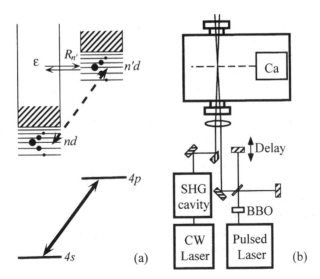

Figure 3.6   (*a*) Energy level diagram describing the modification of the wave packet by the core laser interaction. The Rabi oscillations of the core transition cause the wave packet to oscillate between a bound and autoionizing series (dashed arrow). The superposition of states that comprises the wave packet is depicted by the weighted dots. At the moment shown, the core transition has gone through $\pi/2$ of a Rabi oscillation. The wave packet has equal weight in the two series. (*b*) Optical and atomic beam layout. The pump and probe pulses are created by the Michelson interferometer setup. The pulsed laser path and the cw laser path are collinear.

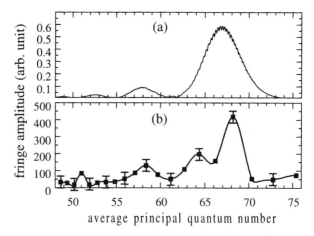

Figure 3.7 Visibility of Ramsey fringes as modified by core laser. In (*a*) a theoretical model is presented for comparison. The experimental results are shown in (*b*). Note that only the synchronized case ($\bar{n} = 67$), which produces a nondecaying wave packet, has significant visibility at the fixed delay time.

Rydberg state. Therefore the general amplitude of the Ramsey fringes was low except when the ratio of the Rabi frequency to the classical frequency of the wave packet was equal to unity. Both cases have similar amplitude of the Ramsey fringes as is expected of a nondecaying wave packet. The small intermediate peak is also of interest. It occurs at one-half the ratio of the Rabi frequency to the Rydberg frequency. In this situation the strong coupling to the autoionization continuum has rapidly reduced the population in the Rydberg states but has also led to the rapid evolution of a different type of nondecaying wave packet. It is characterized by a doubly peaked probability distribution. Neither of the peaks are near the core when the core is excited. A more direct means of observing how the channel interactions affect the coherence stored in the wave packet may be seen in how the phase of the wave packet is modified (in comparison with a zero field wave packet) as a function of the ratio of the Rabi frequency to the classical frequency. Note the rapid change of the phase in the vicinity of submultiples of this ratio. The electron-electron correlations allow the coherence stored in the wave packet to be modified by an independent laser source.

These striking modifications of the behavior of a wave packet are of direct interest to such areas of research as quantum measurement, quantum computing, quantum memories, and the control of decay.

3.3 WAVE PACKET DESIGN SCHEMES (TARGET STATES)

Atomic physics is one of the testing grounds of schemes for preparing special purpose wave packets or target states. Design of such target states has a broad appeal ranging from studies of fundamental questions in quantum mechanics to coherent control of reaction pathways. Some examples of note are the excitation of Schrödinger Cat states

(Noel 1996), the demonstration of wave packet isotope separation (Averbukh 1996), the control of reaction dynamics (Kohler 1995; Bardeen 1995), and ultrafast all-optical switching (Heberle 1995). As might be expected, the preparation of such wave packets has received much attention. The primary means of exciting such states utilizes specially designed laser pulses.

One successful approach to the design problem has come to be known as optimal control theory; see, for example, (Yen 1993). The Hamiltonian of the system is used to define a cost functional. Minimization of that cost functional, within the limits imposed by some constraints, yields an integral equation whose solution defines the optimum optical field to produce a desired target state. A typical way to form the cost functional is to choose a target operator that represents the desired outcome, and then to compute the optical field that maximizes the expectation value of the target operator and meets the imposed constraints. For example, the chosen target operator may be the projection operator $|\psi_{targ}\rangle \langle \psi_{targ}|$ for the representation of the target state ψ_{targ} in phase space. This specific distribution will then occur at some predetermined time after the designed excitation pulse. Typically energy constraints are applied to the electric field; for example, the intensity is required to be nonzero and noninfinite. An excellent application of optimal control theory to the design of Rydberg atomic wave packets may be found in Krause (1997).

For complex multidimensional systems, optimal control theory is, at present, the best available solution to the inverse problem. However, for less complicate systems (ones that can be described by a single quantum number or, perhaps, simple multidimensional systems) it is possible to find an analytical solution for the design problem. Rydberg atomic systems provide access to a physically simple picture of the design of target states. We have developed a pair of analytical solutions for Rydberg atoms. They are based upon two alternative excitation schemes for creating Rydberg atom target states. In this section we will discuss these solutions and the physical intuition they provide.

The two complementary design schemes are based upon complementary time (Chen 1998b) and frequency (Chen 1997) domain views of the problem. Each takes advantage of the relatively smooth character of the potential. This allows the evolutionary behavior of a wave packet to accurately described by a Taylor series expansion of the eigenenergies of the superposed states that form the wave packet. The first-order term of the expansion determines the primary period of the system. While it is the higher-order terms in the expansion that give rise to the dephasing and the revival structures in the wave packet's evolution. All the observed phenomena, to date, can be described by a second-order expansion.

The general idea of each scheme is to excite an initial wave packet superposition with phase terms that exactly compensate the dispersion of the wave packet. Thus, after the initial state has evolved to a specific target time, the desired target state appears. However, there are an infinite number of fields that can produce the required initial state. In optimal control theory the selection from among these possible fields is performed by the applied constraints. Here the selection of the excitation field follows naturally from the analytic formalism.

3.3.1 Frequency Domain Design

A wave packet state is a coherent superposition of several eigenstates (e.g., atomic Rydberg states) which can be expressed as

$$\Psi(\mathbf{r}, t) = \sum_n c_n \psi_n(\mathbf{r}) \exp(-i\omega_n t), \tag{3.1}$$

where w_n is the eigenenergy for the state n. Typically the probabilities $|c_n|^2$ are strongly centered around a mean value $\bar{n}$, in which case only those states with energies near the value $w_{\bar{n}}$ are essential. This permits an expansion of the energy in a Taylor series in n around the centrally excited value $\bar{n}$:

$$\omega_n = \omega_{\bar{n}} + \omega_{\bar{n}}'k + \frac{1}{2}\omega_{\bar{n}}''k^2 + \frac{1}{6}\omega_{\bar{n}}'''k^3 + \dots, \tag{3.2}$$

where each prime on $\omega_{\bar{n}}$ denotes a derivative with respect to $\bar{n}$. The variable k is defined $k = n - \bar{n}$. The derivative terms in (3.2) define distinct time scales that depend on $\bar{n}$. The fundamental period is the classical period, $T_{cl} = 2\pi/|\omega_{\bar{n}}'|$. The second time scale, $T_{rev} = 4\pi/|\omega_{\bar{n}}''|$, defines a revival period, namely the oscillation from a well-formed wave packet to dispersed and back to a well-formed wave packet. Commonly $T_{cl} \ll T_{rev} \ll T_{super}$ so that the T_{rev} defines a slowly varying envelope over the fundamental classical oscillations, and similarly, T_{super} defines such an envelope over the periodic revival oscillations. And so to second order (Averbukh 1989) we find

$$\Psi(\mathbf{r}, t) = \sum_n c_n \psi_n(\mathbf{r}) \exp\left[-2\pi i\left(\frac{k}{T_{cl}} - \frac{k^2}{T_{rev}} + \frac{k^3}{T_{super}}\right)t\right]. \tag{3.3}$$

A simple first case is the design of a pulse that will compensate the quadratic term in k. The short, optical pulse excitation of a wave packet is well understood. It is particularly simple to describe in the weak field limit. In that case it is straightforward to show that the Schrödinger amplitudes of the quantum states, a_n, depend directly upon the Fourier transform of the excitation pulse's shape (Parker 1986a). For example, if the excitation is by a Gaussian pulse, which has a Fourier transform given by $G(\omega) = e^{-(\omega^2 \sigma^2)/2}$, the Schrödinger amplitudes are simply

$$a_n = -\frac{i}{2}\Omega_n G(\Delta_n). \tag{3.4}$$

The Rabi frequencies describing the strength of the interaction between the ground state and excited states of the system are denoted Ω_n. The rotating wave approximation has been employed which results in the appearance of the detuning, $\Delta_n = \omega_n - \omega_{opt}$, in the solution. The optical frequency of the excitation pulse (ω_{opt}) is taken to be equal to $\omega_{\bar{n}}$. Note that the amplitudes a_n are defined by the spectrum of the optical pulse. However, the inverse is not true; the spectrum of the optical pulse needs only to have the required discrete values to produce a given set of a_n. That is, the choice of the Gaussian pulse to produce the initial state is not a unique choice. In a moment we will return to why this choice is a natural one for this system.

In the excitation of the wave packet considered above, it was assumed that σ was real. If it is replaced by a complex parameter, the amplitude of its Fourier transform remains a Gaussian, but a phase term appears. This well-known and compact result for a chirped Gaussian pulse simplifies the present analysis. The evolution structure of the wave packet, within the time scale set by T_{rev}, is primarily determined by the quadratic term of the expansion. Therefore a quadratic phase term arises due to a chirped excitation that can shift this entire evolution structure in time. In particular, a linearly chirped Gaussian pulse may be written,

$$g(t) = A \exp\left[-\frac{t^2}{2\sigma^2(1+ib)}\right],\tag{3.5}$$

where σ is the zero-chirp parameter and b is the linear chirp parameter. The normalization constant is $A = [2\pi\sigma^2(1+ib)]^{-1/2}$. With this excitation the Schrödinger amplitudes are given by

$$a_n = -\frac{i}{2}\,\Omega_n G(\Delta_n) \exp\left(-\frac{i}{2}\Delta_n^2\sigma^2 b\right).\tag{3.6}$$

The phase term may be simplified by expanding Δ_n around the central quantum number $\bar{n}$ and retaining only the first-order term, $\Delta_n = w_n - w_{\bar{n}} \approx w'_{\bar{n}}k$. With this substitution we have

$$a_n = -\frac{i}{2}\,\Omega_n G(\Delta_n) e^{-ib'k^2},\tag{3.7}$$

where $b' = (\omega'_{\bar{n}}\sigma)^2 b/2$ and for the particular case of a Rydberg atom wave packet $b' = \sigma^2 b/2\bar{n}^6$. In the Schrödinger picture, using the second-order expansion for the free evolution of the wave packet state, the probability amplitudes are

$$A_n(t) = -\frac{i}{2}\,\Omega_n G(\Delta_n) \exp\left[-ik\omega'_{\bar{n}}t - ik^2\left(b' + \frac{\omega''_{\bar{n}}}{2}t\right)\right],\tag{3.8}$$

after dropping the common phase factor of $\exp(-i\omega_{\bar{n}}t)$.

The last term (second order in k) determines the structure or envelope of the wave packet's evolution. A commonly desired target state is a well-localized wave packet at a specified point in time. If b is chosen so that this last term is zero at the specified time, then the wave packet formed at this time is identical (to second order) to the well-localized wave packet that is formed at time equal to zero for an unchirped pulse. In essence, the evolution of the unchirped wave packet has been shifted in time by

$$T_s = -\frac{(\omega'_{\bar{n}})^2\sigma^2 b}{\omega''_{\bar{n}}}.\tag{3.9}$$

Since the entire structure of the evolution shifts, it is also possible to shift any structure associated with the wave packet evolution to a specified time. For example, a fractional revival can be moved to a specified target time by adding its known offset in time to T_s. In addition, since the wave packet experiences multiple revivals (Yeazell 1990), there are still many solutions that will result in the creation of the desired target state. Thus the periodic nature of the revivals (Aronstein 1997) limits the necessary time shift to obtain the desired target state. The localized state from any revival (defined by the evolution of the unchirped wave packet) may be shifted to the target time. Therefore the shift need never be larger than $\frac{1}{2}T_{rev}$.

One possible benefit of such a simple target state follows from the recent work on isotope separation (Averbukh 1996). This isotope separation technique requires that the Rydberg atomic wave packets of one isotope be near the nucleus, while at the same time the wave packets of the other isotope are away from the nucleus. In addition both wave packets must be well-localized at this point in time. Unfortunately, the simultaneous satisfaction of both of these conditions is not necessarily simple and may require a prohibitively long evolution time. The shift technique, described above, can remove this difficulty by producing well-localized wave packets at any point in the fundamental oscillatory evolution.

The above example focused upon manipulating the quadratic term of the expansion. It is also possible to compensate the cubic phase term (Chen 1997).

This frequency dependent analytical design scheme offers insight into the general problem of creating exotic wave packets. One of its central ideas is that the states explored by the free evolution of an unchirped wave packet acts as a "shopping list" of quantum states (well-localized wave packets, Schrödinger Cat states, multiply-peaked wave packets, etc.). From this list, a desired target state may be chosen and shifted to the target time. There is flexibility in the design in that the shift may be accomplished from any revival time of the wave packet.

This design precisely defined the parameters of a Gaussian excitation pulse for shifting the appropriate target state to the desired target time. However, the choice of the Gaussian pulse shape appears somewhat arbitrary. This choice does fit well with experimentally available pulse shapes. In the time domain design, which follows, we will show that this choice is also a natural one for these systems.

3.3.2   Time Domain Design

The time domain scheme uses multiple phase-locked pulses to create the target state. The natural basis set to use for this type of excitation is a classical wave packet basis. A classical wave packet is defined as a Gaussian wave packet (Averbukh 1989). These classical wave packets are structures that remain relatively unchanged for a short period of time. Therefore a target state assembled from these classical wave packets will remain in existence for a time on the order of a few classical periods. The number of classical wave packets needed to form the target state gives a measure of its complexity and stability. In addition there is a direct connection between a classical wave packet and the optical field. The connection is that a perturbative, Gaussian pulse creates a wave packet that is nearly a classical wave packet.

The design procedure may be pictured in terms of an analogy with the study of phase conjugation. Phase conjugation corrects phase aberrations on an optical field so that a field propagating through an aberrating medium arrives at the target unaberrated. Here phase conjugation corrects or adjusts the dispersion of a wave packet so that it evolves to desired target state. In fact the corrected initial state is the phase conjugate of the target state. More precisely, the target state is formed of classical wave packets at time equal to zero. It is then evaluated at the target time, and phase is conjugated to give the initial state. This connection between the phase conjugate of the target state is true in general (it is a property of Hamiltonian evolution; (Messiah 1961). Use of the classical basis provides the picture that it is the dispersion of the wave packet that is being corrected. Thus phase conjugation of the target state leads to an initial state that is also constructed of classical wave packets. Since each classical wave packet can be excited by a perturbative, Gaussian pulse, the amplitudes and phases of the classical wave packets forming the initial state give directly the amplitudes and phases of the optical pulse sequence that will generate the initial state. Creation of such a sequence of optical pulses is experimentally straightforward (provided that the number of pulses is not large). In general, this design scheme will only produce excitation fields that are smoothly varying. Such fields are experimentally possible to produce by either a pulse sequence or by use of a pulse shaper (Weiner 1990).

The quantum state produced by a sequence of phase-locked pulses may also be described in terms of superpositions of several wave packets. The pulses are assumed to be Gaussian and perturbative. If j such pulses are used to excite a quantum state, the resulting wavefunction may be expressed as the superposition of several wave packets:

$$\Psi(\mathbf{r}, t) = \psi(\mathbf{r}, t) + \sum_j c_j e^{i\omega\tau_j} \psi(\mathbf{r}, t - \tau_j), \tag{3.10}$$

where $\psi(t)$ is the wave packet excited by each laser pulse in the Schrödinger picture, ω is the optical frequency of the laser field, c_j is the amplitude of the jth delayed pulse, and τ_j is the delay time for each delayed pulse. The $\omega\tau_j$ is the optical phase delay between the jth pulse and the 0th pulse. The ability to vary three parameters for each pulse (amplitude, time delay, and optical phase) offers great flexibility for the generation of a particular quantum state. As mentioned previously, each of the wave packets in the equation above is an excellent approximation of a classical wave packet.

If the target time corresponds to a fractional revival time of the system, the number of pulses need to form the target state is an integer. If we limit ourselves to such times, we can write a formula for the generation of an arbitrary target state. Expressing the target state as a linear superposition of classical wave packets, we have

$$\Psi(\mathbf{r}, t) = \sum_j A_j \psi_{\text{cl}}\left(\mathbf{r}, t + \frac{j}{l} T_{\text{cl}}\right). \tag{3.11}$$

Note that if the target state does not have a classical wave packet within one classical period, then the target state has zero amplitude for all times. For example, it is not possible to create a target state that has a single well-localized wave packet that oscillates at twice the classical period. There is also a fundamental limit on the highest-frequency component of the structure of the target state. The target state structure, precisely the Ramsey fringes associated with the target state, cannot have a frequency component higher than that allowed by the largest energy difference in the superposition of states.

The designed initial wave packet takes the form

$$\Psi(\mathbf{r}, t) = \sum_{s=0}^{l-1} B_s \psi\left(\mathbf{r}, t - \frac{s}{l}T_{\mathrm{cl}}\right). \tag{3.12}$$

The coefficients are

$$B_s = \sum_{j}^{l-1} A_j a_{j+s}^*. \tag{3.13}$$

The amplitudes a_q are given by

$$a_q = \frac{1}{l} \sum_{k=0}^{l-1} \exp\left(-2\pi i \theta_k + 2\pi i k \frac{q}{l}\right), \tag{3.14}$$

where $k = n - \bar{n}$, as defined above.

As an example, consider a novel Schrödinger Cat state at $(\frac{1}{3})T_{\mathrm{rev}}$, which takes the form $\psi_{\mathrm{cl}}(t) + e^{i\phi}\psi_{\mathrm{cl}}(t - \frac{1}{3}T_{\mathrm{cl}})$. This target wave packet consists of two wave packets that are separated by $\frac{1}{3}T_{\mathrm{cl}}$. It is not the state found at the $\frac{1}{4}$ fractional revival of an evolving radial wave packet. In that case the two sub–wave packets are separated by $\frac{1}{2}T_{\mathrm{cl}}$. In fact it is not a state that is found anywhere in the evolution of a radial wave packet (except perhaps at exceedingly long times). Following the above equation, we obtain the coefficients B_s for each wave packet,

$$B_0 = \frac{-i}{\sqrt{3}}\left[1 + \exp\left(i\phi - \frac{4\pi i}{3}\right)\right],$$

$$B_1 = \frac{-i}{\sqrt{3}}(e^{-4\pi i/3} + e^{i\phi}),$$

$$B_2 = \frac{-i}{\sqrt{3}}[e^{-4\pi i/3}(1 + e^{i\phi})]. \tag{3.15}$$

In Figure 3.8 we numerically test our analytical result (the parameter, ϕ, is taken to be zero). The optical pulses occur at $t = 0$, $t = 1/3T_{\mathrm{cl}}$, and $t = 2/3T_{\mathrm{cl}}$. The amplitudes and phase shifts of these pulses are given directly by the amplitudes and phases of B_s. The

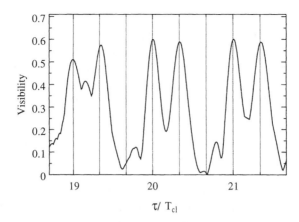

$\tau/\ T_{cl}$

Figure 3.8 Design of Schrödinger Cat target state. The Ramsey fringe visibility indicates the presence of two sub-wave packets separated in time by $1/3T_{cl}$. The system considered is a Rydberg atomic system. The superposition of states has an average principal quantum number of approximately 90 and approximately 7 eigenstates are significantly excited. The target time is $1/3T_{rev} = 20T_{cl}$. The second sub-wave packet follows by $1/3T_{cl}$.

figure shows a spacing of $\frac{1}{3}T_{cl}$ between the Ramsey fringes, in the vicinity of the target time of $(\frac{1}{3})T_{rev}$.

The parameter l sets the upper limit on the number of classical wave packets that can be used to construct the target state. That is, the complexity of the spatial structure of the target state is defined by l. It is also true that l is linked to the target time $t = (m/n)T_{rev}$ (for odd n, $l = n$; for even n, $l = n/2$ provided that $mn/4 = 0(\text{mod } 1)$). However, with a slight shift of the target time, l can be made quite large (e.g., $m/n = 1/2$ vs. 15/31. An upper limit on l is given by the bandwidth of the optical pulse, as discussed above. Note that in turn, l sets an upper limit on the number of pulses needed to excite the initial state.

The central goal of any target state design scheme is to choose an optical field that produces an initial state that evolves into the target state. This time domain scheme chooses optical fields that have a direct one-to-one relationship with the initial wave packet as described in the classical basis. Each classical wave packet can be produced by a single optical pulse of appropriate amplitude and time delay. Although the design was developed in terms of a multiple pulse excitation, note that the optical field that results from the design analysis may be produced by other means. That is, this design is also applicable to pulse-shaping schemes that directly manipulate the spectrum of an optical pulse.

3.4 FINAL COMMENTS

The excitation of atomic wave packets has evolved greatly over the past decade. These studies have gone well beyond the initial thrust of studying the classical-quantum limit. It is now possible to design and create a specific coherent superposition state. The coherence of that state can be subsequently modified by the application of a wide range of fields. That coherence may be measured and the effects of decoherence phenomena studied. Atomic wave packets will continue to have impact on a wide variety of fields ranging from quantum computing, communication, and measurement to the coherent control of reaction dynamics. The atomic system remains a straightforward arena in which to explore these unique phenomena.

ACKNOWLEDGMENTS

This work was supported by the National Science Foundation. Many thanks are owed to X. Chen, P. Lambropoulus, L. Marmet, J. Parker, C. R. Stroud Jr., T. Uzer, and H. Walther for inspiring discussions and contributions of important ideas.

REFERENCES

Aronstein, D. L., and Stroud, C. R. Jr. (1997). *Phys. Rev.* A**55**, 4526.

Averbukh, I. Sh., and Perelman, N. F. (1989). *Phys. Lett.* A**139**, 449.

Averbukh, I. Sh., Vrakking, M. J. J., Villeneuve, D. M., and Stolow, A. (1996). *Phys. Rev. Lett.* **77**, 3518.

Bardeen, C. J., Wang, Q., and Shank, C. V. (1995). *Phys. Rev. Lett.* **75**, 3410.

Campbell, M. B., Bensky, T. J., and Jones, R. R. (1998). *Phys. Rev.* A**57**, 4616.

Chen, X., and Yeazell, J. A. (1997). *Phys. Rev.* A**55**, 3264.

Chen, X., and Yeazell, J. A. (1998). *Phys. Rev.* A**58**, 1267.

Chen, X., and Yeazell, J. A. (1998a). *Phys. Rev. Lett.* **81**, 5772.

Cooke, W. E., Gallagher, T. F., Edelstein, S. A., and Hill, R. M. (1978). *Phys. Rev. Lett.* **40**, 178.

Hanson, L. G., and Lambropoulos, P. (1995). *Phys. Rev. Lett.* **74**, 5009.

Heberle, A. P., Baumberg, J. J., and Kohler, K. (1995). *Phys. Rev. Lett.* **75**, 2598.

Jones, R. R. (1996). *Phys. Rev. Lett.* **76**, 3927.

Kalinski, M., and Eberly, J. H. (1996). *Phys. Rev. Lett.* **77**, 2420.

Kohler, B., Yakovlev, V. V., Che, J., Krause, J. L., Messina, M., K. R., Wilson, R., Schwenter, N., Whitnell, R. M., and Yan, Y. (1995). *Phys. Rev. Lett.* **74**, 3360.

Krause, J. L., Schafer, K. J., Ben-Nun, M., and Wilson, K. R. (1997). *Phys. Rev. Lett.* **79**, 4978.

Lankhuijzen, G. M., and Noordam, L. D. (1996). *Phys. Rev. Lett.* **76**, 1784.

Lyons, B. J., Schumacher, D. W., Duncan, D. I., Jones R. R., and Gallagher, T. F. (1998). *Phys. Rev.* A**57**, 3712.

Marmet, L., Held, H., Raithel, G., Yeazell J. A., and Walther, H. (1994). *Phys. Rev. Lett.* **72**, 3779.

Messiah, A. (1961). *Quantum Mechanics*, Vol. 2, North Holland Amsterdam, p. 667.

Nauenberg, M., Stroud, C. R. Jr., and Yeazell, J. A. (1994). *Sci. Am.* 24 June.

Noel, M. W., and Stroud, C. R. Jr. (1996). *Phys. Rev. Lett.* **77**, 1913.

Parker, J., and Stroud, C. R. Jr. (1986). *Phys. Rev. Lett.* **56**, 716.

Parker, J., and Stroud, C. R. Jr. (1986a). *Physica Scripta* T**12**, 70.

Schumacher, D. W., Lyons, B. J., and Gallagher, T. F. (1997). *Phys. Rev. Lett.* **78**, 4359.

Strehle, M., Weichmann, U., and Gerber, G. (1998). *Phys. Rev.* A**58**, 450.

Weiner, A. M., Laird, D. E., Patel, J. S., and Wullert, J. R. (1990). *Opt. Lett.* **15**, 326.

Wolde, A. ten, Noordam, L. D., Muller, H. G., Lagendijk, A., and van Linden van den Heuvell, H. B. (1988). *Phys. Rev. Lett.* **61**, 2099.

Yan Y., Gillilan, R. E., Whitnell, R. M., and Wilson, K. R. (1993). *J. Phys. Chem.* **97**, 2320.

Yeazell, J. A., and Stroud, C. R. Jr. (1988). *Phys. Rev. Lett.* **60**, 1494.

Yeazell, J. A., and Stroud, C. R. Jr. (1990). *Phys. Rev.* A**40**, 5040.

Yeazell, J. A., Mallalieu, M., and Stroud, C. R. Jr. (1990a). *Phys. Rev. Lett.* **64**, 2007.

Yeazell. J. A., and Stroud, C. R. Jr. (1991). *Phys. Rev.* A**43**, 5153.

CHAPTER 4

From Asteroids to Atoms: Quantum Wave Packets and the Restricted Three-Body Problem of Celestial Mechanics

ERNESTINE LEE, ANDREA F. BRUNELLO, CHARLES CERJAN, TURGAY UZER, and DAVID FARRELLY

> The title "Atomic Mechanics," given to these lectures which I delivered in Göttingen during the session 1923–24, was chosen to correspond to the designation "Celestial Mechanics."
> —Max Born, in the Preface to the German edition of *Mechanics of the Atom*[1]

4.1 INTRODUCTION

The reason why atomic physicists are interested in celestial mechanics is simple: the gravitational and Coulombic potentials are mathematically identical and a one-electron atom is therefore governed by the same Hamiltonian as is the Kepler problem. The classical equations of motion for the Kepler problem are easy to solve: In the solar system, if only the Earth and the Sun were considered, then the Earth would orbit the Sun[2] endlessly, never varying its elliptical course. The presence of the Moon destroys the integrability of the system and the resulting motions become extremely complicated. For millennia astronomers have been fascinated by these motions and a large part of the history of celestial mechanics has been given over to the development of theories of the Moon (Gutzwiller 1998; Neugebauer 1969). To give but one example, Delaunay spent 20 years developing an analytical theory of the

---

[1]The German title was actually "Atommechanik" which corresponds rather well with "Himmelsmechanik," the German name for celestial mechanics.
[2]Strictly speaking, this should be defined with respect to the center of mass of the system.

The Physics and Chemistry of Wave Packets, Edited by John Yeazell and Turgay Uzer
ISBN 0-471-24684-0   © 2000 John Wiley & Sons, Inc.

motion of the Moon which culminated in his publication, in 1860, of an approximate expression for the "disturbing function" that contained 320 terms, each of which was derived by hand   (Danby 1992). This theory was later used as the basis for constructing lunar tables. Remarkably few errors were contained in Delaunay's magnificent expansion which has been thoroughly checked and extended by Deprit and coworkers using symbolic manipulation programs (Deprit, Henrard, and Rom 1971).

When the number of bodies is increased beyond three, the problem becomes even more of a nightmare. In fact, despite an enormous expenditure of effort, no case beyond the two-body problem has been solved completely. This complexity has had its felicitous side, however, in that many of the important developments in applied mathematics and physics, from the publication of Sir Isaac Newton's *Principia* in 1687 to Max Born's *Mechanics of the Atom* in 1925 (see Born 1960), and down to today, have been fueled by progress made in studies of the n-body problem (Danby 1992). In particular, the "old quantum mechanics" described in Born's book includes many concepts and methodologies that were carried over intact from celestial mechanics. Although celestial mechanics and atomic physics temporarily parted ways after the discovery of wave mechanics, the synergism between the two fields left a lasting impression: The notion of a "classical" or "Bohr" atom in which a localized electron revolves around the nucleus in a Kepler orbit continues to persist not only among members of the general public but also among atomic physicists. Indeed, the goal of making a classical atom accounts for a notable fraction of the vast body of contemporary experimental and theoretical research on Rydberg atoms (Gallagher 1994; Connerade 1998). However, the three-body problem itself, arguably the *raison d'être* of celestial mechanics for several centuries, has received relatively less attention from atomic physicists because it does not have a direct quantum counterpart. Three quantum particles cannot mutually attract one another and at the same time interact through a purely Coulombic force law. Nevertheless, in the last five years it has been demonstrated by several research groups that quantum analogues of a particular limit of the three-body problem, *the restricted three-body problem* (Szebehely 1967) not only exist but contain dramatically new physics. This chapter will describe this work and in particular will demonstrate the possibility of producing localized electronic states in atoms that are direct analogues of the coherent states of the harmonic oscillator. No attempt will be made to review all of the work in the field pertaining to the creation of nonstationary, nondispersive wave packets in atoms. Instead the chapter will be limited to a discussion of only those quantum wave packets that are related directly to the restricted three-body problem, this effectively excluding a discussion of work using linearly polarized fields, and so forth.

The wave packets we construct move along a classical orbit without spreading; in other words, the system constitutes a classical atom, although the stability of the wave packets is achieved by the continuous application of external fields. One must admit, of course, that the development of wave mechanics forced the abandonment of the very notion that a genuinely classical atom can exist, since no unique path can be ascribed to an atomic electron. For this reason it is imperative at the outset to state in clear and unambiguous language what is meant by a *classical atom*. Based on ideas that can be traced directly back to Schrödinger, a definition of a classical atom that is consistent with quantum mechanics can be stated as follows:

1. The wave packet representing the electron neither spreads nor disperses as its center moves along a Kepler orbit about the nucleus.

2. The spatial extent of the packet is small compared to the radius of the orbit.

3. The classical orbit along which the center of the wave packet moves is confined to a single plane in space.

Schrödinger was among the first to consider this problem, and his thoughts are collected in his essay *Der Stetige Uebergang von Mikro-zur Makromechanik*[3] (1926, 1968). To set the scene: Schrödinger, having found the stationary states of the harmonic oscillator, realized that they could not represent a harmonically oscillating classical particle since they were smeared out over the available position space. Thus he proceeded to ask whether a superposition of these eigenstates could produce a wave packet that evolved in time harmonically, and exhibited minimal dispersion during this motion, just like a classical particle would. The coherent states of the harmonic oscillator that he thus constructed represent the closest approximation, within the laws of quantum mechanics, to a classical particle and its motion under the influence of a Hooke's law force (Glauber 1963). Schrödinger relayed his discovery of the harmonic oscillator coherent states to Planck in a letter dated May 31, 1926, which concluded: "I believe that it is only a question of computational skill to accomplish the same thing for the electron in a hydrogen atom" (see Prizibram 1967). As before, he sought to superpose stationary states, this time of the hydrogen atom, to form nondispersing wave packets that moved on the elliptic orbits generic to the Kepler-Coulomb problem, thereby creating as classical an atomic electron as quantum mechanics allows. Before the year was out, however, he had concluded that the technical difficulties in the calculation were far greater than in the harmonic oscillator case (this was indicated in a letter to H. A. Lorentz; see Schrödinger 1968).

Ultimately Schrödinger proved unable to construct coherent states of the hydrogen atom, and the prospects for producing such states either in the laboratory or on the computer remained dismal until quite recently. During the past 10 years or so interest in the problem has been rekindled and there are an ever-increasing number of theoretical and experimental attempts to form nonstationary, nondispersive electronic wave packets in Rydberg atoms (Yeazell and Stroud 1990). The harmonic oscillator is, of course, the ideal quantum system in which to carry out studies on truly nondispersive or coherent wave packets due to the constant energy level spacings. Naturally the question arises as to whether it might be possible to prepare locally harmonic regimes in atoms thereby allowing the creation of almost completely nondispersive coherent atomic states that are identical to the coherent states of the harmonic oscillator. One approach (Alber and Zoller 1991) has been to try to form

[3]"The continuous transition from microscopic to macroscopic mechanics."

nondispersive wave packets by manipulating the energy levels of a Rydberg or hydrogen atom so that the level spacings are locally constant or almost constant (i.e., locally harmonic). The energy level spacings of a hydrogen atom are clearly not constant but can be made to be almost so with the use of a perturbation such as an external static electric field (the Stark effect). From perturbation theory, the energy levels of the hydrogen atom in a static electric field (F) are given to first order by

$$E_{nk} \approx -\frac{1}{2n^2} + \frac{3}{2} Fnk, \tag{4.1}$$

where n is the principal quantum number and k is the parabolic quantum number (the Stark effect separates in parabolic coordinates). The electric field thus splits the states of a given principal quantum number manifold into *equally spaced* (locally almost harmonic) Stark states. Thus a common experimental strategy has been to work at very high quantum numbers, possibly but not necessarily in the presence of an external electric field, to create regimes in which the local energy spacings are approximately constant. Laser excitation is then used to form a spatially localized superposition of such atomic states. However, while this wave packet can be localized in a plane and, within that plane, in the radial direction, it cannot be localized angularly. As a result it spreads along its elliptic orbit, and interference effects lead to experimentally detectable recurrences (Nauenberg, Stroud, and Yeazell 1994).

It is the concept of *linearity* or *harmonicity* that provides the point of contact between the restricted three-body problem and the idea of a classical atom which is the foundation of the work described in this chapter. Motion that is governed (if only locally) by a linear force law generates an energy spectrum that is (at least locally) harmonic. Clearly, a classical equilibrium point in a potential energy surface is a good candidate for generating coherent states. However, in a one-electron atom the potential is Coulombic, and the possibility of artificially creating an outer equilibrium point in the total potential by the application of external fields, while attractive, has so far failed to materialize[4] (e.g., the proposal of Clark, Korevaar, and Littman 1985 for creating a "quasi-Penning" trap in which the electron is localized at the Stark saddle point of a hydrogen atom in crossed electric and magnetic fields proved to be not feasible; see Nessman and Reinhardt 1987). An alternative solution was suggested by Iwo Bialynicki-Birula, Maciej Kaliński, and Joe Eberly (Bialynicki-Birula, Kaliński, and Eberly 1994, 1995; Kaliński and Eberly 1995a, 1996a, 1996b, 1997; Bialynicki-Birula and Bialynicki-Birula 1996, 1997a, 1997b) and by the present authors (Farrelly and Uzer 1995; Farrelly, Lee, and Uzer 1994, 1995; Lee, Uzer, and Farrelly 1994; Lee, Farrelly, and Uzer 1997; Lee, Brunello, and Farrelly 1995, 1997; Brunello, Uzer, and Farrelly 1996; Cerjan et al. 1997); the two teams recognized independently (Bialynicki-Birula, Kaliński, and Eberly 1994; Farrelly and Uzer 1995) the possibility of creating so-called Lagrange equilibrium points in an atom by using circularly polarized microwave fields, possibly in combination with a magnetic field (Farrelly, Lee, and Uzer 1995). Both groups also realized that much could be learned

---

[4]Except for a special case described later.

from the work of Lagrange in the 1770s (Szebehely 1967) and of Hill (1878) in the 1890s, each of whom devoted considerable effort to the three-body problem.

The transition from two to three bodies is so fraught with difficulties, in general, that various simplified cases of the three-body problem were developed. Thus Lagrange and Hill directed their efforts to one of these simplified cases, the restricted three-body problem. In the restricted three-body problem all three bodies are assumed to lie in the same plane; two of the bodies have finite mass and revolve around each other in circular orbits, while the third body has infinitesimal mass and moves in their combined field (Szebehely 1967). There are many examples of these requirements being approximately fulfilled in the solar system with the most extreme manifestation being the Sun–Jupiter–asteroid system (see Fig. 4.1). In effect the asteroid moves under the combined influence of (1) a gravitational attraction to the center of mass of the Sun–Jupiter system and (2) the time dependent field produced by Jupiter's circling of the Sun. Despite this simplification the dynamics generated by the restricted three-body problem Hamiltonian is by no means trivial. In the late 1800s Poincaré observed that the restricted three-body problem contained certain motions that were unstable. This behavior is now called *chaotic* motion (Poincaré 1993). The problem contains several other surprises that are germane to the discussion; Lagrange discovered the existence of five equilibrium points, now called the *Lagrange equilibria* (L_1–L_5), two of which were stable despite their being energy *maxima*.[5] For many physicists this is a totally unfamiliar concept. However, the facts are indisputable; two swarms of minor planets called the Trojan asteroids do indeed inhabit the vicinity of these equilibrium points which lie 60° ahead of and behind Jupiter in its orbit.[6] Lagrange was able to demonstrate only linear stability of these equilibrium points. In reality the motion can be rather nonlinear with the oscillations of individual asteroids amounting to as much as 20° of longitude (Danby 1992). Because the asteroids are localized at energy maxima, these large amplitude motions together with dissipative effects (or collisions) can cause asteroids to escape, some of which may ultimately cross the orbits of the inner planets (Levison, Shoemaker, and Shoemaker 1997). This observation has important consequences for the stability of quantum wave packets not to mention the inner planets.[7]

The close analogy between a Rydberg atom in a circularly polarized microwave field and the restricted three-body problem extends to the existence of atomic stable equilibrium points that are directly analogous to the Lagrangian equilibrium points in

[5]In 1764 Lagrange won the Paris Academy of Science's award for the best essay on the librations of the Moon.

[6]Although collectively called "Trojan asteroids," the two individual subgroups are called *The Trojans* (which trail Jupiter) and *The Greeks* (which lead Jupiter). The asteroids being named, for the most part, after the Trojan and Greek heroes in Homer's *Iliad* (Fagles 1990). However, there are heroes from both sides of the war mixed in the two subgroups. In order of discovery, they are Achilles, Patroclus, Hector, Nestor, Priamus, Agamemnon, Odysseus, Aeneas, Anchises, Troilus, and Ajax (Garfinkel 1978, 1982).

[7]Levison, Shoemaker, and Shoemaker (1997) estimate that there are over 200 escaped Trojan asteroids with diameters >1 km currently roaming the solar system, a few of which may be on Earth-crossing orbits.

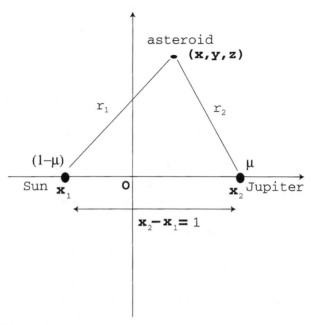

Figure 4.1 Diagram illustrating the relevant parameters used in the restricted three-body problem. The diagram is in the rotating frame and is specific to the Sun–Jupiter–asteroid problem.

celestial mechanics. This led Bialynicki-Birula, Kaliński, and Eberly (1994) to expect that wave packets launched from the atomic equilibrium point analogous to the stable Lagrange points L_4 and L_5 would orbit the nucleus without spreading. The term "Trojan" wave packet seems appropriate for such states. While the analogy between Rydberg atoms and planetary systems turned out to be useful, the finite size of Planck's constant imposes an absolute scale on the atomic problem which soon leads to dispersion. The atomic analogues of these points are stable only over a limited range of parameters, and placing a finite-size minimum uncertainty wave packet at such an equilibrium point may require unrealistically large principal quantum numbers to guarantee stability. In other words, linear stability is not enough to guarantee the actual stability of a particular wave packet.

Still, the discovery of the possibility of forming nonstationary wave packets that are resistant to dispersion quickly led to the growth of a sizable body of literature describing one aspect of the problem or another. For example, Bialynicki-Birula, Kaliński, and Eberly (1995) experimented with differently shaped wave packets in an attempt to minimize dispersion. This followed the suggestion of three of the current authors that Hill's curves of zero velocity provided clues to the size and shape of the stability region (Farrelly, Lee, and Uzer 1995). Following the earlier discovery (Buchleitner 1993; Buchleitner and Delande 1995) of nonTrojan, Floquet states anchored to stable islands in the classical phase space of the linearly polarized microwave problem, Zakrzewski, Delande, and Buchleitner (1995; Delande, Zakrzewski, and Buchleitner 1995) and three of us (Brunello, Uzer, and Farrelly 1996) have shown that it is possible to find eigenstates of the problem in a rotating frame that, being eigenstates, are immune to spreading. In the laboratory frame such eigenstates indeed orbit the nucleus without spreading. However, they are neither wave packets nor coherent states in the sense of Schrödinger and so do not mimic the harmonic oscillator coherent states, they are not minimum uncertainty wave packets. Suggestions for the experimental preparation of these rather exotic and fragile "quantum orchidlike" states can be found in the literature (Kaliński and Eberly 1996b; Delande, Zakrzewski, and Buchleitner 1995). The essential importance of *coherent* wave packets is discussed in Chapter 1 of this book.

The present authors invented a different method in which a homogeneous magnetic field is used in addition to the circularly polarized microwave field to stabilize continuously the packet as it moves around the nucleus; in essence, the states are robust coherent states of an atom dressed by both microwave and magnetic fields. Examination of all of these systems in a rotating frame reveals the key role played by Coriolis forces. In fact a major emphasis of our research is to study the effects of nonconserved Coriolis forces on electronic dynamics. Coriolis effects lead to velocity dependent terms in the Hamiltonian which prevent one from defining a potential energy surface. Hill recognized this problem when he studied the Earth–Moon–Sun problem in a rotating frame (Hill 1878). This led to his introducing the idea of curves of zero relative velocity to visualize the allowed and forbidden dynamical regions as well as the positions of any equilibrium points. This chapter will make extensive use of the concept of a zero-velocity surface (ZVS) as a guide to the synthesis and dynamics of electronic wave packets. The ZVS has been used successfully in a number of recent applications to Rydberg atoms (Farrelly and Uzer 1995; Farrelly, Lee, and Uzer 1994, 1995; Lee, Uzer, and Farrelly 1994; Lee, Brunello, and Farrelly 1995, 1997; Brunello, Uzer, and Farrelly 1997; Lee, Farrelly, and Uzer 1997; Cerjan et al. 1997), Rydberg molecules (Lee, Uzer, and Farrelly 1994), quantum dots (Farrelly, Lee, and Uzer 1998; Lee et al. 1998) and Bose-Einstein condensates of alkalimetal atoms (Brunello, Uzer, and Farrelly n.d.). The authors recommend its use very highly in problems containing nonconserved Coriolis or paramagnetic terms (the latter having the same functional form as Coriol is terms). It is therefore useful at this point to introduce the restricted three-body problem and the idea of a surface of zero relative velocity.

4.2 RESTRICTED THREE-BODY PROBLEM

In the restricted three-body problem two of the bodies (the "primaries") have finite mass and revolve around each other in circular orbits, while a third body, which is assumed not to perturb the primaries, moves in their combined field. All motion occurs in a common plane, so it is useful to do as Hill did and move to a frame that rotates with the Kepler frequency of the primaries. In appropriately scaled units (for details, see Danby 1992) such that the axes are rotating with constant angular velocity

$\omega = 1$ and defining μ and $(1 - \mu)$ to be the masses of the primaries with $\mu \leq \frac{1}{2}$, the Hamiltonian is given by[8]

$$K = \frac{1}{2}(p_x^2 + p_y^2) - (xp_y - yp_x) - \frac{(1 - \mu)}{\rho_1} - \frac{\mu}{\rho_2}, \tag{4.2}$$

where $\rho_1 = \sqrt{(x + \mu)^2 + y^2}$ and $\rho_2 = \sqrt{(x + \mu - 1)^2 + y^2}$. In the case of the Trojan asteroids $\mu \sim 10^{-3}$, and the primaries are the Sun and Jupiter (see Fig. 4.1). In effect the three-body problem has been reduced to that of a two-body potential problem plus a time dependent perturbation which, in the rotating frame, emerges in the form of a Coriolis term. The Coriolis term mixes coordinates and momenta together and so prevents the construction of a potential energy surface. Nevertheless, using the technique invented by Hill to circumvent this difficulty, one can gain considerable insight into the system.

4.2.1 Rotating Axes and the Zero-Velocity Surface

A consequence of Newton's second law is that if a conservative force $\mathbf{P} = -\nabla V$ acts on a particle, then motion with respect to axes that are rotating with constant angular velocity ω about the z-axis will be determined by

$$\mathbf{P} = m[\ddot{\mathbf{r}} + \{2\omega\hat{\mathbf{z}} \times \dot{\mathbf{r}}\} + \{\omega^2\hat{\mathbf{z}} \times (\hat{\mathbf{z}} \times \mathbf{r})\}], \tag{4.3}$$

where the extra terms (as compared to Newton's second law in an inertial frame) in the first and second sets of curly braces are the Coriolis and centrifugal forces, respectively. The following relation has been used

$$\frac{d\mathbf{r}}{dt} = \frac{\partial \mathbf{r}}{\partial t} + \omega \times \mathbf{r}, \tag{4.4}$$

which relates the rate of change of a vector $\mathbf{r}$ in a fixed frame of reference to that in a frame rotating with angular velocity ω. If $\mathbf{r}$ is decomposed into perpendicular and planar components

$$\mathbf{r} = z\hat{\mathbf{z}} + \rho, \tag{4.5}$$

then

$$\mathbf{P} = m[\ddot{\mathbf{r}} + 2\omega\hat{\mathbf{z}} \times \dot{\mathbf{r}} - \omega^2\rho]. \tag{4.6}$$

Using the relation $\rho \cdot \dot{\mathbf{r}} = \rho \cdot \dot{\rho}$ and forming the quantity $\mathbf{P} \cdot \dot{\mathbf{r}}$, one can calculate the work done in going from A to B:

---

[8]The symbol K will be used throughout to denote Hamiltonians in a noninertial rotating frame. This is not to be confused with the "Kamiltonian" introduced in Goldstein (1950).

$$W_{AB} = \int_A^B \mathbf{P} \cdot d\mathbf{r} = \frac{m}{2}(v_B^2 - v_A^2) - \frac{m\omega^2}{2}(\rho_B^2 - \rho_A^2), \tag{4.7}$$

where v_A and v_B are the mechanical velocities. For a conservative field $W_{AB} = V(A) - V(B)$, and we obtain the result

$$K = \frac{1}{2} m\dot{\mathbf{r}}^2 + \frac{1}{2}m\omega^2\rho^2, \tag{4.8}$$

where the last term is called the *rotational potential* (Danby 1992). It is apparent that the motion in the rotating frame is governed by the *modified* potential energy function

$$\Upsilon(x,y,z) = V - \frac{1}{2}m\omega^2(x^2 + y^2), \tag{4.9}$$

which for fixed Υ is the locus of the surfaces of zero velocity (Danby 1992).

4.2.2 Hill's Curves of Zero Velocity

In celestial mechanics the surface defined by $\Upsilon(x, y, z)$ is often termed the *surface of zero relative velocity* or simply the *zero-velocity surface* and was first used by Hill in his studies of the Earth–Moon–Sun problem.

It is important to realize that Υ is not a potential, even though it may share some properties with a regular potential energy surface. Figure 4.2 is a plot of the ZVS for the restricted three-body problem which shows the presence of the five Lagrange points including the stable equilibrium points L_4 and L_5[9] which are clearly energy maxima. They are stabilized by the Coriolis terms in the Hamiltonian, and their existence would not be apparent simply by examining the function V. An excellent discussion is provided by Greenberg and Davis (1978).

The ingredients for creating a quantum analogue of this problem are now clear; one needs a time dependent periodic field (to mimic Jupiter's idealized motion) together with a Coulombic interaction (to represent gravitational interactions). These conditions can be met and made to produce a stable equilibrium point in the problem of a hydrogen atom interacting with a circularly polarized microwave field.

4.3 TROJAN WAVE PACKETS

The problem of a hydrogen atom in a circularly or elliptically polarized microwave field has received considerable attention, mainly from the standpoint of how the polarization of the field affects ionization (Gallagher 1994). In the dipole approximation and atomic units ($\hbar = m_e = e = 1$) the Hamiltonian for a hydrogen atom

---

[9]The L_5-Society (Greenberg and Davis 1978) suggests that a manned space station might reasonably be located at the L_5 point associated with the Earth–Moon system. Perturbations from the Sun complicate the stability analysis.

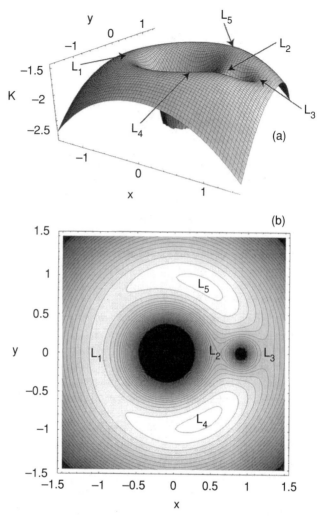

Figure 4.2 (*a*) Isometric and (*b*) contour plot of the ZVS for the restricted three-body problem with $\mu = 0.1$. The five Lagrangian points are labeled L_1 through L_5. The two maxima are the Lagrange points L_4 and L_5. K is the Jacobi constant, which is the energy in the rotating frame.

subjected to a circularly polarized microwave field, and a magnetic field perpendicular to the plane of polarization is

$$H = \frac{\tilde{\mathbf{p}}^2}{2} - \frac{1}{r} - \frac{\omega_c}{2}(xp_y - yp_x) + \frac{\omega_c^2}{8}(x^2 + y^2) + F(x \cos \omega_f t + y \sin \omega_f t). \quad (4.10)$$

The magnetic field is taken to lie along the positive *z*-direction, $\omega_c = eB/m_e c$ is the cyclotron frequency, while ω_f is the microwave field frequency and F is its strength. In a frame rotating with the field frequency ω_f the Hamiltonian becomes

$$H = \frac{\tilde{\mathbf{p}}^2}{2} - \frac{1}{r} - \left(\omega_f + \frac{\omega_c}{2}\right)(xp_y - yp_x) + Fx + \frac{\omega_c^2}{8}(x^2 + y^2), \tag{4.11}$$

where K is called the Jacobi constant[10] in analogy with the restricted three-body problem (Danby 1992), and the coordinates are now interpreted as being in the rotating frame.

Construction of the ZVS is accomplished most easily by a simple short cut: Re-write the Hamiltonian in terms of the velocities using Hamilton's equations of motion, and then set the velocity terms to zero;[11] those terms remaining define the modified potential energy surface, or ZVS:

$$\Upsilon = -\frac{1}{r} + Fx - \frac{\omega_f(\omega_f + \omega_c)}{2}(x^2 + y^2). \tag{4.12}$$

The equilibrium points of the ZVS are found to lie along the x-axis as shown in Figure 4.3. It is L_m that provides the analogy with the Lagrange equilibrium points L_4 and L_5 in the restricted-three body problem. This configuration of equilibria occurs whether or not a magnetic field is added. Bialynicki-Birula (Bialynicki-Birula, Kaliński, and Eberly 1994) originally studied the case in which only a circularly polarized microwave field was used.

For convenience the coordinates and the momenta are scaled as follows:

$$\mathbf{r}' = \omega^{2/3}\mathbf{r}, \tag{4.13}$$

$$\tilde{\mathbf{p}}' = \omega^{-1/3}\mathbf{p}, \tag{4.14}$$

where

$$\omega = \left(\omega_f + \frac{\omega_c}{2}\right). \tag{4.15}$$

Substitution of the scaled quantities into the Hamiltonian and dropping the primes yields the new Hamiltonian

$$\mathcal{K} = \frac{\tilde{\mathbf{p}}^2}{2} - \frac{1}{r} - (xp_y - yp_x) + \varepsilon x + \frac{\omega_s^2}{8}(x^2 + y^2). \tag{4.16}$$

The scaling shows that the *classical* dynamics depends only on the three scaled parameters, K, ω_s, and ε. The stability conditions can now be expressed in terms of the two parameters ω_s and ε. It is now possible to examine the regions of linear stability

---

[10]The terms *Jacobian* and *Jacobin* have, of course, already been taken.

[11]In some cases a more sophisticated analysis is needed. See Lee, Uzer, and Farrelly (1994).

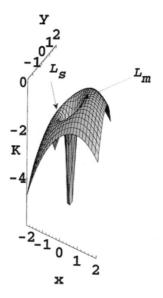

Figure 4.3 Isometric plot of the ZVS for a maximum using scaled units with $F = 0.6$, $\omega c = 0$, and $\omega f = 1$. The saddle L_s and the maximum L_m are shown. K is the Jacobi constant.

for the cases where $\omega_c = 0$ and $\omega_c \neq 0$ using the methods described in Farrelly, Lee, and Uzer (1995). Figure 4.4 shows the regions of stable and unstable motion at the maximum as a function of the parameters ω_s and ε. For $\omega_c \neq 0$ the curve separating stable from unstable motion consists of an upper branch ($\varepsilon^>$) and a lower branch ($\varepsilon^<$) given by

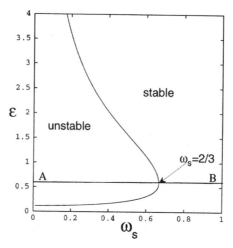

Figure 4.4 Stability regions for the maximum configuration as a function of the scaled parameters ω_s and ε. The area to the left of the curve is unstable, whereas the area to the right is linearly stable. The line AB intersects the point where the two parts of the stability-instability curve coalesce.

$$\varepsilon^{><} = \frac{4 - 5\omega_s^2 \pm (2 \mp \frac{1}{2}\omega_s^2)\sqrt{4 - 9\omega_s^2}}{2^{4/3}\omega_s^{2/3}(2 \pm \sqrt{4 - 9\omega_s^2})^{2/3}},$$ (4.17)

where the upper (lower) sign is taken throughout for $\varepsilon^>$ ($\varepsilon^<$). These two functions become equal to each other exactly at $\omega_s = 2/3$ which is the rightmost point in Figure 4.4. It was something of a surprise to recognize that at this point the planar limit of the problem reduces to the rigorously integrable case discovered by Raković and Chu (1994):

$$K = \frac{(p_x^2 + p_y^2)}{2} - \frac{1}{r} - (xp_y - yp_x) + \left(\frac{2}{3}\right)^{4/3} x + \frac{1}{18}(x^2 + y^2).$$ (4.18)

4.3.1 Coherent States

The next step is to expand the Hamiltonian at the equilibrium point using methods developed in celestial mechanics (Deprit 1966) and nuclear physics (Glas, Mosel, and Zint 1978). Full details and an exhaustive set of references are provided in Lee, Brunello, and Farrelly (1997) and Cerjan et al. (1997). Refs. 34, 35. First a transformation from the original rotating coordinates to the eqolibrium configuration is done using the canonical transformation:

$$x = x_0 + \xi, \quad p_x + p_\xi,$$ (4.19)

$$y = \eta, \quad p_y = \omega x_0 + p_\eta,$$ (4.20)

$$z = \zeta, \quad p_x = p_\zeta.$$ (4.21)

This transforms the Hamiltonian into the form

$$K = \frac{p_\xi^2 + p_\eta^2 + p_\zeta^2}{2} - \omega(\xi p_\eta - \eta p_\xi) + F\xi + \Theta,$$ (4.22)

where the function

$$\Theta = -\frac{1}{r} + Fx_0 + \frac{\omega_c^2}{8}(\xi^2 + \eta^2) - \frac{1}{2}\omega^2 x_0^2 + \frac{1}{8}\omega_c^2 x_0^2 + \left(\frac{\omega_c^2}{4} - \omega^2\right)x_0\xi$$ (4.23)

is expanded around $(\xi, \eta, \zeta) = (0, 0, 0)$. This produces an approximate Hamiltonian describing librations around the equilibrium

$$\mathcal{H} = \frac{p_\xi^2 + p_\eta^2 + p_\zeta^2}{2} + \frac{\omega^2}{2}(a\xi^2 + b\eta^2 + c\zeta^2) - \omega(\xi p_\eta - \eta p_\xi) + K_c$$ (4.24)

with

$$a = \frac{1}{\omega^2}\left(\frac{\omega_c^2}{4} - \frac{2}{x_0^3}\right), \qquad b = \frac{1}{\omega^2}\left(\frac{\omega_c^2}{4} + \frac{1}{x_0^3}\right), \qquad c = \frac{1}{\omega^2 x_0^3}. \tag{4.25}$$

The part of the Hamiltonian containing only constant terms is given by

$$K_c = -\frac{1}{2}\,\omega^2 x_0^2 + F x_0 + \frac{1}{8}\,\omega_c^2 x_0^2 - \frac{1}{x_0}. \tag{4.26}$$

The actual derivation of the conditions for linear stability at the equilibrium point is discussed elsewhere (Farrelly, Lee, and Uzer 1995). In the case where $\omega_c = 0$, the stable region for the problem in hand is extremely limited and results in a very restricted set of values that a and b may take for stable dynamics to be possible: Specifically, the field parameters must be selected so that b lies in the range $8/9 < b < 1$. A transition from stability to instability occurs as one passes outside the stable regime. This transition is known as a Trojan bifurcation (Abraham and Marsden 1987), and it occurs when $\varepsilon = F/\omega_f^{4/3}$ (see Fig. 4.4).

The harmonic approximation to the Hamiltonian to describe librations around the equilibrium point shows the motion in the z (or ζ)-direction to be stable, harmonic, and uncoupled from the planar motion. Therefore we temporarily ignore this degree of freedom.

After a rotation in phase space,

$$\xi' = A\xi + B p_\eta, \tag{4.27}$$

$$\eta' = A\eta + B p_\xi, \tag{4.28}$$

$$p_\xi' = p_\xi + C\eta, \tag{4.29}$$

$$p_\eta' = p_\eta + C\xi, \tag{4.30}$$

with $A - BC = 1$ (to preserve the commutation relations between coordinates and momenta). H can be reduced to the following separable form (Glas, Mosel, and Zint 1978):

$$\mathcal{K} = \frac{1}{2m_\xi}p_\xi'^2 + \frac{1}{2}m_\xi\Omega_\xi^2\xi'^2 + \frac{1}{2m_\eta}p_\eta'^2 + \frac{1}{2}m_\eta\Omega_\eta^2\eta'^2, \tag{4.31}$$

where we will assume (without loss of generality) that $\omega_\eta > \omega_\xi$. The locally harmonic frequencies

$$\Omega_\eta^2 = \omega^2\left(\frac{1}{4}\,\omega_s^2 + 1 - \frac{1}{2}q - \frac{1}{2}\sqrt{9q^2 - 8q + 4\omega_s^2}\right), \tag{4.32}$$

$$\Omega_\xi^2 = \omega^2 \left(\frac{1}{4} \omega_s^2 + 1 - \frac{1}{2} q + \frac{1}{2} \sqrt{9q^2 - 8q + 4\omega_s^2} \right), \tag{4.33}$$

and the "masses"

$$m_\xi = \frac{\sqrt{9q^2 - 8q + 4\omega_s^2}}{(2 + \frac{3}{2} q) + \frac{1}{2} \sqrt{9q^2 - 8q + 4\omega_s^2}}, \tag{4.34}$$

$$m_\eta = \frac{- \sqrt{9q^2 - 8q + 4\omega_s^2}}{(2 - \frac{3}{2} q) - \frac{1}{2} \sqrt{9q^2 - 8q + 4\omega_s^2}}, \tag{4.35}$$

are given here in terms of the dimensionless quantity

$$q = \frac{1}{\omega^2 x_0^3}. \tag{4.36}$$

The "masses" are not physical masses and can be either positive or negative; there is no bound motion if both masses are negative. In the case of motion at a maximum, the masses have opposite signs. Based on the zero-velocity surface, this case occurs when $\omega^2 > \omega_\xi^2, \omega_\eta^2$. When both masses are positive, stable motion at a minimum is indicated. In order to cover the two possibilities for stable motion, we define the index

$$\Lambda = \frac{m_\xi m_\eta}{|m_\xi m_\eta|} \tag{4.37}$$

which is 1 for a minimum and -1 for a maximum.

The energy eigenvalues are given by

$$E = \frac{m_\xi}{|m_\xi|} \left(n_\xi + \frac{1}{2} \right) \hbar |\Omega_\xi| + \frac{m_\eta}{|m_\eta|} \left(n_\eta + \frac{1}{2} \right) \hbar |\Omega_\eta|. \tag{4.38}$$

The magnitude of the ground state energy is defined in terms of an average frequency Ω through

$$E = \frac{\hbar}{2} (|\Omega_\xi| + \Lambda |\Omega_\eta|) = \hbar \Omega \tag{4.39}$$

with this frequency explicitly being given by

$$\Omega = \frac{|\omega|}{2} \sqrt{2 - q + \frac{1}{2} \omega_s^2 + 2\Lambda s(q, \omega_s)}, \tag{4.40}$$

where

$$s(q, \omega_s) = \sqrt{(1 + 2q - \frac{1}{4}\omega_s^2)(1 - q - \frac{1}{4}\omega_s^2)} \, . \tag{4.41}$$

4.3.2 The Initial Wave Packet

The ground-state wavefunction of the (local) electronic Hamiltonian becomes

$$\Psi_{000}(\xi, \eta, \zeta) = N\psi_C(\xi, \eta)\exp\left(-\frac{\omega_\zeta}{2}\zeta^2\right), \tag{4.42}$$

where $\psi_C(\xi, \eta)$ is the normalized ground state wavefunction of the cranked oscillator

$$\psi_C(\xi, \eta) = \left(\frac{\alpha\beta}{\pi^2}\right)^{1/4}\exp\left(-\frac{\alpha}{2}\xi^2 - \frac{\beta}{2}\eta^2 - i\gamma\xi\eta\right). \tag{4.43}$$

The parameters α, β, γ are given by

$$\alpha = \left(\frac{\omega}{3q\hbar}\right)\left(1 + 2q - \frac{1}{4}\omega_s^2 + \Lambda s\right)\sqrt{2 - q + \frac{1}{2}\omega_s^2 + 2\Lambda s(q, \omega_s)} \, , \tag{4.44}$$

$$\beta = \left(\frac{\omega}{3q\hbar}\right)\left(q - 1 + \frac{1}{4}\omega_s^2 - \Lambda s\right)\sqrt{2 - q + \frac{1}{2}\omega_s^2 + 2\Lambda s(q, \omega_s)} \, , \tag{4.45}$$

$$\gamma = \left(\frac{\omega}{3q\hbar}\right)\left(2 + q - \frac{1}{2}\omega_s^2 + 2\Lambda s\right). \tag{4.46}$$

When the stabilizing magnetic field is absent ($\omega_s = 0$), these parameters reduce to the ones used by Bialynicki-Birula, Kaliński, and Eberly (1994; after interchanging x and y and reversing the sign of ω to account for differences in our conventions).

4.3.3 Phase Factors

The transformation to barycentric synodical coordinates requires two shifts, one in coordinate and another in momentum, to reach the equilibrium point from the center of mass. These shifts introduce all-important phase factors. The quantum mechanical consequence of these shifts can be described by a translation operator

$$\mathcal{T}(x_0) = \exp\left(-\frac{ix_0 p_\xi}{\hbar}\right) \tag{4.47}$$

and a boostlike operator

$$\mathcal{B}(\omega x_0) = \exp\left(\frac{i\omega x_0 \eta}{\hbar}\right) \tag{4.48}$$

If $|C>$ is the ket represented by $\psi_C(\xi,\eta)$, then the ket $|I\rangle$ that we need to use as the initial state in the barycentric coordinates is related to it by

$$|I\rangle = \mathcal{T}(x_0)\mathcal{B}(\omega x_0)|C\rangle \tag{4.49}$$

and therefore the wavefunctions are related by

$$\psi_C(x,y) = \left(\frac{\alpha\beta}{\pi^2}\right)^{1/4} \exp(i\nu x_0 y)\exp\left[-\frac{\alpha}{2}(x-x_0)^2 - \frac{\beta}{2}y^2 - i\gamma(x-x_0)y\right], \tag{4.50}$$

where $\nu = \omega/\hbar$.

4.3.4 Wave Packet Propagation

The stability of a Gaussian wave packet launched from the equilibrium depends, in part, on the validity of the locally harmonic approximation to the ZVS at the maximum. If $F = 0$, the ZVS is flat (i.e., not harmonic at all) transverse to the field direction in the rotating frame but becomes increasingly harmonic with increasing F. However, for $\omega_c = 0$, a transition to instability (the Trojan bifurcation; Abraham and Marsden 1987) occurs when $\varepsilon_c = F_c/\omega^{4/3} \approx 0.1156$ which limits the range of linear dynamics. Figure 4.5a shows a Gaussian wave packet defined as in Bialynicki-Birula, Kaliński, and Eberly (1994) ($\omega_c = 0$) and with the microwave field chosen such that the maximum is located at $x_0 = 10^4$ a.u. (the value suggested in Bialynicki-Birula, Kaliński, and Eberly 1994), with $b = 0.9562$ and $\varepsilon = 0.0444$ (Bialynicki-Birula, Kaliński, and Eberly 1994) . It is apparent that much of the packet spills out of the harmonic regime. The classical simulations shown in Figure 4.6 reveal that a sea of classical chaos surrounds the maximum even in the "stable" regime, and thus any leakage of the packet into the chaotic zone will lead to spreading. In fact the wave packet originally reported in Bialynicki-Birula, Kaliński, and Eberly (1994) did start to spread after only a few periods. If $x_0 \geq 10^7$ a.u., however, most of the packet can nest quite comfortably atop the harmonic part of the maximum—see Figure 4.5b but this represents an inordinately large atom. Based partly on observations using the ZVS (Farrelly, Lee, and Uzer 1994), Bialynicki-Birula, Kaliński, and Eberly (1995) constructed a "curved wave packet" whose stability properties were considerably improved; this packet fitted more closely the contours of the ZVS. A different approach is needed to create a genuinely coherent state.

The addition of a magnetic field ($\omega_c \neq 0$) changes the situation dramatically, since it is now possible to adjust the relative sizes of the coefficients a and b in order to enlarge the stable domain. In particular, it allows one to increase F beyond F_c, thereby increasing the size of the harmonic regime at the maximum. Quantum propagation using fast Fourier techniques as described in Cerjan et al. (1997) confirm that

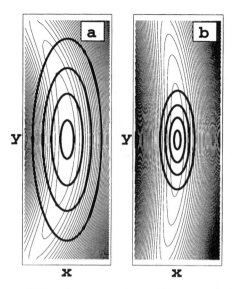

Figure 4.5 Level curves of Gaussian wave packets (thick contours) superimposed on those of the ZVS (thin contours) at the maximum with $b = 0.9562$. Thick lines are contours at 0.25, 0.5, 0.75, and 0.95 of the Gaussian probability density centered at (in a.u.): (a) $x_0 = 10^4$; (b) $x_0 = 10^7$.

expectations based on direct examination of the classical ZVS are borne out quantum mechanically in the correspondence principle limit.[12]

We use two measures to diagnose quantum mechanical state localization. One is the autocorrelation function which is defined as

$$\sigma_a(t) = \int_{-\infty}^{\infty} \psi^*(x, y, t)_{\text{init}} \psi(x, y, t) dx dy, \qquad (4.51)$$

where $\psi(x, y, t)_{\text{init}}$ is the initial wave packet. Another reasonable measure of the deviation from stationary behavior is the time dependence of the virial commutator

$$\sigma_v(t) = \int_{-\infty}^{\infty} \psi^*(x, y, t)[H(x, y), p_x x + p_y y] \psi(x, y, t) dx dy. \qquad (4.52)$$

The classical virial theorem states that if the motion is damped or periodic, then the time average of the virial expression establishes a relation between the kinetic and

---

[12]The ZVS need have no quantum mechanical counterpart, since velocities and not momenta are used in its definition. However, in the limit of large quantum numbers, as our simulations show, the ZVS pro vides an excellent replacement for a potential energy surface.

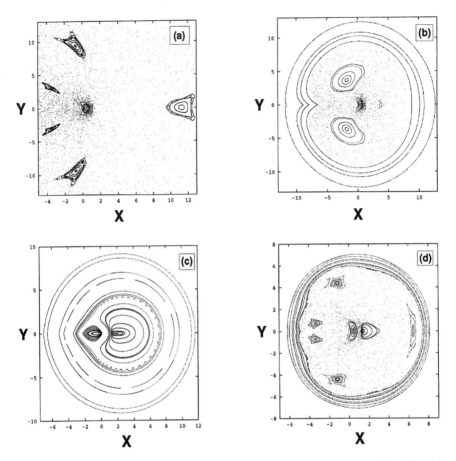

Figure 4.6 Poincaré surfaces of a section along the line AB. In each case $\varepsilon = 0.6$, while ω_s is changed. In scaled units, (a) $\omega_s = 0.2$, $x_0 = 1.25$; (b) $\omega_s = 0.6$, $x_p = 1.30$; (c) $\omega_s = 2/3$, $x_0 = 1.32$; and (d) $\omega_s = 0.8$, $x_0 = 1.36$. Frames (a) and (b) are in the unstable region; (c) corresponds to the integrable limit, while (d) is in the stable region. Note that even in the stable regime the maximum is surrounded by an ocean of chaos.

potential energy contributions (Goldstein 1950). The quantal analogue is provided by the Heisenberg equations of motion. As is well-known, if a stationary state is used for the averaging, then the virial expression vanishes, again providing a relation between the different energy contributions. For the specific case of the Hamiltonian operators above, the commutator of the virial operator, $p_x x + p_y y$, with the Hamiltonian operator produces

$$[H, p_x x + p_y y] = (p_x^2 + p_y^2) - V(x - x_0). \tag{4.53}$$

Thus, if the expectation value of the right-hand side is constant or periodic, the quantal motion is localized.

The results of the calculations are summarized in Figure 4.7. The figure corresponds to the case where the initial wave packet is placed at a maximum of the ZVS which has been 'flattened' by the external magnetic field. The time is in units of ω_c with a total time of 12 field cycles for this calculation, which corresponds to 362 picoseconds. This is a somewhat counterintuitive arrangement for those who are used

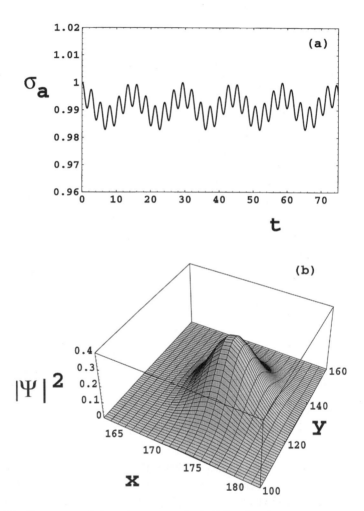

Figure 4.7   Quantum evolution of a wave packet initially placed at the maximum. In atomic units the field parameters are $F = 5.0 \times 10^{-8}$, $\omega_c = 5.0 \times 10^{-6}$, and $\omega_f = 1.0 \times 10^{-6}$. The maximum lies at $x_0 = 1.0 \times 10^4$ bohr. The plots are in scaled units as defined in the text, and the nucleus is denoted by the symbol $\oplus$. (a) Autocorrelation function as a function of time; (b) contours of the probability density at $t = 0$ on a section of the 256×256 grid used in the propagation. The ranges along the abscissa and ordinate are $(-64, 64)$ and $(-19, 19)$ scaled units respectively. (c) Contour diagram of the wave packet at $t = 0$ on the FFT grid. Also shown is a contour plot of the ZVS: (d) as (b) at $t = 12$ cycles; (e) as (c) at $t = 12$ cycles; (f) time dependence of the virial commutator.

to thinking of stable motions being confined to minima of a potential. However, the initial phase conditions ensure that the centrifugal terms will dominate, effectively preventing the transfer of potential to kinetic energy. The autocorrelation function is highly periodic and regular. For the present conditions the time dependence is nearly periodic with no perceptible growth over the 12 cycle propagation time. The results

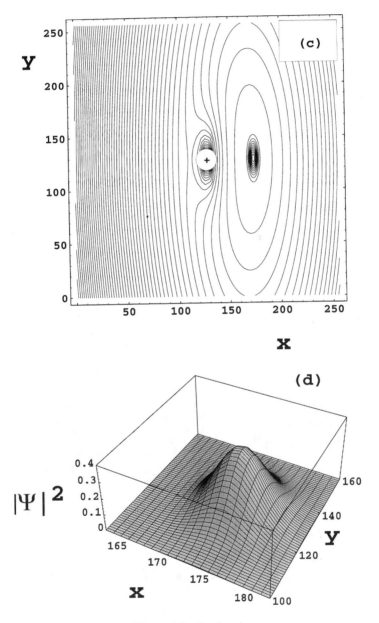

Figure 4.7 Continued

in Figure 4.7 should be compared to those reported in the original calculations by Bialynicki-Birula, Kaliński, and Eberly (1994) in which only a circularly polarized microwave field was applied.

Decay of the wave packet can occur through a variety of mechanisms including chaos assisted tunneling (Zakrzewski, Delande, and Buchleitner 1998), simple

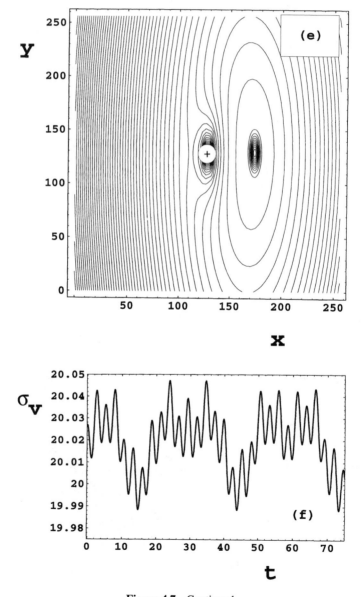

Figure 4.7 Continued

overlap of parts of the packet with the surrounding chaotic sea (Farrelly, Lee, and Uzer 1994; Brunello, Uzer, and Farrelly 1996; Lee, Brunello, and Farrelly 1996), and spontaneous emission (Bialynicki-Birula and Bialynicki-Birula 1997a; Hornberger and Buchleitner 1998; Delande and Zakrzewski 1998). Each of these mechanisms has been studied in the context of Trojan wave packets. The next section describes how it is possible, by changing the field configuration slightly, to minimize dramatically all of these decay mechanisms.

4.4 A SATURNIAN ATOM

In this section we show theoretically how coherent states that are localized at an external energy *minimum* rather than a maximum can be synthesized in a Rydberg atom. Doing this considerably improves the stability of the packet. The connection with celestial mechanics is maintained, since the resulting dynamics resembles the motion of a charged dust grain in one of the ethereal rings of a giant planet such as Saturn's E, F, and G rings or the gossamer ring of Jupiter.[13] In this sense the system may be considered to be a one electron Saturnian atom (Lee, Farrelly, and Uzer 1997).

By reversing the direction of the magnetic field as compared to the previous section, the following Hamiltonian is produced (the sign of F, which is immaterial, has also been changed for convenience to keep the stable equilibrium along the positive x-axis).

$$K = \frac{\widetilde{\mathbf{p}}^2}{2} - \frac{1}{r} - \left(\omega_f - \frac{\omega_c}{2}\right)(xp_y - yp_x) + \frac{\omega_c^2}{8}(x^2 + y^2) - Fx. \qquad (4.54)$$

A key point is that this configuration of fields allows the coefficient of the paramagnetic term to be reduced or even eliminated. By mixing coordinates and momenta the paramagnetic term prevents the normal separation of the Hamiltonian into potential and (positive definite) kinetic parts: However, a true potential energy surface can be defined if $\omega_f = \omega_c/2$, thereby eliminating the paramagnetic term. A typical section through the resulting surface is shown in Figure 4.8 for experimental parameters that are consistent with those that are currently (if not routinely) achievable (Gallagher 1994). Note particularly the existence of a saddle point and an outer harmonic minimum in the potential. In the laboratory frame the equilibrium at the minimum corresponds to a circular orbit in the plane, and localization of the electron in this well thus produces a giant atomic dipole rotating at the microwave frequency in the x-y plane.

It is again convenient to scale coordinates and momenta:

$$\mathbf{r}' = \omega_c^{2/3}\mathbf{r}, \qquad \mathbf{p}' = \omega_c^{-1/3}\mathbf{p}. \qquad (4.55)$$

---

[13]It is of some historical interest that in his paper discussing the scattering of α particles by atoms, Rutherford (1911) cited Nagaoka's earlier paper in which was proposed the idea of a "Saturnian atom" consisting of rings of rotating electrons (Nagaoka 1904). See also Born (1960) and Bohr (1913).

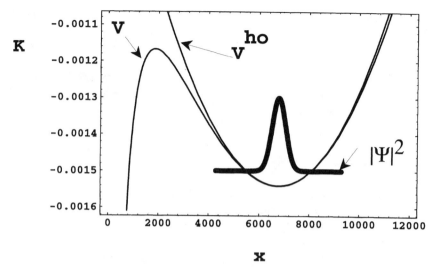

Figure 4.8   ZVS for the Saturnian atom configuration with $\omega_c = 3.46$ T, $\omega_f = 50$ GHz, and F = 2000 V/cm. A section ($y = z = 0$) through the potential is shown. Also plotted is the harmonic approximation to the potential V^{ho}, and the Gaussian probability density of the ground state $|\Psi|^2$.

Dropping the primes and separating the motion along the z-direction (as before) the resulting Hamiltonian is given by

$$\mathcal{K} = \frac{1}{2}\,(p_c^2 + p_y^2) - \frac{1}{r} - \left(\Omega - \frac{1}{2}\right)(xp_y - yp_x) + \frac{1}{8}\,(x^2 + y^2) - \varepsilon x, \qquad (4.56)$$

where $K = K/\omega_c^{2/3}$, $\Omega = \omega_f/\omega_c$, and $\varepsilon = F/\omega_c^{4/3}$. It is significant that if the particle is initially confined to the plane $z = 0$, with no component of velocity in the z-direction, then it is guaranteed to remain in that plane (below the zero field ionization limit the motion in the z-direction is essentially harmonic around the minimum and uncouples from planar motion). Thereby the system can be made to meet the criteria of a planetary atom—an electronic wave packet prepared in the well will move in a nondispersive fashion along circular Keplerian orbits that lie in a given plane. Since the motion is harmonic in the direction perpendicular to this plane, then truly three-dimensional localized coherent states can be prepared. This is also borne out in our quantum simulations and those of Zakrzewski, Delande, and Buchleitner (1995). Poincaré surfaces of section for a number of trajectories and $\Omega = \frac{1}{2}$ (published elsewhere; Lee, Brunello, and Farrelly 1997) reveal no surprises; they simply show sets of elliptical, foliated Kolmogorov-Arnold-Moser (KAM) curves which is a clear signature of stable (i.e., nondispersive) harmonic motion. In fact the planar limit of the problem is integrable at $\Omega = \frac{1}{2}$ (Raković, Uzer, and Farrelly 1998).

Apart from the special case that $\Omega = \frac{1}{2}$, it is not possible to define a potential energy surface but it is still possible to construct a ZVS, given by

$$\Upsilon = -\frac{1}{r} + \frac{\omega_f(\omega_c - \omega_f)}{2}(x^2 + y^2) - Fx. \tag{4.57}$$

Examination of the ZVS reveals that provided that $\Omega < 1$, it is possible to produce an outer well at large distances from the nucleus. In this case the motion may be strongly chaotic, but provided that tunneling is unimportant, which it will be for $x_0 \sim 10^4$ a.u., the electron will be confined within the well by the curves of zero velocity for all values of K below the ionization threshold (Lee, Brunello, and Farrelly 1995). This leads to the somewhat exotic scenario of a wave packet executing totally periodic motion as it revolves around the nucleus while at the same time behaving fully chaotically in the vicinity of the equilibrium point that supports it. Chaotic motion, level spacing distributions, and state scarring within the outer well have recently been studied in detail (Brunello, Uzer, and Farrelly 1996).

Figure 4.9 represents the culmination of our efforts. The Larmor precession due to the external magnetic field and the precession due to the circularly polarized field

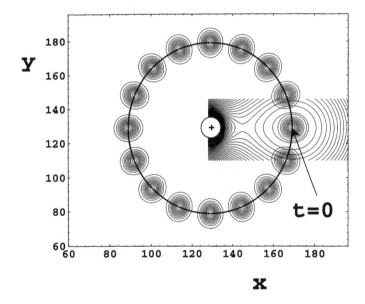

Figure 4.9 Quantum evolution of a wave packet initially placed at the minimum of the Saturnian atom configuration with vanishing paramagnetic term. This is an excellent approximation to a rigorously coherent state. The parameters are, in atomic units: $F = 3.899 \times 10^{-7}$, $\omega_c = 1.6887 \times 10^{-5}$, and $\omega_f = 8.4435 \times 10^{-6}$. The minimum is at $x_0 = 4880.0$. The axes are labeled by the grid reference point used in the FFT. The parameters for the wave packet are as follows: $\alpha = 7.354 \times 10^{-6}$, $\beta = 8.9385 \times 10^{-6}$, and $\gamma = 0$. The evolution is shown in the nonrotating frame for one cycle of the microwave field. The wave packet is propagated in the rotating frame, and the result is transformed to this frame. The propagation begins on the right-hand side at $t = 0$, and scaled units are used throughout. The contours on the right-hand side show the ZVS at $t = 0$.

cancel out the paramagnetic term exactly. The minimum is almost exactly harmonic, and its vacuum state is therefore almost perfectly coherent: While in the rotating frame, the initial wave packet does nothing for 12 field cycles amounting to 107 picoseconds; in the nonrotating frame it is revolving around the nucleus on a large circular orbit without spreading—like a classical electron traveling on the circular orbits of the Bohr atom. Since this calculation was performed, we have propagated fully three-dimensional Trojan and Saturnian atom wave packets; they fully confirm the two-dimensional quantum propagations presented here.

4.5 RELATED WORK, STATE PREPARATION, AND STATE DETECTION

The discovery of the possibility of forming nonstationary nondispersive states in Rydberg atoms, whether at a maximum or at a minimum, has led to a number of related developments. Kaliński and Eberly (1996a) extended their work to predict the existence in the hydrogen atom of stable nonspreading wave packets that have no classical counterparts, although this is the subject of some discussion (Delande, Zakrzewski, and Buchleitner 1997; Kaliński and Eberly 1997). Like the Trojans these states are both angularly and radially localized yet are not associated with a Lagrange equilibrium point. In fact they are exactly π out of phase with the corresponding Trojan wave packet. Their stabilization results from quantum interference effects and these states have been called "anti-Trojans."[14] These same researchers further showed that Trojan-like states can also be supported by a linearly polarized field (Kaliński and Eberly 1995b), which can be thought of as being a superposition of a left and a right circularly polarized field. Kaliński and Eberly (1996b) also developed an analytic Mathieu theory of three-dimensional Trojan wave packets that is valid beyond the harmonic approximation. They demonstrated that electronic Trojan wave packets can be generated from an angularly completely delocalized state (a circular Rydberg state) by adiabatic switching of a circularly polarized electric field; a similar idea was also proposed in Delande, Zakrzewski, and Buchleitner (1995). More recently Kaliński studied the effect of a weak magnetic field on the time evolution of a quantum state that is a coherent superposition of two counterrotating Trojan wave packets in a linearly polarized field. The interference pattern of the electron probability density and of the differential cross section for half-cycle pulse ionization was found to show periodic modulation as a function of magnetic flux cutting the plane of the motion. Kaliński noted that the proportionality of the shift to the magnetic flux was reminiscent of the Aharonov-Bohm effect (Kaliński 1998).

Bialynicki-Birula and Bialynicki-Birula (1997a) studied the spontaneous emission of Trojan wave packets as did Delande and Zakrzewski (1998). The former authors claim to have improved on the treatment in Bialynicki-Birula and Bialynicki-Birula (1997a) which they say underestimates the classical and quantum decay rates by a

---

[14]Perhaps a designation more in keeping with the analogy to the Trojan asteroids would be to refer to these anti-Trojan wave packets simply as "Greek" wave packets.

factor of n_0. Their estimate provides excellent agreement with exact numerical calculations. Delande and his coworkers have emphasized an alternative Floquet picture of the Trojan eigenstates that is particularly helpful in discussing spontaneous emission. Bialynicki-Birula and Bialynicki-Birula (1997b) also introduced the idea of a rotational frequency shift that is an analogue of the Doppler shift. The new frequency shift was proposed to occur in atomic systems that lack rotational invariance but have stationary states in a rotating frame such as the Trojan wave packets. In our group, using numerically exact three-dimensional quantum propagation methods, we have demonstrated the stability of both the Trojan and the Saturnian states to sizable deviations from exact circularity of the field, for example, in the presence of elliptically polarized fields (Griffiths and Farrelly 1992). The states produced behave like driven harmonic oscillator coherent states (Glauber 1963; Brunello, Lee, and Farrelly n.d.). Schweizer, Jaub, and Uzer (1998) have investigated methods for the optimal localization of wave packets on invariant structures such as periodic orbits. As already noted we have also extended the concept of a Trojan state to alkali vapor Bose-Einstein condensates in a TOP trap (Brunello, Uzer, and Farrelly n.d.).

There is considerable theoretical and experimental interest in Rydberg molecules, and Bialynicki-Birula and Bialynicki-Birula (1996) have proposed the existence of Trojan states in Rydberg molecules where the rotation of the core (taken to be a point dipole) plays the role of the rotating electric field. This work is similar to earlier work using the ZVS in the context of Rydberg molecules (Lee, Uzer, and Farrelly 1994).

The wave packets described have not yet been prepared experimentally, although schemes for making them have been proposed, such as by adiabatic switching of circular states (which can be produced) (Kaliński and Eberly 1996b; Delande, Zakrzewski, and Buchleitner 1995; Delande and Zakrzewski 1997). Newly developed half-cycle pulses show promise in the detection of these states (Jones, You, and Bucksbaum 1993; Reinhold et al. 1996).

4.6 ELECTRON LOCALIZATION IN ARTIFICIAL ATOMS

The discussion above has centered on the dynamics of atomic and molecular Rydberg electrons in the presence of Coriolis or Coriolis-like forces (nonconserved paramagnetic interactions). This section describes the treatment of nonconserved Coriolis-like forces that may arise in what have come to be known as quantum dots, or artificial atoms, specifically *quantum dot helium*. As different as the two classes of problem may seem, a single electron in a quantum dot and an electron in a Rydberg atom or molecule share a vital similarity: They are each bound by highly symmetric central field potentials that can be solved exactly in the semiclassical limit: the Coulomb and harmonic oscillator potentials. Quantum dots are sometimes called *artificial atoms*. They are deemed such because they are similar to natural atoms in that the electrons are attracted to a central location and occupy discrete orbitals. In a natural atom, this central location is a positively charged nucleus; in an artificial atom, the electrons are confined to a bowl-shaped parabolic potential well that plays the role of the Coulombic potential of natural atoms. However, because of the relatively large size of an artificial atom, its physical characteristics differ considerably from those of a natural atom. As the size of an atom is increased, the repulsive Coulomb energy between electrons decreases due to the increased average distance between electrons. Because the differences in the orbital energies decrease faster than the Coulomb energy, however, electron–electron interactions become much more important in large atoms than in small ones (Pang and Louie 1990; Maksym 1996; Ashoori 1996). Also at low temperatures electrons fall into distinct quantum-mechanical energy levels of the artificial atom, and the large Coulomb energy exerts a profound influence. In a quantum dot the large Coulomb repulsion may cause electrons to act lethargically, reluctant to displace themselves to make way for another electron (this gives the impression that the electrons actually attract one another). As a result, transitions not easily observed in the spectra of natural atoms are readily seen for artificial ones. For example, singlet-triplet transitions in the ground state of quantum dot helium can be induced by an external field on the order of *only* 1 to 2 T as compared to approximately 4×10^5 T in a natural atom. Increased sensitivity to perturbations, together with an enhancement in the importance of electron–electron interactions, means that one can conduct quantum physics experiments in a quantum dot that might be impossible in an atom. For this reason, investigating the response of these systems to strong perturbations that break the symmetry of the system, such as an application of a magnetic field, is a fundamental problem in quantum physics with potential applications in device physics. This is connected to the increasing interest in the electronic structure and properties of donor centers in semiconductors and quantum wells in strong magnetic fields (Pang and Louie 1990; Maksym 1996). Many novel effects in these systems, such as the quantum Hall effect, metal-insulator transitions, and electron localization, are intimately related to the presence of high magnetic fields and impurity states (Ashoori 1996).

The actual problem studied in this chapter is closely connected to a recent series of experiments performed by Ashoori et al. (1992) using single-electron capacitance spectroscopy. There the effect of an applied magnetic field on the discrete quantum levels of a quantum-dot system containing only a few electrons was investigated. During their investigation Ashoori et al. (1992) unexpectedly detected pairs of electrons tunneling *together* onto a GaAs quantum dot. Usually electrons tunnel one by one onto a quantum dot due to the Coulomb blockade effect. The mechanism by which these electron pairs formed remains unclear, although Ashoori et al. (1992) were empirically able to associate their discovery with the presence of Si impurity atoms that had migrated from the backing layer (insulating layer) creating an off-axis impurity center. This led Wan, Ortiz, and Phillips (1995) to propose a model involving an off-axis impurity center in which coupling to phonons mediates the pairing state through bipolaron formation. However, the model proposed by Wan, Ortiz, and Phillips (1995) neglected nonconserved paramagnetic interactions that arise from an off-axis impurity center in the presence of a magnetic field. In the case of Rydberg atoms and molecules, this interaction was shown earlier to be extremely important in the underlying dynamics of the electron. Similarly paramagnetic interactions play a vital role in the dynamics of quantum-dot electrons; as already noted the presence of the paramagnetic term complicates the analysis but enriches the dynamics. This difficulty prompted Wan, Ortiz, and Phillips (1995) to drop the paramagnetic term altogether in order to allow them to perform routine diffusion quantum Monte Carlo (DMC) calculations. This assumption admittedly simplifies the DMC calculation, since DMC cannot be applied directly if the ground state is nonbosonic. However, it also introduces significant errors (Farrelly, Lee, and Uzer 1998; Lee et al. 1998). Elsewhere we have developed an approach that allows the computation of electron interactions accurately, for the first time, without excluding paramagnetic interactions, using DMC (Lee et al. 1998). The concept of the ZVS, which has been previously discussed in detail, was used as a tool to direct the quantum computations of the electron dynamics. In this section we demonstrate that coherent electronic states can be formed in quantum-dot helium at Lagrange-like equilibrium points.

4.6.1 The Hamiltonian

In the effective mass approximation, the Hamiltonian for two electrons simultaneously subject to (1) a parabolic confinement potential, (2) a hydrogenic (Coulombic) impurity center displaced (by a distance x_0) from the center of the confinement potential, and (3) an applied magnetic field (cyclotron frequency ω_c) along the z-axis is given by Pang and Louie (1990)

$$H = H(1) + H(2) + \frac{1}{r_{12}}, \qquad (4.58)$$

where

$$H(i) = \frac{p_i^2}{2} - \frac{1}{r_i} + \frac{\omega_c}{2} L_z^{(i)} + \frac{\omega_c^2}{8} (x_i^2 + y_i^2) + \frac{\omega_0^2}{2} [(x_i - x_0)^2 + y_i^2] + V_Q(z_i) \quad (4.59)$$

and

$$L_z^{(i)} = x_i p_{y_i} - y_i p_{x_i}, \qquad (4.60)$$

$$r_i = \sqrt{x_i^2 + y_i^2 + z_i^2}, \qquad (4.61)$$

$$r_{12} = \sqrt{(x_2 - x_1)^2 + (y_2 - y_1)^2 + (z_2 - z_1)^2}. \qquad (4.62)$$

The term $V_Q(z)$ is a square well of height V_0 and width L which confines the electrons laterally and takes on the following values:

$$V_Q(z_i) = \begin{cases} V_0 & \text{if } |z| < \dfrac{L}{2}, \\ 0 & \text{if } |z| \geq \dfrac{L}{2}. \end{cases} \qquad (4.63)$$

Units have been chosen such that $\hbar = m^* = a_0^* = 1$, where m^* is the effective mass for the electron and values appropriate to GaAs have been used (Pang and Louie 1990; Wan, Ortiz, and Phillips 1995).

This Hamiltonian is identical (apart from the presence of the square well and some reinterpretation of parameters) to that for a He atom interacting with circularly polarized microwave and magnetic fields. In a He atom under such conditions, the states we prepare are two electron wave packets with the electrons orbiting the nucleus synchronously, although these states must be distinguished from those of Schlagheck and Buchleitner (1998) which emerge from frozen planetary configurations in a linearly polarized field. The states we construct have a totally different origin that is unrelated to the planetary states of the He atom.

Zero-Velocity Surface

Paramagnetic interactions complicate the analysis because they mix coordinates and momenta, and this in turn means that a potential energy surface cannot be defined. As stated earlier, the ZVS was invented precisely to deal with this problem and nonconserved Coriolis-like interactions in the three-body problem of celestial mechanics.

Knowledge of the topology of configuration space is also a considerable advantage when using quantum Monte Carlo methods because it aids in the construction of trial variational quantum Monte Carlo wave functions which may then be used to guide, using importance sampling, the DMC computations (Lee et al. 1998). In the present problem the absence of a potential energy surface and the multidimensional nature of the configuration space means that one is working essentially in the dark. A way out is provided by the ZVS in a similar way to that described in previous sections for nondispersive atomic and molecular wave packets.

The ZVS is constructed as usual by rewriting the Hamiltonian in terms of the mechanical velocities:

$$H = \sum_{i=1}^{2} \frac{1}{2} (\dot{x}_i^2 + \dot{y}_i^2 + \dot{z}_i^2) + \Upsilon. \tag{4.64}$$

The Hamiltonian is now cleanly split into a part that is positive definite in velocities and a part, Υ, that depends solely on coordinates. Curves of zero velocity that bound the classical motion are level curves of the function

$$\Upsilon = \frac{1}{\varepsilon r_{12}} + \sum_{i=1}^{2} \left(-\frac{1}{\varepsilon r_i} + \frac{\omega_0^2}{2} [(x_i - x_0)^2 + y_i^2] + V_Q(z_i) \right). \tag{4.65}$$

Note particularly that the ZVS *does not* depend explicitly on the magnetic field: At an equilibrium point the Lorentz force clearly vanishes, and therefore the form of the ZVS correctly predicts that any equilibria will be unaffected by an applied magnetic field.

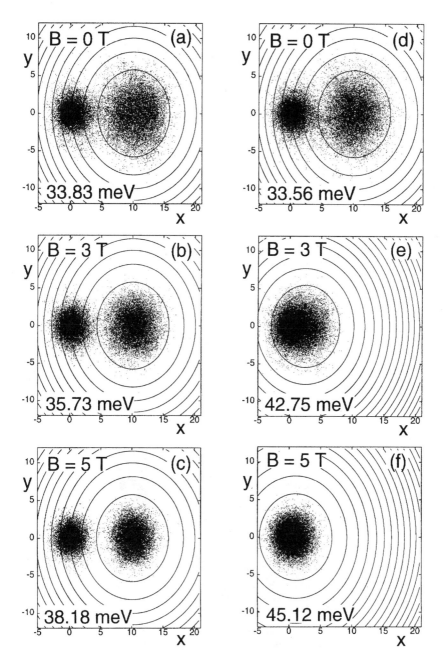

Figure 4.10 Final distribution of walkers as obtained from DMC with parameters as in the text and $L = 1.515a_0^*$, $V_0 = 24.113$ a.u. Results in the left-hand column, (a)–(c), were obtained including the paramagnetic term, and those in (d)–(f) were by omitting the paramagnetic interaction. Each frame shows the final DMC energy and the magnetic field used. Also shown are levels curves of the ZVS (left-hand column) and the potential energy surface that results from dropping the paramagnetic term (right-hand column).

Figure 4.10 shows the "electron clouds" computed by DMC for several magnetic field strengths. The standard deviations are within 0.10 meV for all cases. The most striking observation is the contrast between keeping and omitting paramagnetic interactions. The results in Figure 4.10a–c account for paramagnetic interactions using the procedure just described. The two electron clouds are localized at distinct, spatially separated sites associated with Lagrange-like equilibrium points. With increasing magnetic field, the two clouds become progressively tighter and more localized about their individual centers. Also shown in Figure 4.10a–c are level curves of the multidimensional ZVS, obtained by projecting the $x_1 y_1$ part of the ZVS onto the xy-plane while taking $x_2 \approx 0$, $y_2 = z_1 = z_2 = 0$. These plots clearly reveal the existence of an outer well that is centered roughly at x_0 for *all* B values. Remarkably, without further calculation, the ZVS encapsulates both the zero and infinite magnetic field limits in a single picture. Dropping the paramagnetic term in equation (4.1), on the other hand, leads to a counterintuitive and unphysical migration of the outer electron to the impurity center as is apparent in the sequence Figure 4.10d–f, with a substantial increase in the ground-state energy. Superimposed are level curves of the potential energy surface (PES) obtained by dropping the paramagnetic term. The outer potential well, prominent in Figure 4.10d, quickly shrinks and moves toward the impurity center, taking the electron with it, as the magnetic field is increased. Succinctly, neglect of the paramagnetic term leads to a strongly gauge dependent PES.

Another, and quite remarkable pair of Lagrange-like equilibria also exist in this problem (and consequently in a He atom in combined circularly polarized microwave and magnetic fields). Figure 4.11 is a projection of the ZVS onto the $x_1 x_2$-plane

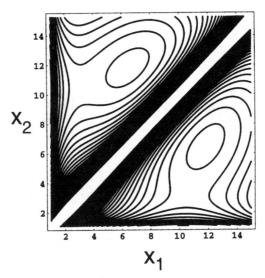

Figure 4.11 ZVS in the $x_1 x_2$ plane showing the two global minima that occur, for the parameters in the text, at $(x_i, x_j) \approx (6.394 a_0^*, 11.679 a_0^*)$, $\hbar \omega_0 = 1.5$ meV, $x_0 = 10$ a_0^*, and $y_i = z_i = 0$. The singular line, when $r_{12} = 0$, is apparent along the diagonal $x_1 = x_2$.

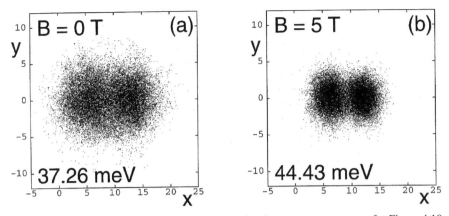

Figure 4.12 Final distribution of DMC walkers, for the same parameters as for Figure 4.10 but for the second type of electron configuration. The classical equilibria in the ZVS are the same as in Figure 4.11.

showing two global minima that lie along the x-axis. The results of the DMC calculations for this kind of configuration are presented in Figure 4.12. For the case where $B = 0$, there is considerable overlap of the two electron clouds, although, nevertheless, the electrons do exhibit noticeable spatial localization. This is shown in Figure 4.12a. Upon increasing B, the picture rapidly sharpens and the electrons become more localized in their individual wells as shown in Figure 4.12b. The situation is similar to two-electron planetary states of the He atom (Richter and Wintgen 1990) but much more open to the control of the experimenter. The classical support for nondispersive two-electron configurations in a helium atom subjected to linearly polarized microwaves has been investigated by Schlagheck and Buchleitner (1998, 1999).

4.7 CONCLUSIONS

This chapter has demonstrated that coherent states in Rydberg atoms can be produced by a combination of circularly polarized microwave and magnetic fields. Such states neither spread nor disperse as they execute their revolutions around the nucleus, although a Trojan wave packet will slowly decay due to tunneling. A more significant source of dispersion will arise if the tails of the wave packet penetrate appreciably into the nonlinear or chaotic parts of phase space. This source of spreading can be anticipated by examining the contours of the ZVS, a classical-mechanical construct. In the laboratory frame, if these dispersive factors can be minimized, the electronic wave packet will travel along a circular Kepler orbit while remaining localized radially and angularly for a finite (but possibly very large) number of Kepler periods. An important point in our study is that the stability of such a packet can be enhanced considerably by using a magnetic field in addition to the circularly polarized field. Indeed, the initial wave packet displayed remarkable localization with the addition of a static magnetic field, whereas the absence of this field can lead to rapid delocalization.

Two distinct types of electron configuration in quantum dot helium with an off-axis impurity center were also identified in which the electrons are spatially localized around classical equilibrium points. This localization is enhanced by an applied magnetic field in which nonconservative paramagnetic interactions are central to the mechanism by which localization is maintained. The two systems are really examples of "Wigner" molecules (Maksym 1996). They may differ in energy by less than perturbation theory estimates of the bipolaron stabilization energy (≈ 6 meV; Wan, Ortiz, and Phillips 1995), which suggests that it should be possible, in an appropriate experiment, to detect electron pair formation in either of the two different arrangements exemplified by Figures 4.10 and 4.12.

The applications of Hill's methods and the importance of Lagrange equilibria in quantum physics are not exhausted (Lee 1997). Applications have also been made to Rydberg molecules (important in zero-electron-kinetic energy or ZEKE spectroscopy) and to the stability of Bose-Einstein condensates in the time-averaged orbiting or TOP trap. Without doubt the legacies of Lagrange and Hill as well as other giants of celestial mechanics continue to influence the direction of many important areas of physics and astronomy.

ACKNOWLEDGMENTS

We have benefited from many conversations with Professor Iwo Bialynicki-Birula, Professor Joe Eberly, and Dr. Martin Gutzwiller which are gratefully acknowledged. One of the authors (D. F.) thanks Dr. André Deprit for useful insights into various aspects of celestial mechanics and Alan Feuerbacher for useful discussions about mathematics and astronomy in antiquity. This work was supported by NSF grant CHE-9633671. Part of it was also performed under the auspices of the United States Department of Energy by the Lawrence Livermore National Laboratory under Contract W-7405-Eng-48.

REFERENCES

Abraham, R., and Marsden, J. E. (1987). *Foundations of Mechanics*, 2nd ed. Addison-Wesley, Redwood City, CA, pp. 675–688.

Alber, G., and Zoller, P. (1991). *Phys. Rep.* **199**, 231.

Ashoori, R. C. (1996). *Nature* **379**, 413.

Ashoori, R. C., Stormer, H. L., Weiner, J. S., Pfeiffer, L. N., Pearton, S. J., Baldwin, K. W., and West, K. W. (1992). *Phys. Rev. Lett.* **68**, 3088.

Bialynicki-Birula, I., and Bialynicki-Birula, Z. (1996). *Phys. Rev. Lett.* **77**, 4298.

Bialynicki-Birula, I., and Bialynicki-Birula, Z. (1997a). *Phys. Rev.* **A56**, 3623.

Bialynicki-Birula, Z., and Bialynicki-Birula, I. (1997b). *Phys. Rev. Lett.* **78**, 2539.

Bialynicki-Birula, I., Kaliński, M., and Eberly, J. H. (1994). *Phys. Rev. Lett.* **73**, 1777.

Bialynicki-Birula, I., Kaliński, M., and Eberly, J. H. (1995). *Phys. Rev. Lett.* **75**, 973.

Bohr, N. (1913). *Phil. Mag.* **26**, 1, 476, and 857.

Born, M. (1960). *Mechanics of the Atom*. Trans. J. W. Fisher. Reprint F. Ungar, New York.

Brunello, A. F., Uzer, T., and Farrelly, D. (1996). *Phys. Rev. Lett.* **76**, 2874.

Brunello, A. F., Uzer, T., and Farrelly, D. Unpublished manuscript.

Brunello, A.F., Lee, E., and Farrelly, D. Unpublished manuscript.

Buchleitner, A. (1993). Doctoral dissertation. Université Pierre et Marie Curie, Paris.

Buchleitner, A., and Delande, D. (1995). *Phys. Rev. Lett.* **75**, 1487.

Cerjan, C., Lee, E., Farrelly, D., and Uzer, T. (1997). *Phys. Rev.* A**55**, 2222.

Clark, C. W., Korevaar, E., and Littman, M. G. (1985). *Phys. Rev. Lett.* **54**, 320.

Connerade, J.P. (1998). *Highly Excited Atoms*. Cambridge University Press, Cambridge.

Danby, J. M. A. (1992). *Fundamentals of Celestial Mechanics*, 2nd ed. Willman-Bell, Richmond, VA, pp. 384–385.

Delande, D., and Zakrzewski, J. (1997). *J. Phys.* B**30**, L87.

Delande, D., and Zakrzewski, J. (1998). *Phys. Rev.* A**58**, 466.

Delande, D., Zakrzewski, J., and Buchleitner, A. (1995). *Europhys. Lett.* **32**, 107.

Delande, D., Zakrzewski, J., and Buchleitner, A. (1997). *Phys. Rev. Lett.* **79**, 3541.

Deprit, A. (1966). *Astron. J.* **71**, 77.

Deprit, A., Henrard, J., and Rom, A. (1971). *Astron. J.* **76**, 269. Pavelle, R., Rothstein, M., and Fitch, J. (1981). *Sci. Am.* **245**, 136.

Fagles, R. trans. (1990). *Homer, The Iliad*. Viking Press, New York.

Farrelly, D., Lee, E., and Uzer, T. (1994). *Phys. Rev. Lett.* **75**, 972.

Farrelly, D., Lee, E., and Uzer, T. (1995). *Phys. Lett.* A**204**, 359.

Farrelly, D., Lee, E., and Uzer, T. (1998). *Phys. Rev. Lett.* **80**, 3884.

Farrelly, D., and Uzer, T. (1995). *Phys. Rev. Lett.* **74**, 1720.

Gallagher, T. (1994). *Rydberg Atoms*. Cambridge University Press, Cambridge.

Garfinkel, B. (1978). *Cel. Mech.* **18**, 259.

Garfinkel, B. (1982). *Astron. J.* **82**, 368.

Glas, D., Mosel, U., and Zint, P. G. (1978). *Z. Phys.* A**285**, 83.

Glauber, R. J. (1963). *Phys. Rev.* **131**, 2766.

Goldstein, H. (1950). *Classical Mechanics*. Addison-Wesley, New York, pp. 69ff.

Greenberg, R., and Davis, D. R. (1978). *Am. J. Phys.* **46**, 1068.

Griffiths, J. A., and Farrelly, D. (1992). *Phys. Rev.* A**45**, 2678.

Gutzwiller, M. (1998). *Rev. Mod. Phys.* **70**, 589.

Hill, G. W. (1878). *Am. J. Math.* **1**, 5.

Hornberger, K., and Buchleitner, A. (1998). *Europhys. Lett.* **41**, 383.

Jones, R. R., You, D., and Bucksbaum, P. (1993). *Phys. Rev. Lett.* **70**, 1236.

Kaliński, M. (1998). *Phys. Rev.* A**57**, 2239.

Kaliński, M., and Eberly, J. H. (1995a). *Phys. Rev.* A**52**, 4285.

Kaliński, M., and Eberly, J. H. (1995b). *Phys. Rev.* A**52**, 2460.

Kaliński, M., and Eberly, J. H. (1996a). *Phys. Rev. Lett.* **77**, 2420.

Kaliński, M., and Eberly, J. H. (1996b). *Phys. Rev.* A**53**, 1715.

Kaliński, M., and Eberly, J. H. (1997). *Phys. Rev. Lett.* **79**, 3542.

Lee, E. (1997). Doctoral dissertation, Utah State University.

Lee, E., Brunello, A. F., and Farrelly, D. (1995). *Phys. Rev. Lett.* **75**, 3641.

Lee, E., Brunello, A. F., and Farrelly, D. (1997). *Phys. Rev.* A**55**, 2203.

Lee, E., Farrelly, D., and Uzer, T. (1997). *Optics Exp.* **1**, 221.

Lee, E., Puzder, A., Chou, M. Y., Uzer, T., and Farrelly, D. (1998). *Phys. Rev.* B**57**, 12 281.

Lee, E., Uzer, T., and Farrelly, D. (1994). *Chem. Phys. Lett.* **231**, 241.

Levison, H., Shoemaker, E. M., and Shoemaker, C. S. (1997). *Nature* **385**, 6611.

Maksym, P. A. (1996). *Phys. Rev.* B**53**, 871.

Nagaoka, E. (1904). *Phil. Mag.* **7**, 445.

Nauenberg, M., Stroud, C. R., and Yeazell, J. A. (1994). *Sci. Am.* **270**, 44.

Nessman, C., and Reinhardt, W. P. (1987). *Phys. Rev.* A**35**, 3269.

Neugebauer, O. (1969). *The Exact Sciences in Antiquity.* Dover, New York.

Pang, T., and Louie, S. G. (1990). *Phys. Rev. Lett.* **65**, 1635.

Poincaré, H. (1993). *New Methods of Celestial Mechanics*, ed. D. L. Goroff American Institute of Physics, New York, part 2, ch. 11. (Originally published as *Les Méthodes nouvelles de la mécanique céleste* in 1892–1899).

Prizibram, K., ed., (1967). In *Letters in Wave Mechanics.* Philosophical Library, New York.

Raković, M. J., and Chu, S. I. (1994). *Phys. Rev.* A**50**, 5077.

Raković, M. J., Uzer, T., and Farrelly, D. (1994). *Phys. Rev.* A**57**, 2814.

Reinhold, C. O., Burgdörfer, J., Frey, M. T., and Dunning, F. B. (1996). *Phys. Rev.* A**54**, 33.

Richter, K., and Wintgen, D. (1990). **65**, 1965.

Rutherford, E. (1911). *Phil. Mag.* **21**, 669.

Schlagheck, P., and Buchleitner, A. (1998). *J. Phys.* B**31**, 489.

Schlagheck, P., and Buchleitner, A. (1998). *J. Phys.* B**31**, L489.

Schlagheck, P., and Buchleitner, A. (1999). *Europhys. Lett.* **46**, 24.

Schrödinger, E. (1926). *Naturwissenschaften* **14**, 664.

Schrödinger, E. (1968). In *Sources of Quantum Mechanics*, ed. B. L. van der Waerden. Dover, New York.

Schweizer, W., Jans, W., and Uzer, T. (1998). *Phys. Rev.* A**58**, 1382.

Szebehely, V. (1967). *Theory of Orbits: The Restricted Problem of Three Bodies.* Academic, New York.

Wan, Y., Ortiz, G., and Phillips, P. (1995). *Phys. Rev. Lett.* **75**, 2879.

Yeazell, J. A., and Stroud C.R., Jr. (1990). *Act. Phys. Pol.* A**78**, 253.

Zakrzewski, J., Delande, D., and Buchleitner, A. (1995). *Phys. Rev. Lett.*, **75**, 4015.

Zakrzewski, J., Delande, D., and Buchleitner, A. (1998). *Phys. Rev.* E**57**, 1458.

CHAPTER 5

Destruction of Quantum Coherence and Wave Packet Dynamics

GERNOT ALBER

5.1 INTRODUCTION

The development of short, powerful laser pulses and of sophisticated trapping techniques within the last few years has stimulated numerous theoretical and experimental investigations on the dynamics of wave packets in elementary, material quantum systems. These wave packets are nonstationary, spatially localized quantum states that are situated on the border between the microscopic and macroscopic domain. A detailed understanding of their dynamics is essential for our conception of quantum mechanics and of its connection with classical mechanics. So far the interplay between classical and quantum mechanical aspects of their dynamics have been investigated in Rydberg systems (Alber and Zoller 1991), in molecules (Garraway and Suominen 1995; Sepulveda and Grossmann 1996), in clusters (Knospe and Schmidt 1996), and in nano structures (Koch et al. 1996). These studies have concentrated mainly on semiclassical aspects which may be attributed to the smallness of the relevant de Broglie wave lengths. Thereby quantum aspects still manifest themselves in interferences between probability amplitudes that are associated with various families of classical trajectories. However, for a comprehensive understanding of the emergence of classical behavior, a detailed understanding of the destruction of quantum coherence is also required. Typically this destruction of coherence arises from external stochastic forces or environmental influences which cannot be suppressed. Though by now many aspects of the coherent dynamics of these wave packets are understood to a satisfactory degree, still scarcely anything is known about the influence of destruction of quantum coherence.

The main aim of this chapter is to discuss characteristic physical phenomena that govern the destruction of quantum coherence of material wave packets. For systematic investigations on this problem, it is advantageous to deal with physical systems in which wave packets can be prepared and detected in a controlled way and in which the mechanisms causing the destruction of quantum coherence can be influenced to a large extent. Rydberg atoms (Seaton 1983; Fano and Rau 1986) are paradigms of

The Physics and Chemistry of Wave Packets, Edited by John Yeazell and Turgay Uzer
ISBN 0-471-24684-0 © 2000 John Wiley & Sons, Inc.

elementary quantum systems that meet these requirements. The high-level density of Rydberg states close to an ionization threshold is particularly convenient for the experimental preparation of spatially localized electronic wave packets by coherent superposition of energy eigenstates (Alber and Zoller 1991). Furthermore the dynamics of electronic Rydberg wave packets exhibits universal features that apply to atomic and molecular Rydberg wave packets as well as to Rydberg wave packets in more complex systems such as clusters. This dynamical universality might be traced back to the fact that almost over its whole classically accessible range, the dynamics of a Rydberg electron is governed by the Coulomb potential of the positively charged ionic core. This universality, together with the fact that Rydberg systems are amenable to a systematic theoretical description with the help of semiclassical methods, makes them attractive for theoretical investigations. In recent years many detailed investigations have been performed concerning various fundamental aspects of the coherent dynamics of Rydberg wave packets, such as the influence of core scattering processes (Alber 1989; Dando et al. 1995; Hüpper et al. 1995), the connection between classical bifurcation phenomena and quantum dynamics (Beims and Alber 1993, 1996; Main et al. 1994), the influence of the stimulated light force on the atomic center of mass motion (Alber 1992; Alber and Strunz 1994), or the influence of electron correlations on wave packet dynamics in laser-induced two-electron excitation processes (Hanson and Lambropoulos 1995; Zobay and Alber 1995; van Druten and Muller 1995).

The dynamics of Rydberg electrons is governed by characteristic features that greatly influence the way in which they can be affected by external stochastic forces or environments. Notably Rydberg electrons can be influenced by laser fields of moderate intensities and by their statistical properties only in a small region around the atomic nucleus (Giusti and Zoller 1987). Furthermore Rydberg systems are characterized by unique threshold phenomena that result from the infinitely many bound states and from the continuum states converging toward an ionization threshold. In addition radiative decay rates of Rydberg states are so small that in typical situations of current experimental interest the direct influence of radiative damping can be neglected. However, the dissipative influence of radiative decay might become significant, if Rydberg systems interact with intense laser fields. In order to demonstrate characteristic physical phenomena governing the destruction of quantum coherence of electronic Rydberg wave packets, in the subsequent discussion two stochastic mechanisms will be considered in detail, namely radiative damping which is mediated by electron correlations between a Rydberg wave packet and a resonantly excited, tightly bound core electron and fluctuations of laser fields.

The investigation of radiative damping mediated by electron correlation effects is motivated by the recently revived interest in laser-induced two-electron excitation processes (Jones and Bucksbaum 1991; Stapelfeldt et al. 1991; Robicheaux 1993; Grobe and Eberly 1993; Hanson and Lambropoulos 1995; Zobay and Alber 1995; van Druten and Muller 1995). Nonresonant laser-induced excitation processes in which two valence electrons of an atom, such as an alkaline earth atom, are excited simultaneously have already been playing an important role in spectroscopy for a long time (Gallagher 1994). Typically one of the valence electrons is excited into a Rydberg state and the other one into a tightly bound core state. Due to the availability of intense laser light sources, recently other cases have become accessible experimentally in which both of these electrons are excited resonantly so that the influence of the laser field can no longer be treated with the help of perturbation theory. The resulting strong modifications of the electron correlations may give rise to interesting novel phenomena. If the Rydberg electron is prepared in a wave packet state, these coherent laser-modified electron correlations may even lead to an almost complete suppression of autoionization (Hanson and Lambropoulos 1995). In the subsequent discussion it will be demonstrated that these coherent effects are rather sensitive to the destruction of coherence which is caused by radiative decay of the tightly bound, excited core electron.

Due to the inherent stochastic nature of laser light the investigation of optical excitations of atoms or molecules by fluctuating laser fields is one of the central problems of laser spectroscopy. So far research on this problem has concentrated predominantly on laser-induced excitation of isolated energy eigenstates (Agarwal 1976; Dixit et al. 1980; Vemuri et al. 1991). By now this special class of excitation processes is understood to a satisfactory degree. Despite these successes so far scarcely anything is known about the effect of laser fluctuations on optical excitation processes in which the level density of the resonantly excited states is large and in which wave packets are prepared. A paradigm in this respect is the laser-induced excitation of Rydberg and continuum states close to an ionization threshold which typically leads to the preparation of an electronic Rydberg wave packet. This physical system is well suited for investigating fundamental aspects of the destruction of quantum coherence on wave packet dynamics. In the subsequent discussion it will be demonstrated that this fluctuation-induced destruction of quantum coherence, together with the peculiar threshold phenomena of Rydberg systems, leads to a variety of novel phenomena. One of these generic effects is stochastic ionization which manifests itself in a characteristic scenario of nonexponential decays (Alber and Eggers 1997).

5.2 COHERENT DYNAMICS OF RYDBERG ELECTRONS: GENERAL THEORETICAL CONCEPTS

In this section a brief review of general theoretical concepts is presented that are useful for the description of the dynamics of Rydberg electrons. These concepts have already been used successfully to describe various aspects of the coherent dynamics of electronic Rydberg wave packets. Thereby we will concentrate mainly on cases of recent experimental and theoretical interest in which a weakly bound Rydberg electron interacts with a laser field and additional weak electric and/or magnetic fields (Alber 1989; Dando et al. 1995; Hüpper et al. 1995; Moser et al. 1997). Throughout this chapter Hartree atomic units will be used for which $e = \hbar = m_e = 1$ (e and m_e are the electronic charge and mass, respectively).

Rydberg electrons are atomic or molecular electrons whose dynamics is dominated by highly excited energy eigenstates close to an ionization threshold. In the simplest possible case the energies of these Rydberg states are given by the well-known relation $\varepsilon_n = -1/[2(n - \alpha)^2]$ (Seaton 1983; Fano and Rau 1986). Thereby the quantum defect α is approximately energy independent for energies sufficiently close to the ionization threshold at energy $\varepsilon = 0$. In typical optical excitation processes only Rydberg states with small values of the angular momentum l are excited, $l \ll n$. These Rydberg states of low angular momenta are essentially de-localized over the whole space which is classically accessible to them, $(l + \frac{1}{2})^2 < r < 1/|\varepsilon_n|$. ($r$ denotes the radial distance of the electron from the nucleus measured in units of the Bohr radius $a_0 = 5.29 \times 10^{-11}$m.)

If a Rydberg electron interacts with a laser field of moderate intensity and with a weak, static electric and/or magnetic field one can distinguish three characteristic spatial regimes:

1. *The core region* $(0 < r < O(1))$. It extends a few Bohr radii around the atomic nucleus. Inside this core region, Rydberg electrons of low angular momenta which are able to penetrate this core region interact with all other atomic core electrons. These interactions lead to characteristic electron correlations effects such as autoionization and channel coupling. Quantitatively these effects can be described by quantum defect parameters which are approximately energy independent close to an ionization threshold (Seaton 1983; Fano 1986; Aymar et al. 1996).

If a Rydberg electron of low angular momentum interacts with a laser field of moderate intensity, whose electric field strength is given by

$$\mathbf{E}(t) = \mathbf{E}_0 e^{-i\omega t} + c.c., \tag{5.1}$$

two major effects take place. First, the Rydberg electron experiences an intensity dependent ponderomotive energy shift of magnitude $\delta\omega_p = |\mathbf{E}_0|^2/\omega^2$. This energy shift is independent of the energy of the Rydberg electron and may thus be interpreted as an energy shift of the ionization threshold. Second, all other dominant energy exchange processes between a Rydberg electron and the laser field are localized within a region typically extending a few Bohr radii around the atomic nucleus. This localization of the electron-laser coupling inside the core region relies on two sufficient conditions, namely moderate laser intensities and sufficiently high laser frequencies preferably in the optical frequency domain (Giusti and Zoller 1987). Thereby laser intensities are considered to be moderate provided that the stationary oscillation amplitude α_{osc} of an electron in the laser field (in the absence of the Coulomb potential of the ionic core) is significantly less than the extension of the core region,

$$\alpha_{osc} = |\mathbf{E}_0|/\omega^2 \ll 1. \tag{5.2}$$

Furthermore in this context laser frequencies ω are considered to be high, if they are much larger than the inverse classical Kepler period T_n of the Rydberg electron,

$\omega T_n \gg 1$ with $T_n = 2\pi(n - \alpha)^3$. Classically speaking at these high laser frequencies, it is only in a region close to the nucleus that the acceleration of a Rydberg electron is sufficiently large that an appreciable energy exchange of the order of $\Delta\varepsilon \approx \omega$ can take place between the laser field and the Rydberg electron (compare also with (5.5)). As a consequence the interaction of a Rydberg electron with a laser field of moderate intensity and sufficiently high frequency is completely different from its interaction with a microwave field whose frequency is comparable with its classical Kepler frequency $1/T_n$. Even if the field strength of such a microwave field is small in the sense that $\alpha_{osc} \ll 1$, the small frequency of the microwave field implies that energy can be exchanged with the microwave field essentially at any distance of the Rydberg electron from the atomic nucleus.

2. *The Coulomb region* $(O(1) < r < a)$. Outside the core region the dynamics of a highly excited Rydberg electron is dominated by the $1/r$ Coulomb potential of the positively charged ionic core. If the Rydberg electron is influenced by a weak external electric or magnetic field, this is only valid for distances of the Rydberg electron from the nucleus which are smaller than the critical distance $a \gg 1$ at which the external potentials are no longer negligible. If this critical distance is located inside the classically accessible region, $a < 1/|\varepsilon_n|$, these external fields influence the dynamics of the Rydberg electron significantly.

3. *The asymptotic region* $(1 \ll a < r)$. In the asymptotic region the influence of weak external fields is as important as the Coulomb force originating from the positively charged ionic core. In general, in this region the resulting dynamics of the Rydberg electron is complicated by the fact that its classical dynamics is no longer integrable and exhibits signatures of chaos.

In each of these characteristic spatial regimes different approximations can be applied for the dynamical description of the Rydberg electron. All photon absorption and emission processes and all electron correlation effects that take place inside the core region have to be described quantum mechanically. Since the Bohr radius is small in comparison with the extension of the Coulomb region and of the asymptotic region, outside the core region the dynamics of a Rydberg electron can be described with the help of semiclassical methods.

Starting from these elementary considerations a systematic theoretical description of Rydberg electrons can be developed that is based on a synthesis of semiclassical methods and of concepts of quantum defect theory (Alber 1989; Alber and Zoller 1991). Thereby solutions of the Schrödinger equation that are valid inside the core region and at the boundary to the Coulomb region have to be matched to semiclassical wavefunctions that are valid in the Coulomb region and in the asymptotic region. The values of the wavefunction at the border between the core region and the Coulomb region are determined by the solution of the Schrödinger equation inside the core region. Within the framework of quantum defect theory, these values are determined by approximately energy independent quantum defect parameters. These quantum defect parameters originate from two different types of interactions, namely electron correlation effects and laser-induced photon absorption and emission processes. For moderate laser intensities and sufficiently high frequencies, these latter type of processes give rise to intensity dependent quantum defects. Thus, in the simplest case of a one-channel approximation, for example, these interactions inside the core region can be characterized by a complex quantum defect of the form (Alber and Zoller 1988)

$$\mu = \alpha + i\beta. \tag{5.3}$$

The real part of this quantum defect defines the energies of the Rydberg electron in the absence of the laser field, $\varepsilon_n = -1/[2(n - \alpha)^2]$. The imaginary part β describes the influence of laser-induced transitions of the Rydberg electron into continuum states well above threshold. In lowest order of perturbation theory, it is given by

$$\beta = \pi|\langle\varepsilon = \omega|d \cdot E_0|\varepsilon = 0\rangle|^2, \tag{5.4}$$

with d denoting the atomic dipole operator. For hydrogen and linearly polarized laser light, for example, this imaginary part of the quantum defect can be evaluated approximately with the help of the Bohr correspondence principle. According to this principle the dipole matrix element entering (5.4) is approximated by a Fourier coefficient of the classical trajectory of a Rydberg electron of energy $\varepsilon = 0$ (Landau and Lifshitz 1975), namely

$$\langle\varepsilon = \omega|d \cdot E_0|\varepsilon = 0\rangle = \frac{1}{2\pi} \int\limits_{-\infty}^{\infty} dt e^{i\omega t} x(t) \cdot E_0 = \frac{6^{2/3}}{2\pi\sqrt{3}} \Gamma\left(\frac{2}{3}\right)\omega^{-5/3}|E_0|. \tag{5.5}$$

($\Gamma(x) = \int_0^\infty du u^{x-1}e^{-u}$ denotes the Euler gamma function.) Thereby $x(t)$ describes the parabolic classical trajectory of an electron that moves in the Coulomb field of the nucleus with energy $\varepsilon = 0$. Consistent with the previous qualitative discussion, the $\omega^{-5/3}$ dependence in (5.5) demonstrates that the dominant contribution to this dipole matrix element originates from a spatial region around the nucleus with a size of the order of $r_c \approx \omega^{-2/3}$. This characteristic size r_c is the distance a classical electron of (asymptotic) energy $\varepsilon = 0$ can depart from the nucleus during the relevant photon absorption time $t_{photon} \approx 1/\omega$.

In the Coulomb and asymptotic region the quantum mechanical state can be determined semiclassically. To make these ideas more precise, let us consider the general form of the semiclassical solution of the time independent Schrödinger equation which is valid in the Coulomb and asymptotic region. It has the general form (Maslov and Fedoriuk 1981; Delos 1986)

$$\psi(\varepsilon,|x) = \sum_j \varphi(\varepsilon, y_j)\sqrt{\frac{J(0, y_j)}{|J(t_j, y_j)|}} \ e^{i[S_j(t_j,y_j)-\mu_j(t_j)\pi/2]}. \tag{5.6}$$

This wavefunction is determined by two different types of quantities, namely the probability amplitude $\varphi(\varepsilon, y)$ of finding the electron at position y on the boundary between the core region and the Coulomb region and by quantities that describe the classical motion of the Rydberg electron outside the core region (compare with Fig. 5.1). The probability amplitude $\varphi(\varepsilon, \mathbf{y})$ is determined by the quantum defect parameters that describe the electron correlations and the electron-laser interaction inside the core region. According to (5.6) the probability amplitude $\psi(\varepsilon, \mathbf{x})$ of finding the electron at position $\mathbf{x}$ outside the core region is also determined by properties of all those classical trajectories j which start at the boundary between the core region and the Coulomb region at position $\mathbf{y}$ and reach the final point $\mathbf{x}$ at any "time" t. In this context the variable t represents a curve parameter and not a physical time. Together with the initial positions $\mathbf{y}$, the curve parameter t constitutes a global coordinate system for the family of classical trajectories that leave the core region and form a Lagrangian manifold (Maslov and Fedoriuk 1981; Delos 1986). The important classical properties of trajectory j that determine $\psi(\varepsilon, \mathbf{x})$ are as follows:

1. Its classical action (eikonal) $S_j(t_j, \mathbf{y}_j)$.

2. The determinant of its Jacobi field

$$J(t_j, \mathbf{y}_j) = \frac{dx_1 \wedge dx_2 \wedge dx_3}{dt \wedge dy_1 \wedge dy_2} \Big|_j$$

which characterizes its stability properties.

3. Its Maslov index $\mu_j(t_j)$ which characterizes the number of conjugate points and their multiplicity.

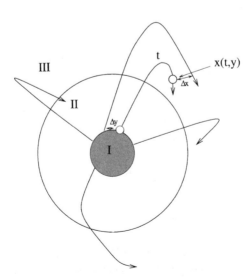

Figure 5.1 Schematic representation of the characteristic spatial regions that determine the dynamics of a Rydberg electron. Some relevant classical trajectories for the semiclassical wave function are also indicated.

According to this general theoretical approach it is apparent that Rydberg systems differ from one another only as far as their dynamics inside the core region is concerned. This part of the dynamics can be described generally by a few quantum defect parameters. Thus Rydberg systems exhibit universal behavior, and the quantum defect parameters characterize the associated universality classes. Furthermore the semiclassical analysis of the dynamics of the Rydberg electron in the Coulomb region and in the asymptotic region implies that probability amplitudes describing atomic transitions between an initial and a final state can be represented as a sum of contributions associated with all possible classical paths (including their multiple returns) that connect the regions of support of the initial and the final state. In particular, if the dominant contribution of a transition amplitude originates from the core region, for example, it is all classical paths that start and end inside the core region and are relevant for the theoretical description. On the basis of this combination of methods of quantum defect theory with semiclassical path representations for relevant quantum mechanical transition amplitudes, many aspects of the coherent dynamics of electronic Rydberg wave packets have already been described successfully (Beims and Alber 1993, 1996; Alber et al. 1994; Zobay and Alber 1998).

5.3 DISSIPATIVE DYNAMICS OF ELECTRONIC RYDBERG WAVE PACKETS

So far in the context of wave packet dynamics of material particles, the investigation of dissipative and stochastic influences that destroy quantum coherence has not received much attention. Definitely to some extent this may be attributed to the complications arising from the high-level densities that have to be taken into account for a proper theoretical description of wave packet dynamics. In general, they turn the solution of master equations for the relevant density operator into a difficult mathematical and numerical problem. Electronic wave packets in Rydberg systems are an extreme example of this kind due to their almost macroscopic size and the infinitely high-level density of Rydberg states at an ionization threshold. In the subsequent discussion it will be demonstrated that a combination of the semiclassical methods discussed in Section 5.2 together with stochastic simulation methods constitutes a powerful theoretical approach for describing many aspects of the destruction of quantum coherence in wave packet dynamics. In addition this theoretical approach offers insight into the intricate interplay between the semiclassical aspects of the dynamics of a Rydberg electron outside the core region and its coupling to the radiation field inside the core region. In the subsequent sections two types of physical processes will be discussed in detail by which this coupling to the radiation field can destroy the quantum coherence of an electronic wave packet, namely spontaneous emission of photons and the intrinsic fluctuations of a laser field. Motivated by the recent interest in laser-induced two-electron excitation processes characteristic effects of radiative damping are explored first; they are mediated by the correlation between an electronic Rydberg wave packet and a resonantly excited, tightly bound core electron. Then it is demonstrated that as a result of the peculiar threshold properties of Rydberg systems the destruction of quantum coherence which is brought about by a fluctuating laser field gives rise to a variety of novel phenomena.

5.3.1 Radiative Damping Mediated by Electron Correlations

Due to the long radiative life times of Rydberg states (radiative life times scale as $(n - \alpha)^3$; Gallagher 1994) the direct influence of spontaneously emitted photons is negligible under typical laboratory situations. However, destruction of quantum coherence originating from radiative damping might become significant in cases in which more than one atomic or molecular electron is excited resonantly by a laser field. In such cases the influence of a photon that is emitted spontaneously by one of these excited electrons can influence another excited Rydberg electron via electron correlation effects. Isolated core excitation (ICE) processes (Cooke et al. 1978) are a particular class of laser-induced two-electron excitation processes that has received considerable attention recently. In the following it is demonstrated that in these types of excitation processes the dissipative influence of radiative damping mediated by electron correlations may influence the dynamics of electronic wave packets significantly.

ICE excitation processes have been studied extensively in the alkaline earth elements as the corresponding singly-charged ions are excited easily with laser fields in the optical or near-UV regime. In Figure 5.2 a typical laser-induced ICE process is shown schematically for a magnesium atom. In a first step, the atom is excited from its $|3s^2\rangle$ ground state to a Rydberg state $|3snd\rangle$ by two-photon excitation. After this excitation process the atom consists of the $Mg^+(3s)$ ionic core and the nd-Rydberg electron which tends to be located at large distances from the core. By applying a second laser pulse tuned to a resonance of the Mg^+ ion, the remaining core electron is excited, for example, to the $3p$-state of the ionic core. The direct influence of the laser

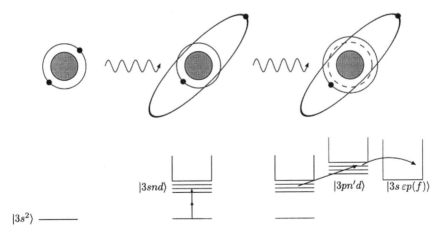

Figure 5.2 Schematic representation of a laser-induced isolated core excitation process in Mg. After initial preparation in a $|3snd\rangle$ Rydberg state, a second laser pulse excites the core $3s \rightarrow 3p$ transition. The Rydberg states of the excited core autoionize.

field on the highly excited Rydberg electron is usually negligible in comparison to its interaction with the second, tightly bound valence electron. But the laser field influences the Rydberg electron indirectly by electron correlation effects. Immediately after the core transition the Rydberg electron experiences a "shakeup" by the different short-range core potential to which it has to accommodate. A quantitative measure for the degree of this shakeup is given by the difference between the quantum defects of the two channels associated with the $3s$ and the $3p$-states of the ionic core. The early work on ICE spectroscopy of alkaline earth elements has concentrated on nonresonant core transitions which can be described in the lowest order of perturbation theory with respect to the laser field (Gallagher 1994). Nonperturbative effects of laser fields have become of interest only recently in connection with the development of powerful tunable laser sources (Jones and Bucksbaum 1991; Stapelfeldt et al. 1991; Robicheaux 1993; Grobe and Eberly 1993). They are particularly important in resonant core excitation processes in which one of the laser fields induces Rabi oscillations of the ionic core. A variety of new coherent effects have been predicted theoretically in this context (Robicheaux 1993; Hanson and Lambropoulos 1995; Zobay and Alber 1995; van Druten, and Muller 1995); they rely on the coherent interplay between the Rabi oscillations of the ionic core and the dynamics of an electronic Rydberg wave packet which is influenced by these Rabi oscillations through the resulting shakeup processes (for a review on these theoretical developments, see Zobay and Alber 1998; see also Chapter 3 of this book for experimental developments). However, due to the possibility of spontaneous emission of photons by the resonantly excited core electron, all these effects are expected to be particularly sensitive to the resulting destruction of quantum coherence.

In order to investigate these dissipative effects in detail, let us consider a typical laser-induced two-electron excitation process in an alkaline earth atom as represented in Figure 5.3. It is assumed that the atom is prepared initially in its ground state $|g\rangle$. The atom is situated in a cw-laser field whose electric field strength is given by $\mathbf{E}(t) = \mathcal{E}\mathbf{e}e^{-i\omega t} + c.c.$; it is tuned near resonance with a transition of the positively charged ionic core. Typically electron correlations imply that as long as the atom

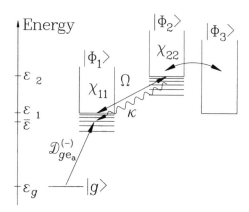

Figure 5.3 Three-channel excitation scheme including spontaneous emission process and autoionization.

remains in its initial state $|g\rangle$, this laser field is well detuned from any atomic transition. Thus the laser field has negligible effect on the atomic dynamics. But as soon as an outer valence electron is excited to Rydberg state close to an ionization threshold the cw-laser field starts to induce transitions between the two resonantly coupled states of the ionic core which have energies ε_1 and ε_2, respectively. Let us concentrate on a case where one of the valence electrons is excited coherently to Rydberg states by a short and weak laser pulse with electric field strength $\mathbf{E}_a(t) = \mathcal{E}_a(t)\mathbf{e}_a e^{-i\omega_a t} + c.c.$ (Typically the pulse envelope $\mathcal{E}_a(t)$ will be modeled by a Gaussian shape centered around time t_a with pulse duration τ_a.) Thus a radial electronic Rydberg wave packet is prepared by this short laser pulse (Alber and Zoller 1991). This wave packet moves in the Coulomb field of the positively charged ionic core. Whenever it penetrates the core region, it is shaken up by the Rabi oscillations of the resonantly driven core. Furthermore, whenever the core emits a photon spontaneously, this emission process will disrupt the relative phases of the electronic wave packet and will thus destroy quantum coherence. The dynamics of this electronic wave packet under the influence of the Rabi oscillations of the ionic core can be investigated by typical pump-probe experiments, for example.

For the theoretical description of the resulting destruction of quantum coherence one has to solve the corresponding optical Bloch equation for the density operator of the two atomic valence electrons. In the case depicted in Figure 5.3, for example, the optical Bloch equation is given by (Zobay and Alber 1996)

$$\dot{\rho}(t) = -i[H, \rho(t)] + \frac{1}{2}\{[L, \rho(t)L^\dagger] + [L\rho(t), L^\dagger]\}. \tag{5.7}$$

Thereby the Hamiltonian

$$H = \sum_{i,j=1,\ldots,3} H_{i,j} + V_{\text{ICE}} \tag{5.8}$$

characterizes the coherent part of the dynamics. The dynamics of the valence electrons are described by the Hamiltonian

$$H_{i,j} = (\mathbf{h}_{jj}\delta_{ij} + \mathbf{V}_{ij} + \varepsilon_{cj}\delta_{ij})|\Phi_i\rangle\langle\Phi_j| \tag{5.9}$$

with

$$\mathbf{h}_{jj} = -\frac{1}{2}\frac{d^2}{dr^2} + \frac{l_j(l_j+1)}{2r^2} - \frac{1}{r}. \tag{5.10}$$

The short-range potential $\mathbf{V}_{ij}$ describes electron-correlation effects originating from the residual core electrons (Aymar et al. 1996). In ICE transitions the angular momentum l of the excited Rydberg electron is conserved to a good degree of approximation, $l_1 = l_2 = l$ (Gallagher 1994). In the rotating wave approximation the channel thresholds ε_{cj} are given by $\varepsilon_{c1} = \varepsilon_1, \varepsilon_{c2} = \varepsilon_2 - \omega, \varepsilon_{c3} = \varepsilon_3 - \omega$. The operator

$$V_{\text{ICE}} = -\frac{1}{2}\Omega(|\Phi_2\rangle\langle\Phi_1| + |\Phi_1\rangle\langle\Phi_2|)\otimes\mathbf{1}_r \tag{5.11}$$

describes the laser-induced core transitions between the core states $|\Phi_1\rangle$ and $|\Phi_2\rangle$, and Ω is the Rabi frequency originating from the cw-laser field. The operator $\mathbf{1}_r$ denotes the identity operator for the radial coordinate of the Rydberg electron. Thus the role of the Rydberg electron as a spectator becomes obvious from (5.11).

The stochastic part of the dynamics of the density operator $\rho(t)$ is described by the Lindblad operator

$$L = \sqrt{\kappa}|\Phi_1\rangle\langle\Phi_2| \otimes \mathbf{1}_r \tag{5.12}$$

which characterizes the radiative decay of the ionic core from its excited state to its ground state by spontaneous emission of photons with rate κ.

Due to the high-level density of Rydberg states close to an ionization threshold and due to the presence of the adjacent electron continuum, usually severe problems arise if one tries to solve the optical Bloch equation (5.7) numerically by expanding the density operator $\rho(t)$ into a basis set of atomic energy eigenfunctions. Many of these problems can be circumvented successfully by combining the semiclassical methods as discussed in Section 5.2 with stochastic simulation methods (Zobay and Alber 1996). Besides numerical advantages this approach gives direct insight into the classical aspects of the dynamics of the Rydberg electron and the destruction of quantum coherence caused by the radiative decay of the core. Thereby the density operator is represented by a (fictitious) ensemble of pure states that are associated with definite numbers of spontaneously emitted photons (Mollow 1975), namely

$$\rho(t) = \sum_{N=0}^{\infty} \rho^{(N)}(t), \tag{5.13}$$

with the N-photon contributions

$$\rho^{(N)}(t) = \int_0^t dt_N \int_0^{t_N} dt_{N-1} \ldots \int_0^{t_2} dt_1 |\psi(t|t_N, \ldots, t_1)\rangle \langle\psi(t|t_N, \ldots, t_1)|.$$

The time evolution of the N-photon states $|\psi(t|t_N, \ldots, t_1)\rangle$ is given by

$$|\psi(t|t_N, \ldots, t_1)\rangle = e^{-iH_{\text{eff}}(t-t_N)}\Theta(t - t_N)Le^{-iH_{\text{eff}}(t_N-t_{N-1})}\Theta(t_N - t_{N-1})L \ldots$$

$$Le^{-iH_{\text{eff}}t_1}\Theta(t_1)|\psi(t = 0)\rangle \tag{5.14}$$

with the effective (non-Hermitian) Hamiltonian

$$H_{\text{eff}} = H - \frac{i}{2}L^\dagger L. \tag{5.15}$$

($\Theta(x)$ is the unit step function.) The physical interpretation of (5.14) is straightforward. With each emission of a photon at one of the N random emission times $t_1 \leq t_2 \leq \ldots \leq t_N$, the quantum state "jumps" into a new state by application of the Lindblad operator of (5.12). Between two successive jumps the state evolves according to the Hamiltonian of (5.15). Thus the decomposition of (5.13) may also be interpreted as an unraveling of the density operator into contributions associated with all possible quantum jumps. This decomposition of the density operator $\rho(t)$ offers significant advantages in cases where the number of spontaneously emitted photons is small or where the evaluation of the relevant pure states can be simplified by the application of semiclassical methods. In particular, it is possible to derive general semiclassical path representations for the N-photon states of the optical Bloch equation (5.7). Thus all physical observables of interest can be expressed as a sum of probability amplitudes which are associated with repeated returns of a Rydberg electron to the ionic core. During its motion under the influence of the Coulomb potential of the ionic core photons may be emitted spontaneously by the laser-excited core at any position of the Rydberg electron along its path. These photon emission processes disrupt the coherent quantum mechanical time evolution of the Rydberg electron.

As an example let us consider a coherent process which has received considerable attention recently, namely laser-induced stabilization against autoionization (Hanson and Lambropoulos 1995). This effect is based on a synchronization between the dynamics of the ionic core, which performs Rabi oscillations, and the dynamics of a laser-prepared electronic wave packet. This effect may be understood as follows: At the time of the preparation of the electronic Rydberg wave packet by the short laser pulse, the core is in its ground state. If the mean Kepler period $T_{orb} = 2\pi(-2\bar{\varepsilon})^{-3/2}$ ($\bar{\varepsilon}$ the mean excited energy of the Rydberg electron) of this wave packet is chosen equal to a multiple of the Rabi period $T_{Rabi} = 2\pi/\Omega$ of the core, the Rydberg electron will encounter the core in the ground state at each of its subsequent returns to the nucleus. Since autoionization of a Rydberg electron can take place only inside the core region (Seaton 1983; Fano and Rau 1986; Aymar et al. 1996), this implies that the effective autoionization rate of the electronic wave packet will become much smaller than the autoionization rate of the mean excited Rydberg state $\Gamma_{\bar{n}}$ in the absence of the laser field. In addition it has been demonstrated (Hanson and Lambropoulos 1995) that this suppression of autoionization is accompanied by a reduction of dispersion of the electronic wave packet. This suppression of dispersion is brought about by the Rabi-oscillating core which acts like a quantum-mechanical shutter and effectively cuts off the tails of the wave packet which arrive at the nucleus out of phase with small probability. Since this stabilization against autoionization is based on the coherent interplay between electron correlations and laser-induced Rabi oscillations, it is expected to be particularly sensitive against the destruction of quantum coherence due to spontaneous emission of photons by the ionic core.

In the presence of radiative decay of the ionic core the physical picture is changed significantly. In the simplest case of synchronization, namely for $T_{orb} = T_{Rabi}$, the first photon will be emitted spontaneously by the ionic core most probably at a time $(M + \frac{1}{2})T_{Rabi}$ (with M denoting any integer) because then the core is in its excited state with high probability. Due to the synchronization at these times, the electronic

Rydberg wave packet is close to the outer turning point of its Kepler orbit. The spontaneous emission of a photon reduces the excited core to its ground state. Therefore, at the subsequent return of the electronic wave packet to the core at time $(M + 1)T_{orb}$, the ionic core will be in its excited state so that the Rydberg electron will autoionize on a time scale of the order of $1/\Gamma_{\bar{n}}$. Thus the laser-induced stabilization against autoionization will be destroyed. Typically $\Gamma_{\bar{n}} \gg \kappa$, so the Rydberg electron will autoionize with high probability long before the core can emit a second photon spontaneously. Consequently it is expected that the influence of the radiative damping on this coherent stabilization phenomenon can be described approximately by taking into account only the zero- and one-photon contributions of the density operator $\rho(t)$.

The influence of radiative damping described above manifests itself clearly in the time dependent autoionization rate $\gamma(t)$ into channel three, for example, which results from the dynamics of the electronic Rydberg wave packet. An experimental technique for measuring $\gamma(t)$ has been developed recently (Lankhuijzen and Noordam 1996). This time dependent ionization rate $\gamma(t)$ can be decomposed into N-photon contributions with the help of semiclassical path representations, namely

$$\gamma(t) = \sum_{N=0}^{\infty} \int_0^t dt_N \ldots \int_0^{t_2} dt_1 \gamma^{(N)}(t). \tag{5.16}$$

It is expected that the zero- and one-photon contributions (Zobay and Alber 1996)

$$\gamma^{(0)}(t) = \frac{1}{2\pi}(1 - e^{-4\pi \mathrm{Im}\mu_2}) \mid \int_{-\infty+i0}^{\infty+i0} d\varepsilon_1 e^{-i\varepsilon_1 t}(0, 1, 0)\mathbf{O}\sum_{M_1=0}^{\infty}(e^{i2\pi\tilde{\nu}_1}\tilde{\chi})^{M_1}$$

$$\times e^{i2\pi\tilde{\nu}_1}\underset{ge_a}{\tilde{\mathcal{D}}}(-)\tilde{E}_a(\varepsilon_1 - \varepsilon)\mid^2,$$

$$\gamma^{(1)}(t) = \left(\frac{1}{2\pi}\right)^3(1 - e^{-4\pi\mathrm{Im}\mu_2}) \mid \int_{-\infty+i0}^{\infty+i0} d\varepsilon_1 d\varepsilon_2 e^{-i\varepsilon_2(t-t_1)}e^{-i\varepsilon_1 t_1}(0, 1, 0)\mathbf{O}$$

$$\times \sum_{M_2=0}^{\infty}(e^{i2\pi\tilde{\nu}_2}\tilde{\chi})^{M_2}\tilde{S}_{2,1}^{(M_2,M_1)}\sum_{M_1=0}^{\infty}(\tilde{\chi}e^{i2\pi\tilde{\nu}_1})^{M_1}\underset{gea}{\tilde{\mathcal{D}}}(-)\tilde{E}_a(\varepsilon_1 - \varepsilon)\mid^2 \tag{5.17}$$

are dominant. In (5.17) the laser-induced excitation by the short laser pulse is characterized by the Fourier transform of the pulse envelope

$$\tilde{E}_a(\Delta\varepsilon) = \int_{-\infty}^{\infty} dt' E_a(t)e^{i\Delta\varepsilon(t-t_a)} \tag{5.18}$$

and by the (3×1)-column vector $\widetilde{\mathcal{D}}_{ge_a}^{(-)}$ whose components are the energy normalized photoionization dipole matrix elements (Seaton 1983) into channels one, two, and three. The dynamics of the Rydberg electron under the influence of the Rabi oscillations of the ionic core are described by the (3×3) scattering matrix $\widetilde{\chi}$ and by the (3×3) diagonal matrix $e^{i2\pi\widetilde{\nu}}$ with matrix elements $(e^{i2\pi\widetilde{\nu}})_{jj} = e^{2i\pi[2(\widetilde{\mathcal{E}}_{cj}-\varepsilon)]^{-1/2}} \Theta(\widetilde{\mathcal{E}}_{cj} - \varepsilon)$ $(j = 1, 2, 4)$. All matrices and column vectors with a tilde refer to the basis of photon-dressed core states $|\widetilde{\Phi}_j\rangle$ $(j = 1, 2, 3)$ (Robicheaux 1993; Zobay and Alber 1995). These dressed channel states are related to the corresponding bare states $|\Phi_j\rangle$ by the orthogonal transformation $\mathbf{O}$ which diagonalizes the laser-induced core coupling:

$$\mathbf{O}^T[\varepsilon_c - \frac{i\kappa}{2|\Phi_2\rangle\langle\Phi_2|} - \frac{1}{2}\Omega(|\Phi_2\rangle\langle\Phi_1| + |\Phi_1\rangle\langle\Phi_2|)]\mathbf{O} = \widetilde{\varepsilon}_c. \tag{5.19}$$

Thereby the diagonal matrix $\widetilde{\varepsilon}_c(\varepsilon_c)$ contains the energies of the dressed (bare) core states. Thus the relations $\widetilde{\mathcal{D}}_{ge_a}^{(-)} = \mathbf{O}^T\mathcal{D}_{ge_a}^{(-)}$ and $\widetilde{\chi} = \mathbf{O}^T\chi\mathbf{O}$ hold with the bare photo-ionization dipole matrix elements $\mathcal{D}_{ge_a}^{(-)}$ and with the bare scattering matrix

$$\chi = \begin{pmatrix} e^{2\pi i\mu_1} & 0 & 0 \\ 0 & e^{2\pi i\mu_2}\chi_{23} & \\ 0 & \chi_{32} & \chi_{33} \end{pmatrix}. \tag{5.20}$$

The quantum defects of the bare channels one and two are denoted μ_j. These channels have opposite parity and cannot be coupled by electron correlation effects. The matrix elements χ_{23} and χ_{32} characterize the configuration interaction between channels 2 and 3 which results in autoionization of channel 2. The autoionization rate of a Rydberg state of channel 2 with principal quantum number n is related to the imaginary part of the quantum defect μ_2 by $\Gamma_n = 2\text{Im}(\mu_2)/[n - \text{Re}(\mu_2)]^3$.

Equations (5.17) are examples of semiclassical path representations for the zero- and one-photon ionization rates $\gamma^{(0)}(t)$ and $\gamma^{(1)}(t)$. Their physical interpretation is straightforward: After the initial excitation by the short laser pulse, those fractions of the electronic Rydberg wave packet that are excited into closed photon-dressed core channels return to the core region periodically. The integers M_1 and M_2 count the numbers of these returns. Between two successive returns the Rydberg electron acquires a phase of magnitude $(2\pi\widetilde{\nu})_{jj}$ while moving in the photon-dressed core channel j. This phase equals the classical action of motion along a purely radial Kepler orbit with zero angular momentum and energy $\varepsilon - \widetilde{\varepsilon}_{cj} < 0$. Entering the core region the Rydberg electron is scattered into other photon-dressed core channels by laser-modified electron correlation effects which are described by the scattering matrix $\widetilde{\chi}$. The ionic core can emit a photon spontaneously at any time during the motion of the Rydberg electron. Quantitatively this photon emission process is described by the quantity

$$\widetilde{\mathbf{S}}_{2,1}^{(M_2,M_2)} = \int\limits_{0}^{T_{M_1,M_2}} d\tau e^{2i\pi\widetilde{\nu}_2(1-\tau/T_{M_1,M_2})}(e^{-i\pi/2}\widetilde{\mathbf{L}})e^{2i\pi\widetilde{\nu}_1\tau/T_{M_1,M_2}} \tag{5.21}$$

in (5.17) with $T_{M_1, M_2} = t/(M_1 + M_2 + 1)$. According to (5.21) this spontaneous photon emission by the ionic core can take place at any time τ between two successive returns of the Rydberg electron to the core region. At time τ the Rydberg electron has acquired a phase of magnitude $(2\pi\tilde{\nu})_{jj}\tau/T_{M_1, M_2}$ in channel j. The disruption of the phase of the Rydberg electron by this spontaneous emission process is described by the action of the Lindblad operator $\tilde{\mathbf{L}} = \mathbf{O}^T\mathbf{L}\mathbf{O}$. It also leads to a phase change of magnitude $(\pi/2)$. After the completion of the photon emission process the Rydberg electron acquires an additional phase of magnitude $(2\pi\tilde{\nu})_{jj}(1 - \tau/T_{M_1, M_2})$ in the photon-dressed core channel j until it reaches the core region again.

A representative time evolution of the autoionization rate $\gamma(t)$ is shown in Figure 5.4. The full curve in Figure 5.4a has been obtained by numerical solution of the optical Bloch equation (5.7) with the help of a conventional basis expansion in atomic energy eigenstates. The corresponding zero- and one-photon contributions are also presented in Figures 5.4b and c. In Figure 5.4 the sum of zero- and one-photon contributions are not plotted, since they are indistinguishable from the numerical result (full curve in Figure 5.4a). The chosen parameters represent typical values realizable in alkaline earth experiments. The comparison of $\gamma(t)$ (full curve of Figure 5.4a) with the corresponding result in the absence of radiative damping (dotted curve in Figure 5.4a) demonstrates that the influence of radiative damping is already significant at interaction times of the order of T_{orb}. With the help of the zero- and one-photon contributions of (5.17), the dissipative influence of radiative damping can be analyzed in detail. As apparent from Figure 5.4b, the zero-photon rate vanishes at integer multiples of the mean Kepler period T_{orb} because at these times the core is in its ground state. The maxima of Figure 5.4b at times $(M + \frac{1}{2})T_{\mathrm{orb}}$ originate from fractions of the electronic wave packet which are close to the core at times when the core is in its excited state. Also visible are typical revival effects at times of the order of $25T_{\mathrm{orb}}$. The one-photon rate of Figure 5.4c exhibits maxima and minima at times MT_{orb} and $(M + \frac{1}{2})T_{\mathrm{orb}}$. These maxima indicate that the photon is emitted by the ionic core most probably whenever the Rydberg electron is close to the outer turning point of its classical Kepler orbit. Thus the core will be in its excited state when the Rydberg electron returns to the nucleus, so autoionization will take place with a high probability.

5.3.2 Electronic Wave Packets in Fluctuating Laser Fields

The main aim of this section is to discuss characteristic effects that govern the dynamics of a Rydberg electron in an intense and fluctuating laser field. It is demonstrated that for moderate laser intensities (compare with (5.2)) a variety of novel, nonperturbative effects appear that influence the long-time behavior of Rydberg electrons significantly. A generic consequence of the interplay between the peculiar threshold phenomena of Rydberg systems and the destruction of quantum coherence due to laser fluctuations is stochastic ionization (Alber and Eggers 1997). It is demonstrated that this process also implies an upper time limit on the applicability of two-level approximations even in cases where all characteristic frequencies,

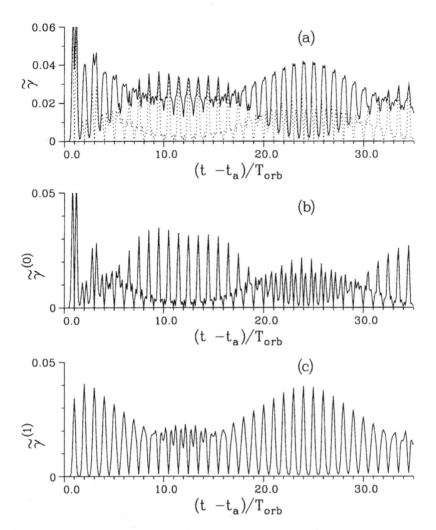

Figure 5.4 Autoionization and resonant excitation of the core under the condition of period matching, $T_{orb} = T_{Rabi}$. The parameters are $\kappa^{-1} = 7$ ns, $\bar{\nu}_1 = [-2(\varepsilon_2 - \varepsilon_4)]^{-1/2} = 73$ ($T_{orb} = 59$ ps), $\mu_1 = 0.0$, $\mu_2 = 0.5 + i0.1$, $\tau_a = 0.4 T_{orb}$ with $\mathcal{E}_a(t) = \mathcal{E}_a^{(0)} e^{-4(t-t_a)^2 \ln/\tau_a^2}$. (a) Scaled ionization rate $\tilde{\gamma}(t) = \gamma(t) T_{orb} \tau_a / |\mathcal{D}_{ge}^{(-)} \mathcal{E}_a^{(0)}|^2$ as obtained from the optical Bloch equations (full curve); (b) and (c) Scaled zero- and one-photon contributions $\tilde{\gamma}^{(0)}(t)$ and $\tilde{\gamma}^{(1)}(t)$. (Reprinted from Zobay and Alber 1996; Copyright © 1998 by the American Physical Society.)

namely Rabi frequencies and laser bandwidths, are small in comparison with the Kepler frequency of a resonantly excited Rydberg electron.

Nowadays laser fluctuations can be controlled to such a degree that it is possible to realize various theoretical models of laser radiation in the laboratory (Vemuri et al. 1991). One of the most elementary theoretical models of laser radiation is the phase diffusion model (PDM) (Haken 1970). It describes approximately the electric field produced by an ideal single mode laser that is operated well above the laser threshold. Thereby the electric field of a laser is represented by a classical, stochastic process with well stabilized amplitude and a fluctuating phase, namely

$$\mathbf{E}(t) = \mathbf{E}_0 e^{i\Phi(t)} e^{-i\omega t} + c.c. \tag{5.22}$$

The fluctuations of the phase $\Phi(t)$ are modeled by a real-valued Wiener process (Klöden and Platen 1992), namely

$$M d\Phi(t) = 0, \ [d\Phi(t)]^2 = 2b \, dt. \tag{5.23}$$

Thereby M indicates the mean over the statistical ensemble. The PDM implies a Lorentzian spectrum of the laser radiation with bandwidth b.

In order to investigate the influence of laser fluctuations on the optical excitation of Rydberg states close to an ionization threshold, let us consider the simplest possible case, namely one-photon excitation from a tightly bound initial state $|g\rangle$ with energy ε_g. In the dipole and rotating wave approximation, the Hamiltonian which describes this excitation process is given by

$$H(\Phi(t)) = \varepsilon_g |g\rangle \langle g| + \sum_n \varepsilon_n |n\rangle \langle n|$$

$$- \sum_n (|n\rangle \langle g| \langle n|\mathbf{d}|g\rangle \cdot \mathbf{E}_0 e^{i\Phi(t)} e^{-i\omega t} + \text{h.c.}). \tag{5.24}$$

In (5.24) the index n refers to Rydberg and continuum states. The energies of the excited Rydberg states are denoted ε_n, and $\mathbf{d}$ is the atomic dipole operator. Let us assume for the sake of simplicity that the excited Rydberg and continuum states can be described with the help of quantum defect theory in a one-channel approximation (Seaton 1983). Thus they are characterized by an approximately energy independent quantum defect $\mu = \alpha + i\beta$. As was explained in Section 5.2 (equation (5.4)), the imaginary part β describes photon absorption from the highly excited Rydberg states to continuum states well above threshold.

For the description of nonperturbative aspects of this laser excitation process, one has to solve the time dependent Schrödinger equation with the stochastic Hamiltonian (5.24) (interpreted as a stochastic differential equation of the Ito type; Klöden and Platen 1992) together with the stochastic differential equation for the phase (5.23). It is the simultaneous presence of the Coulomb threshold with its infinitely many bound states and the continuum, on the one hand, and the laser fluctuations, on the other hand, that makes this solution a highly nontrivial task. Nevertheless, for the case of the PDM the resulting mathematical and numerical problems can be circumvented successfully (Alber and Eggers 1997). Thus even analytical results can be derived in the limit of long interaction times which is dominated by stochastic ionization of the Rydberg electron. Thus let us start from the equation of motion for the mean values $\rho_{nn'}(t) = M\langle n|\psi(t)\rangle \langle \psi(t)|n'\rangle$, $\rho_{ng}(t) = [\rho_{gn}(t)]^* = M e^{-i\Phi(t)} \langle n|\psi(t)\rangle \langle \psi(t)|g\rangle$ and $\rho_{gg}(t) =$

$M|\langle g|\psi(t)\rangle|^2$ which can be combined to form a density operator $\rho(t)$ (Agarwal 1976). From equations (5.23) and (5.24) it can be shown that this density operator fulfills the master equation

$$\frac{d}{dt}\rho(t) = -i[H_{mod}, \rho(t)] + \frac{1}{2}\{[L, \rho(t)L^\dagger] + [L\rho(t), L^\dagger]\}. \tag{5.25}$$

Thereby the modified Hamiltonian $H_{mod} \equiv H(\Phi(t) \equiv 0)$ describes laser induced excitation of Rydberg states close to threshold in the absence of phase fluctuations. The destruction of quantum coherence that is brought about by the laser fluctuations is characterized by the Lindblad operator

$$L = \sqrt{2b}|g\rangle\langle g|. \tag{5.26}$$

On the basis of this master equation Fourier representations can be developed for the density matrix elements whose kernels can be determined explicitly with the help of quantum defect theory. Thus all complications arising from the Coulomb threshold are taken into account properly. These Fourier representations are useful for numerical calculations of averaged transition probabilities which are highly accurate even in the limit of long interaction times. Furthermore these representations are convenient starting points for the derivation of analytical results. Thus the averaged initial state probability $\rho_{gg}(t)$, for example, is given by (Alber and Eggers 1997)

$$\rho_{gg}(t) = \sum_{N=0}^{\infty} \frac{1}{2\pi} \int_{-\infty+i0}^{\infty+i0} dz\, e^{-izt} A_{gg}(z)]2bA_{gg}(z)]^N$$

$$\tag{5.27}$$

$$= \frac{1}{2\pi} \int_{-\infty+i0}^{\infty+i0} dz\, e^{-izt} A_{gg}(z)[1 - 2bA_{gg}(z)]^{-1}$$

with

$$A_{gg}(z) = U(z) + U^*(-z),$$

$$U(z) = \{-C_1(z) + C_2(z)$$

$$+ i\sum_{Re\tilde{\varepsilon}_n < 0}[1 - \frac{d}{dz}\sum{}^*(z_1 - z)]^{-1}[z_1 - \bar{\varepsilon} + ib - \sum(z_1)]^{-1}|_{z_1 = z + \tilde{\varepsilon}_n}\}\Theta(z) \tag{5.28}$$

and with

$$C_1(z) = \frac{1}{2\pi(z + 2ib)} \ln \frac{z - \bar{\varepsilon} + i(b + \gamma/2)}{-\bar{\varepsilon} + i(\gamma/2 - b)},$$

$$C_2(z) = \frac{1}{2\pi[z + i(\gamma + 2b)]} \ln \frac{z - \bar{\varepsilon} + i(b + \gamma/2)}{-\bar{\varepsilon} + i(\gamma/2 + b)}. \tag{5.29}$$

In the spirit of the discussion of Section 5.3 (compare with equation (5.13)), $\rho_{gg}(t)$ is represented as a sum of contributions of all possible quantum jumps N which can be induced by the Lindblad operator of (5.26). According to (5.27) these contributions give rise to a geometric series which can be summed easily. The sum appearing in (5.28) extends over all dressed states $\tilde{\varepsilon}_n$ of the effective Hamiltonian $H_{\text{eff}} = H_{\text{mod}} - iL^\dagger L/2$. The mean excited energy is given by $\bar{\varepsilon} = \varepsilon_g + \omega + \delta\omega$, with $\delta\omega$ denoting the relative quadratic Stark shift between the initial state $|g\rangle$ and the ponderomotive shift of the excited Rydberg states (compare with the general discussion in Section 5.2). Besides the threshold contributions $C_1(z)$ and $C_2(z)$ the characteristic kernel $A_{gg}(z)$ is determined by the (resonant part of the) self energy of the initial state $|g\rangle$, namely

$$\Sigma(z) = \sum_n \frac{|\langle n|\mathbf{d} \cdot \mathbf{E}_0|g\rangle|^2}{z - \varepsilon_n} = -i\frac{\gamma}{2} - i\gamma \sum_{M=1}^{\infty} (e^{i2\pi(-2z)^{-1/2}} \chi)^M. \tag{5.30}$$

This self-energy is characterized by the laser-induced depletion rate

$$\gamma = 2\pi|\langle \varepsilon = 0|\mathbf{d} \cdot \mathbf{E}_0|g\rangle|^2 \tag{5.31}$$

of the initial state $|g\rangle$ and by the scattering matrix element

$$\chi = e^{i2\pi\mu} \tag{5.32}$$

which describes all effects arising from scattering of the Rydberg electron by the ionic core and from photon absorption (compare with equation (5.3)). The sum over M in (5.30) originates from the multiple returns of the Rydberg electron to the core region where the dominant contribution to the self-energy comes from. With each of these returns the Rydberg electron of energy $z < 0$ accumulates a phase of magnitude $2\pi(-2z)^{-1/2}$, and with each traversal of the core region it accumulates a (complex) phase of magnitude $2\pi\mu$ due to scattering by the core and due to photon absorption. The laser-induced depletion rate γ, the imaginary part of the quantum defect β, and the second order Stark shift $\delta\omega$ describe the influence of the laser field on the Rydberg electron. Since these quantities depend on the laser intensity, they are not affected by phase fluctuations of the laser field.

Master equations of the form of (5.25) with a self adjoint Lindblad operator are of general interest as phenomenological models of continuous quantum measurement processes (Braginsky and Khalili 1992). In this context (5.25) would model excitation of Rydberg and continuum states close to an ionization threshold by a classical, deterministic laser field in the presence of continuous measurement of the initial state $|g\rangle$. Thereby the inverse bandwidth $1/b$ would determine the mean time between successive measurements.

Some qualitative aspects of the time evolution of an excited Rydberg electron under the influence of a fluctuating laser field are apparent from the contour plots of Figures 5.5 and 5.6 which refer to one-photon excitation of a hydrogen atom by linearly polarized laser light with $|g\rangle = |2s\rangle$. It is assumed that Rydberg states around $\bar{n} = (-2\bar{\varepsilon})^{-1/2} = 80$ are excited. According to the general discussion in Section 5.2 (compare with equation (5.4)) the laser-induced transitions from the excited Rydberg states to continuum states well above threshold are described by an imaginary quantum defect with $\beta = 0.00375\gamma$.

In Figure 5.5a both the bandwidth of the laser field b and the field-induced depletion rate γ of state $|g\rangle$ are assumed to be small in comparison with the mean level spacing of the excited Rydberg states, $b, \gamma \ll \bar{n}^{-3}$. Thus one may be tempted to think that this excitation process can be described well within the framework of a two-level approximation in which only states $|2s\rangle$ and $|80p\rangle$ are taken into account. However, Figure 5.5a demonstrates that this expectation is only valid for sufficiently small interaction times. Indeed, the early stages of the excitation process are dominated by Rabi oscillations of the electron between the initial and the resonantly excited state.

These Rabi oscillations are damped by the fluctuating laser field. An equilibrium is attained for interaction times $t \geq 1/b$ for which all coherence between the two resonantly coupled states is negligibly small and for which $\rho_{gg}(t) \approx \rho_{\bar{n}\bar{n}}(t) \approx \frac{1}{2}$. This characteristic, well-known two-level behavior is exemplified in Figure 5.5a by the stationary probability distribution of the excited Rydberg state. (The probability distribution of state $|g\rangle$ which is localized in a region of a few Bohr radii around the nucleus is not visible on the radial scale of Figure 5.5a.) Figure 5.5a indicates that for interaction times that are larger than a critical time t_1, this simple picture of the two-level approximation breaks down and the probability distribution of the excited Rydberg electron starts to spread toward larger distances from the core. (here $t_1 \approx 5 \times 10^5 T$ with $T = 2\pi\bar{n}^3$ denoting the mean classical orbit time). Simultaneously the probability distribution becomes more and more spatially delocalized with all nodes disappearing. In order to obtain a more detailed understanding of this diffusionlike process the time evolutions of the initial state probability and of the ionization probability are shown in Figure 5.5b. From Figure 5.5b it is apparent that this spatial spreading of the Rydberg electron is connected with a diffusion in energy space toward the ionization threshold. At interaction times $t \geq t_c \approx 7 \times 10^9 T$, the Rydberg electron has reached the ionization threshold and the ionization probability $P_{\text{ion}}(t)$ rises significantly from a negligibly small value to a value close to unity. Simultaneously the initial state probability $P_{gg}(t)$ starts to decrease faster. This stochastic diffusion of the Rydberg electron which eventually leads to ionization is a characteristic phenomenon brought about by the fluctuations of the exciting laser field. With the help of the theoretical approach presented above, this characteristic stochastic ionization process can be analyzed in detail. Thus it can be shown (Alber and Eggers 1997) that the diffusion of the Rydberg electron towards the ionization threshold starts at time

$$t_1 = \frac{8}{\pi b \gamma T} \tag{5.33}$$

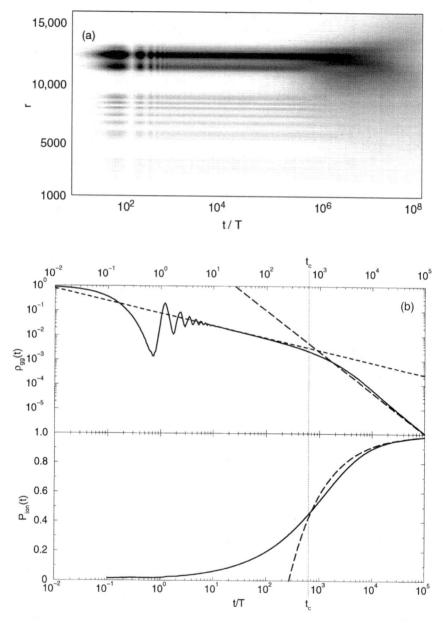

Figure 5.5   Excitation of an isolated Rydberg state: Radial contour plot (*a*) and $P_{gg}(t)$, $P_{ion}(t)$ (*b*) as a function of the interaction time t in units of the mean Kepler period T. The parameters are $\bar{n} = (-2\bar{\varepsilon})^{-1/2} = 80$ ($T = 78$ ps), $\gamma T = 0.1$, and $bT = 0.01$. Various approximate asymptotic time dependences are also indicated, namely equation (5.35) (short dashed) and equations (5.36) and (5.37) (long dashed).

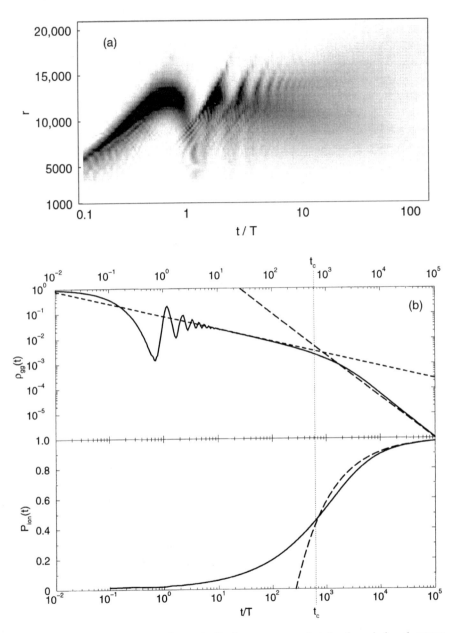

Figure 5.6 Excitation of an electronic Rydberg wave packet by laser-induced power broadening: Radial contour plot (*a*) and $P_{gg}(t)$, $P_{ion}(t)$ (*b*) as a function of the interaction time t in units of the mean Kepler period T. The parameters are $\bar{n} = (-2\bar{\varepsilon})^{-1/2} = 80$ ($T = 78$ ps), $\gamma T = 10.0$, and $bT = 10.0$. Various approximate asymptotic time dependences are also indicated, namely equation (5.35) (short dashed) and equations (5.36) and (5.37) (long dashed); $\gamma t = 10.0$ and $bT = 10.0$.

and eventually leads to stochastic ionization at interaction times $t \geq t_c$ with

$$t_c = \frac{4\pi}{\sqrt{27}\gamma b} \left[\frac{(\bar{\varepsilon}^2 + 3(b^2 + \gamma^2/4)/4)^{3/2}}{\bar{\varepsilon}^2 + b^2 + \gamma^2/4} \right]^{1/2}. \tag{5.34}$$

The time evolution of $P_{gg}(t)$ is approximately given by

$$P_{gg}(t) = \frac{2}{\sqrt{\pi}} [2b\gamma T]^{-1/2} t^{-1/2} \tag{5.35}$$

for $t_1 < t < t_c$ and crosses over to the power law

$$P_{gg}(t) = \frac{(\gamma + 2b)^2}{(2b\gamma\varphi/\pi)^2} \left[\frac{\gamma b \Gamma^3(5/3)}{27\pi(\bar{\varepsilon}^2 + b^2 + \gamma^2/4)} \right]^{1/3} t^{-5/3} \tag{5.36}$$

for interaction times $t > t_c$. The variable φ characterizes the distance of the mean excited energy $\bar{\varepsilon}$ from the ionization threshold and is determined by the relation $-\bar{\varepsilon} + i(b + \gamma/2) = Re^{i\varphi}(0 \leq \varphi < \pi)$. At times $t > t_c$ the ionization probability rises according to the power law

$$P_{\text{ion}}(t) = 1 - \frac{\pi\Gamma(2/3)(\gamma + 2b)}{6b\gamma\varphi} \left[\frac{\gamma b}{\pi(\bar{\varepsilon}^2 + b^2 + \gamma^2/4)} \right]^{1/3} t^{-2/3}. \tag{5.37}$$

These approximate time evolutions are indicated by the dashed curves in Figure 5.5b. The analytical results of (5.33) and (5.34) explicitly show how the critical times t_1 and t_c for the breakdown of the two-level approximation and for stochastic ionization depend on the characteristic parameters of the problem, namely the mean excited energy $\bar{\varepsilon}$, the laser bandwidth b, and the laser-induced depletion rate of the initial state γ.

In Figure 5.6 both the laser bandwidth and the laser-induced depletion rate of the initial state $|g\rangle$ are larger than the mean level spacing $\bar{n}^{-3}$ of the excited Rydberg states. In this case, the initial state is depleted by the laser field in a time that is small in comparison with the mean Kepler period of the excited Rydberg states, that is, $1/\gamma << T = 2\pi\bar{n}^3$, and so the electronic Rydberg wave packet is prepared by power broadening (Alber and Zoller 1988). The initial stage of the preparation of this electronic wave packet by power broadening manifests itself in an approximately exponential decay of $P_{gg}(t)$ with rate γ. The repeated returns of fractions of this wave packet to the core region give rise to recombination maxima of $P_{gg}(t)$ which occur roughly at multiples of the mean Kepler period T. In the absence of laser fluctuations, the nonperturbative time evolution of such an electronic wave packet under the influence of a laser field is already well understood. In the completely coherent case with each return to the core region, a fraction of the electronic wave packet can be scattered resonantly in the presence of the laser field by stimulated emission and reabsorption of a laser photon accompanied by an electronic transition to the initial state $|g\rangle$ and back again. This emission and reabsorption process of a laser photon causes a time delay of the electronic wave packet of the order of $1/\gamma$ with respect to unscattered fractions of the electronic wave packet. These repeated scattering processes lead to a splitting of the original wave packet into many partially overlapping fractions. In the completely coherent case the interference of these overlapping fractions inside the core region eventually give rise to a complicated time dependence of $P_{gg}(t)$ (Alber and Zoller 1991).

Characteristic qualitative aspects of the time evolution of an electronic wave packet in the presence of laser fluctuations are apparent from Figure 5.6a. Clearly, the initial stages of the time evolution are dominated by the preparation of the electronic wave packet and by its repeated returns to the core region. However, at sufficiently long interaction times eventually the spatially localized electronic wave packet starts to spread out uniformly over the whole classically accessible region. Furthermore this classical region starts to grow monotonically with increasing interaction time. Characteristic quantitative details of this time evolution are apparent from Figure 5.6b. For sufficiently small interaction times the familiar recombination maxima of the repeated returns of the electronic wave packet to the core region are clearly visible. However, since the coherence time of the laser field is small in comparison with the mean Kepler period, $1/b \ll T$, interferences between probability amplitudes associated with repeated returns to the core region are destroyed. Thus the details of the early stages of the time evolution of this electronic wave packet appear to be much simpler than in the completely coherent case. As a consequence of the diffusion of the electronic wave packet, at longer interaction times the recombination maxima of $P_{gg}(t)$ disappear and merge into the power law of (5.35). At interaction times larger than t_c, stochastic ionization of the Rydberg electron becomes significant, and the power law decay of $P_{gg}(t)$ crosses over to the decay law of (5.36). Simultaneously the ionization probability rises to a value close to unity according to the approximate power law of (5.37).

In general, stochastic ionization originating from laser fluctuations will compete with other coherent ionization mechanisms such as autoionization. As a consequence a number of new interesting phenomena are expected to arise that are not yet explored. In order to obtain first insights into basic aspects of this competition, let us generalize our previous model to one-photon excitation of an autoionizing Rydberg series (Eggers and Alber 1998). Thus it will be assumed that the laser-excited autoionizing Rydberg series can be described within the framework of quantum defect theory in a two-channel approximation. In particular, let us concentrate on a case where the fluctuating laser field excites Rydberg states close to an ionization threshold of an excited state of the ionic core (channel one) which can autoionize into channel two. For simplicity let us assume that direct excitation of channel two from the initial state $|g\rangle$ is not possible and that the effectively excited energy interval $(\overline{\varepsilon} - b, \overline{\varepsilon} + b)$ also covers continuum states of channel one. The early stages of this ionization process will be governed by an exponential decay of the initial state $|g\rangle$ with the laser-induced depletion rate γ, by autoionization of the excited Rydberg states of channel one into channel two, and by direct laser-induced ionization into the continuum states of

channel one. As long as stochastic ionization is negligible, that is, for interaction times
t with $1/\gamma < t < t_c$, this ionization process will reach a metastable regime. Thereby the
probability of ionizing into channel one is simply determined by the part of the
effectively excited energy interval $(\bar{\varepsilon} - b, \bar{\varepsilon} + b)$ which is located above the ionization
threshold, ε_1, of channel one. However, as soon as $t > t_c$, it is expected that the
branching ratio between channels one and two is changed. For interaction times
$t > t_c$, all Rydberg states whose autoionization lifetimes exceed the stochastic
ionization time, $1/\Gamma_n > t_c$ (Γ_n is the autoionization rate of Rydberg state $|n, 1\rangle$), will no
longer autoionize into channel two but will eventually ionize stochastically into
channel one. Thus for interaction times $t > t_c$ it is expected that the probability of
ionizing into channel one is determined by the part of the effectively excited energy
interval $(\bar{\varepsilon} - b, \bar{\varepsilon} + b)$ which is located above an energy of the order of $\varepsilon_1 - 1/t_c$. Thus
stochastic ionization is expected to lead to an effective lowering of the ionization
threshold ε_1 of channel one. This manifestation of the competition between
autoionization and stochastic ionization is clearly apparent from Figure 5.7 where the
time evolution of $P_{gg}(t)$ is depicted together with the corresponding time evolutions
of $P_{\text{ion-ch1}}(t)$ and $P_{\text{ion-ch2}}(t)$ In the case depicted in Figure 5.7 the laser-induced
depletion rate γ is so small that no electronic Rydberg wave packet is prepared by

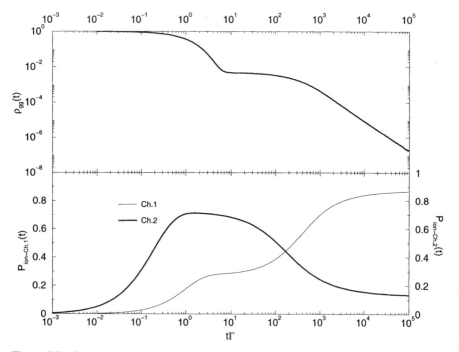

Figure 5.7 Competition between autoionization and stochastic ionization: Time evolution of
$P_{gg}(t)$ and of the ionization probabilities into channels one and two $P_{ion-ch1}(t)$ and $P_{ion-ch2}(t)$.
The parameters are $\bar{n} = \alpha_1 + (-2\bar{\varepsilon})^{-1/2} = 80$, $\alpha_1 = 0.1$, $\gamma T = 1.0$, $bT = 300.0$, and
$\Gamma_n = 2\tau(n - \alpha_1)^{-3}/\pi$ with $\tau = 10^{-5}$ a.u.

power broadening. However, due to the large laser bandwidth, $bT \gg 1$, many Rydberg states are involved in the excitation process. This implies that to a good degree of approximation initially state $|g\rangle$ decays exponentially with rate γ.

ACKNOWLEDGMENTS

Financial support is acknowledged by the Deutsche Forschungsgemeinschaft within the *Schwerpunktprogramm* "Zeitabhängige Phänomene in Quantensystemen der Physik und Chemie."

REFERENCES

Agarwal, G. S. (1976). *Phys. Rev. Lett.* **37**, 1383.

Alber, G. (1989). *Z. Phys.* D**14**, 307.

Alber, G. (1992). *Phys. Rev. Lett.* **69**, 3045.

Alber, G., and Eggers, B. (1997). *Phys. Rev.* A**56**, 820.

Alber, G., and Strunz, W. T. (1994). *Phys. Rev.* A**50**, R3577.

Alber, G., Strunz, W. T., and Zobay, O. (1994). *Mod. Phys. Lett.* B**8**, 1461.

Alber, G., and Zoller, P. (1988). *Phys. Rev.* A**37**, 377.

Alber, G., and Zoller, P. (1991). *Phys. Rep.* **199**, 231.

Aymar, M., Greene, C. H., and Luc-Koenig, E. (1996). *Rev. Mod. Phys.* **68**, 1015.

Beims, M. W., and Alber, G. (1993). *Phys. Rev.* A**48**, 3123.

Beims, M. W., and Alber, G. (1996). *J. Phys.* B**29**, 4139.

Braginsky, V. B., and Khalili, F. Ya. (1992). *Quantum Measurement.* Cambridge University Press, Cambridge.

Cooke, W. E., Gallagher, T. F., Edelstein, S. A., and Hill, R. M. (1978). *Phys. Rev. Lett.* **40**, 178.

Dando, P. A., Monteiro, T. S., Delande, D., and Taylor, K. T. (1995).*Phys. Rev. Lett.* **74**, 1099.

Delos, J. B. (1986). *Adv. Chem. Phys.* **65**, 161.

Dixit, S. N., Zoller, P., and Lambropoulos, P. (1980). *Phys. Rev.* A.**21**, 1289.

van Druten, N. J., and Muller, H. G. (1995). *Phys. Rev.* A**52**, 3047.

Eggers, B., and Alber, G. (1999). *J. Phys.* B**32**, 1019.

Fano, U., and Rau, A. R. P. (1986). *Atomic Collision and Spectra.* Academic Press, New York.

Gallagher, T. (1994). *Rydberg Atoms.* Cambridge University Press, Cambridge.

Garraway, B. M., and Suominen, K. A. (1995). *Rep. Prog. Phys.* **58**, 365.

Giusti-Suzor, A., and Zoller, P. (1987). *Phys. Rev.* A**36**, 5178.

Grobe, R., and Eberly, J. H. (1993). *Phys. Rev.* A**48**, 623.

Haken, H. (1970). In *Handuch der Physik*, ed., S. Flügge, vol. 25, Springer, Berlin.

Hanson, L. G., and Lambropoulos, P. (1995). *Phys. Rev. Lett.* **74**, 5009.

Hüpper, B., Main, J., and Wunner, G. (1995). *Phys. Rev. Lett.* **74**, 2650.

Jones, R. R., and Bucksbaum, P. H. (1991). *Phys. Rev. Lett.* **67**, 3215.

Klöden, P. E., and Platen, E. (1992). *Numerical Solution of Stochastic Differential Equations.* Springer, Berlin.

Knospe, O., and Schmidt, R. (1996). *Phys. Rev.* A**54**, 1154.

Koch, M., von Plessen, G., Feldmann, J., and Goebel, E. O. (1996). *J. Chem. Phys.* **120**, 367.

Landau, L. D., and Lifshitz, E. M. (1975). *The Classical Theory of Fields.* Pergamon, Oxford, p. 181ff.

Lankhuijzen, G. M., and Noordam, L. D. (1996). *Phys. Rev. Lett.* **76**, 1784.

Main, J., Wiebusch, G., Welge, K. H., Shaw, J., and Delos, J. B. (1994). *Phys. Rev.* A**49**, 847.

Moser, I., Mota- Furtado, F., O'Mahony, P. F., and dos Santos, J. P. (1997). *Phys. Rev.* A**55**, 3724.

Maslov, V. P., and Fedoriuk, M. V. (1981). *Semiclassical Approximation in Quantum Mechanics.* Reidel, Boston.

Mollow, B. R. (1975). *Phys. Rev.* A**12**, 1919.

Robicheaux, F. (1993). *Phys. Rev.* A**47**, 1391.

Seaton, M. J. (1983). *Rep. Prog. Phys.* **46**, 167.

Sepulveda, M. A., and Grossmann, F. (1996). *Adv. Chem. Phys.* **96**, 191.

Stapelfeldt, H., Papaioannou, D. G., Noordam, L. D., and Gallagher, T. F. (1991). *Phys. Rev. Lett.* **67**, 3223.

Vemuri, G., Anderson, M. H., Cooper, J., and Smith, S. J. (1991). *Phys. Rev.* A**44**, 7635.

Zobay, O., and Alber, G. (1995). *Phys. Rev.* A**52**, 541.

Zobay, O., and Alber, G. (1996). *Phys. Rev.* A**54**, 5361.

Zobay, O., and Alber, G. (1998). *Prog. Phys.* **46**, 3.

CHAPTER 6

Wave Packets and Half-Cycle Pulses

PHILIP BUCKSBAUM

6.1 INTRODUCTION

This chapter describes the interaction of wave packets with a new type of electromagnetic radiation: the half-cycle pulse. A half-cycle pulse is nothing more than an electric field that is rapidly switched on and then switched off again. When the duration of the field is smaller than one picosecond, it is no longer possible to produce the pulse with a simple pair of conducting plates and a mechanical or electronic switch. The pulse moves through space like a propagating electromagnetic wave with a very special property, namely that the time-integral of the field is not zero! This means that the pulse can impart an impulse to a free electron, or to a weakly bound Rydberg electron in an atom or molecule. These "kicked" electrons, the techniques for kicking them, and their subsequent motion are the subject of this chapter. Most of the discussion is devoted to a summary of the experiments and theories involving half-cycle pulse interactions with wave packets. To introduce the subject and place it in the context of the rest of this book, we will begin with a short recapitulation of the dynamics of Rydberg wave packets and their ionization. Then we have a section about the production and propagation of half-cycle pulses. The behavior of this radiation differs from laser light because of the very broad bandwidth and presence of a dc component to the field. Finally we will introduce and summarize the physics of kicked atomic electrons.

6.2 HOW DOES A HALF-CYCLE PULSE VIEW A RYDBERG WAVE PACKET?

Half-cycle pulses can affect the motion and phase of Rydberg wave packets. Rydberg eigenstates of an atom or molecule are stationary and therefore are relatively real, with only a uniform unobservable global phase. Wave packets move, and therefore the Schrödinger wave function associated with a wave packet state acquires position-dependent phase, which can be observed in interference experiments. Figure 6.1 shows such a wave packet produced from six p-states in atomic cesium, and

The Physics and Chemistry of Wave Packets, Edited by John Yeazell and Turgay Uzer
ISBN 0-471-24684-0 © 2000 John Wiley & Sons, Inc.

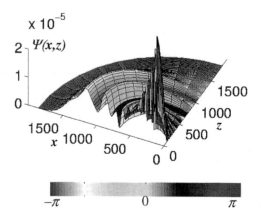

Figure 6.1   Measured quantum wave packet in cesium, made from 26p–31p states. (Reprinted with permission from Weinacht 1998.)

measured using a recently developed interference technique, which takes a "snapshot" of the quantum wave function (Weinacht 1998).

The motion of wave packets can also be quite beautiful. Figure 6.2 shows the radial motion of a similar wave packet in cesium (Raman 1997a). The probability density sloshes back and forth in its potential well, alternately picking up speed and slowing down, spreading, splitting, and coming back together, a true quantum sculpture.

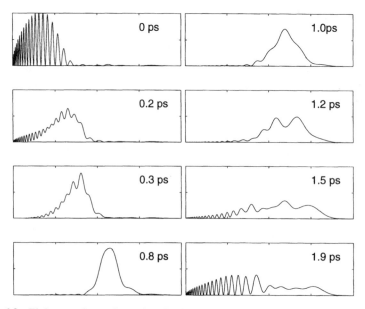

Figure 6.2   Eight snapshots of a cesium Rydberg wave packet moving through one Kepler cycle. (Reprinted with permission from Raman 1997a.)

One of the great frustrations of experimental work on Rydberg atom wave packets is that these beautiful features of motion cannot be observed directly by optical means. We have no camera that can record such delicate features evolving so rapidly. Stroboscopic techniques, particularly pump-probe spectroscopy, have solved the time part of the problem: subpicosecond pulses can interact with a Rydberg wave packets much more rapidly than their orbit time. The pump-probe experiments developed by the editor of this volume and by others have shown the collapse and revival pattern of wave packets in time-delayed ionization by short optical probe.

Unfortunately, these experiments only show the wave packet as it passes near the ion core. It is as though we placed a mask on the whole atom, except for a small hole near the core, and were observing the wave packet through this hole. The signal in this case looks like Figure 6.3 (Yeazell 1990). Although the wiggles convey some basic information about the motion of the wave packet, it is clear that much is missing. For example, the breathing motion of the wave packet as it splits in two during some parts of its orbit appears here only as a change in the signal period.

This severe restriction of photoionization for visualizing wave packets is because the process depends heavily on the presence of a nearby massive object (the ion) in order to conserve momentum. A free electron cannot of course absorb a photon because of the conservation of momentum: If the electron is at rest, for example, absorbing $\hbar\omega$ of energy will give the electron a momentum of $\sqrt{2m\hbar\omega}$, whereas the initial state has momentum $\hbar\omega/c$, which for visible (~$2eV$) photons means a

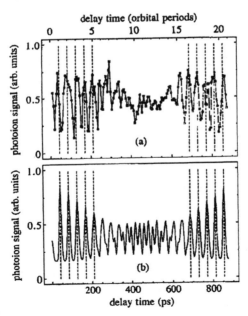

Figure 6.3 Rydberg wave packet probed by ionization. Oscillations in the ionization signal show the motion of the wave packet, and its collapse and revival due to dispersion. (Reprinted with permission from Yeazell 1990.)

momentum conservation mismatch of $\approx 500\hbar k$. When wave packets are away from the ion core, they act very much like free electrons and will not absorb light, and they therefore will not photoionize. We can quantify this on simple physical grounds: Consider an electron on an elliptical cometlike orbit around an ion. Photoionization is most likely if the electron is close enough to the ion to complete a transit in one optical period. In that case the momentum of the electron reverses as it passes around the ion just as the field direction reverses, so the field can accelerate the electron over both ingoing and outgoing halves of its orbit. We can use Kepler's laws to convert this to an approximate condition on the critical radius for photoionization: $r_{crit} \sim \tau^{2/3}$, where τ is the optical period. Inserting atomic units, we find for an optical wavelength of *600* nm, $\tau = 2$ fs or 80 atomic units, so $r_{crit} \approx 19$ bohr radii, or about 5% of the orbital radius of an $n = 20$ state. The cross section for photoionization drops quickly to zero beyond this radius.

According to this argument, we might better seek longer wavelengths, or more to the point, longer optical periods for the photoionizing probe. However, although this increases the range of the interaction, it also increases the time, so the resolution is severely degraded. It looks like the idea of the Rydberg wave packet camera using photoionization ultimately fails!

One way out of this problem is a radical and simple new type of radiation, the half-cycle pulse (HCP). The unipolarity gives this field a peculiar property not shared by optical ultrafast pulses: Its bandwidth is actually greater than its central frequency. In other words, there is a dc component to the field. This means that the usual dictum that a free electron cannot absorb energy from a light pulse does not apply for an HCP. Since the time-integrated electric field $F(t)$ is nonzero, so is the impulse imparted to an electron, even if it is free: $p_e \rightarrow p_e' = p_e + \int eF(t)dt$. If the momentum increases enough, bound electrons can overcome their binding energy and escape, that is, become ionized. The impulse needed to overcome the binding potential depends only on the momentum of the electron and its potential energy. Therefore we can map out the momentum distribution of an eigenstate, or measure how the momentum distribution moves as a wave packet moves around in the Coulomb potential.

6.3   WHAT AN ELECTRIC FIELD DOES TO A RYDBERG ATOM

A half-cycle pulse is a dc electric field that is switched on, and then switched off. Two important parameters are its duration τ and the peak field amplitude F_0. The physics of Rydberg state ionization is quite different if τ is greater than or shorter than the characteristic Kepler orbit time.

Slow half-cycle pulses are easy to make with a couple of field plates and a mechanical or electronic switch. In Rydberg physics, slow pulsed fields are mainly used to resolve Rydberg states through their threshold for field-ionization. This kind of field ionization is "polarizing the atom to death," that is to say, adiabatically transforming the bound state into the continuum by slowly altering the potential.

The physics of field ionization is described very well in the recent book by Gallagher (1994), so we will only summarize some points here for the purpose of making this chapter self-contained. A simple model gives approximately the correct estimate for the field required to ionize a Rydberg state with low angular momentum: The atom-plus field potential for a field polarized along $\hat{z}$ is $V(r) = e^2/r - eFz$. There is a saddle point on the $\hat{z}$-axis with energy $-2e^{3/2}F^{1/2}$ (see Fig. 6.4), and therefore states with energy above this can escape through the saddle. Stated the other way, this means that the critical field for ionizing a state with energy E is $F_{crit} = E^2/4e^3$. Let's relate this to the Rydberg principal quantum number n. If we assume that the energy eigenvalue does not shift during the polarization process prior to reaching the critical field, then we can simply equate E with the Rydberg formula. Then

$$F_{crit} \approx \frac{1}{16n^4}\left[\frac{m^2e^5}{\hbar^2}\right] = 2000 \text{ V} cm^{-1} \times \left[\frac{20}{n}\right]^4 . \tag{6.1}$$

The two main flaws in this theory are (1) it ignores the energy shift (Stark shift) of the atoms prior to ionization and (2) it ignores the spatial distribution of the wavefunction. These are not unrelated, since the polarization-induced energy shift of the Rydberg state as the field turns on depends on the orientation of its dipole moment, or in classical terms on the orientation of the Runge-Lenz vector, which lies along the major axis of its elliptical orbit. In a pure $1/r$ potential like that of hydrogen, the classical orbit does not precess, and the quantum eigenstates have permanent dipole moments, which increase linearly with the applied field. States oriented "downhill" become more tightly bound as the field turns on, and therefore they have a higher critical field. The Rydberg state with the largest negative Stark shift has an additional binding energy of $\Delta E \approx 3/2n^2F[\hbar^2/me]$, in first-order perturbation theory. This means that it will not ionize until the field reaches a value of $F_{crit} = 1/(9n^4)$, 78% higher than the simple one-dimensional estimate. Hydrogen Rydberg states with positive Stark shifts become less bound as the field increases, and they reach the saddle point energy much earlier. They do not ionize, however, because their wavefunction is located mostly on the "uphill" side of the atom, where the overlap with the downhill continuum nearly vanishes. These states survive to much higher fields than the downhill Stark states. Finally states with large projections of their angular momentum

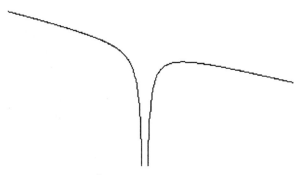

Figure 6.4 The function $V(\mathbf{r}) = e^2/r - eFz$, representing a Coulomb potential in a uniform electric field.

along the field axis, for example, circular orbiting states with orbital planes perpendicular to the field, cannot ionize until the saddle point tunnel opens wide enough to overlap their orbits. So we see that the expression given above is actually just a lower limit for field ionization.

In atoms other than hydrogen the presence of electrons in the ion core changes the potential from a pure Coulomb $-1/r$ form. This breaks the Runge-Lenz symmetry for those states with low enough angular momentum to penetrate inside the ion core: These Rydberg states have corresponding elliptical orbits that can precess rapidly between the uphill to the downhill sides of the ion. Since their trajectories can find the saddle point, these states field-ionize near the value $F_{crit} = (2E)^{-2}$. The eigenvalues of the nonhydrogenic quantum system do not have a simple linear stark shift because low angular momentum states are not degenerate with high L-states. As the field increases, these states follow parabolic trajectories on a field–energy plot, until they intersect a manifold of high L-, low m-states. The manifolds of low m-states from neighboring n-values suffer avoided crossings when they meet, so the eigenvalues wiggle up and down within a narrow range, as shown in Figure 6.4. The result is ionization near the classical value.

The complicated Stark energy diagram in Figure 6.4 is much easier to understand from the vantage point of the different dynamical time scales in the problem. The orbital motion of a bound atomic electron in an electric field has three different kinds of motion, each with its own orbital period. First, there is the Keplerian orbital motion, which for low l-states is characterized by radial oscillations between the core and some outer turning point. If there is an electric field present, this orbit may not be closed, but it is still easy to define by radial motion, which reverses periodically. The period of motion is equal to 2π divided by the energy splitting between two neighboring n-states, as can be seen easily from the following argument: The radial eigenfunctions of two adjacent n-states differ by one node. If they interfere constructively near $r = 0$, then they must have different signs, and therefore interfere destructively, as $r \to \infty$. Each state has a time dependent phase given by $\exp(-iE_n t/\hbar) = \exp[-i(Ry/\hbar n^{*2})t]$. If the states n and n' with effective quantum numbers n^* and n^*+1, respectively, interfere constructively at $r = 0$ and $t = 0$, then they will interfere constructively at $r = 0$ again at $\tau = 2\pi/(1/n^{*2} - 1/n'^{*2}) \sim 2\pi n^3 = 1.26[n/20]^3$ps.

A second, slower type of motion is the precession of the angular momentum of the electron due to the torque exerted by the electric field. A typical trajectory of an electron in the combined Coulomb and static uniform field is shown in Figure 6.5. The precession from a linear orbit through orbits of lower ellipticity, to a circular orbit, and then back again, has a period approximately equal to 2π divided by the energy splitting between neighboring Stark states in the stark manifold, or $\tau = 2\pi/(3nF) = 13.5[n/20][F/(1 \text{ kV/cm})]$ ps. Of course the magnitude of the torque depends on the inclination of the plane of the electron's orbit with respect to the electric field direction. Circular orbits perpendicular to the electric field do not evolve in shape at all. These correspond to orbits with $m = l_{max} = n - 1$. The period of the precession, however, is independent of the inclination.

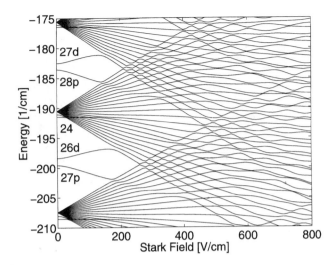

Figure 6.5 Some eigenvalues of Cs $m_1 = 1$ states in an electric field.

The third type of classical motion to consider is the precession of the semi-major axis of the elliptical orbit of the electron. This precession does not occur in a pure $-1/r$ potential, so it is absent for hydrogen, or for any atom in a state of high angular momentum. Low angular momentum multielectron atoms have deviations from a pure Coulomb potential due to the electrostatic interaction of the Rydberg electron with the atomic core. This causes a phase shift in the orbit as the electron passes inside the orbits of the electrons in the ion core. The electron experiences a higher effective charge for the Coulomb field because the charges of the outermost core electrons no longer completely screen the nucleus. The average potential energy is more negative, the average kinetic energy is higher, and the total energy of the state is lower than for the corresponding state in hydrogen. The effect in an alkali spectrum is a small decrease, or *quantum defect*, in the effective principal quantum number. The quantum defect precession is not due to a constant force on the electron but rather on short kicks received once per Kepler period. Therefore there is no characteristic time scale but a phase shift $\delta = n - n^*$, where $n^* = \sqrt{Ry/E}$ is the "effective" quantum number.

When a static electric field is applied to a Rydberg atom, the precession of the orbit and the precession through angular momentum space can result in some complex behavior (see Fig. 6.6). The situation is particularly interesting if the angular momentum precession time and Kepler orbit times are commensurate; then the electron can find itself near perion (the point of closest approach) just when its angular momentum is low, permitting core penetration. The value of the field where this commensurate condition occurs can be calculated by equating the precession time with an integer s times the Kepler orbit time. In atomic units,

$$\frac{2\pi}{3(-2E)^{-1/2}F} = s2\pi(-2E)^{-3/2}, \quad \text{or} \quad F = \frac{4E^2}{3s}. \tag{6.2}$$

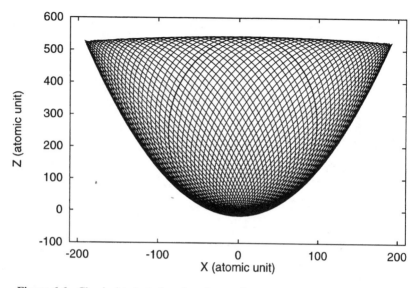

Figure 6.6   Classical trajectories of an electron in an atom in a static electric field.

We can derive a similar expression using the quantum eigenvalues of the Stark spectrum. Consider a Stark spectrum in hydrogen in the region where the eigenvalues from the states Ψ_{nkm} are just beginning to overlap states from the next higher n-manifold $\Psi_{(n+1)k'm}$. Here k, k' are the parabolic quantum numbers, which take the values $(n - m - 1)$, $(n - m - 3)$, . . . , $(1 + m - n)$. The Stark shift, in first order, is given by $(3/2)nkF$. For simplicity let us consider the manifold of Stark states with $m = 0$. For these states the first crossing occurs when the maximum eigenvalue of the n-manifold crosses the minimum eigenvalue of the $(n + 1)$-manifold:

$$\frac{-1}{2n^2} + \frac{3}{2}n(n-1)F = \frac{-1}{2(n+1)^2} + \frac{3}{2}(n+1)(1-n-1)F,$$

or

$$F = \frac{1}{3n^5} + O(1/n^6) \approx \frac{4E^2}{3n}. \tag{6.3}$$

This is just the field where a classical electron executes exactly n Kepler orbits during its full precession through angular momentum. Each successive crossing involves an integer ratio with one fewer Kepler orbits during the angular momentum precession. Close to these crossings, core scattering effectively couples these two types of motion. This coupling breaks the eigenvalue degeneracy and leads to "avoided crossings." When wave packets are constructed near avoided crossings, their eigenvalues are no longer regularly spaced. This leads to modifications in the way the wave packet disperses. This is the quantum analogue of core scattering.

The preceding simple dynamical picture is useful if the electric field is static, or nearly so, while the electron precesses through its orbit. But what happens if the field changes substantially during an orbit? Periodic rapid oscillations of the field magnitude can transfer population among different eigenstates and ultimately ionize the atom. The interplay between classical motion and quantum interference in periodically driven Rydberg systems has been a subject of intense investigation, but we will not discuss it here and rather refer the reader to extensive reviews (Casati 1987; Jensen 1991; Koch 1995).

6.4   THREE WAYS TO IONIZE A RYDBERG ELECTRON

Thus far we have discussed three physical mechanisms for ionizing a Rydberg electron by an electromagnetic field: direct field ionization by a static or nearly static electric field, photoionization by an optical-frequency field, and ionization by a half-cycle pulse. Let us summarize the general properties that differentiate these processes through their dependence on field strength, pulse duration, and state energy.

1. In optical photoionization, the presence of a "critical radius" as described in the introduction means that photoionization matrix elements only depend on the part of the wave function close to the atomic nucleus. In a Rydberg series we have $\psi(r = 0) \sim 1/n^{3/2}$, so the cross section falls as $1/n^3$. The rate also increases linearly with intensity; namely it is quadratic in the electric field so long as the photon energy exceeds the binding energy of the atom. For a given field strength the rate also increases linearly with pulse duration.

2. A static field will also ionize the atom once the field strength reaches a critical threshold value, which scales with the inverse-square of the state energy, or $1/n^4$. Below this threshold, the tunnel ionization rate is exponential in the field strength. Above the threshold, the ionization probability is essentially unity for times longer than the characteristic dynamical time scale associated with the state. For Rydberg states this is the Kepler orbit time: $\tau_{Kepler} = 2\pi n^3$.

3. Midway between these two quite different behaviors is the regime of the half-cycle pulse. The field of the HCP turns on and off too fast to recover static field behavior. In the extreme limit where the HCP is on for much less than one Kepler orbit, the HCP imparts an impulse that only depends on the time-integrated electric field. The energy gained is then proportional to the local momentum in different parts of the atomic potential. For Kepler orbits longer than a nanosecond ($n \cong 200$), the HCP can be made easily using conventional electronics to drive a parallel plate capacitor. These states are very difficult to study, however, because they are so fragile: an electric field of only a fraction of a volt will ionize them (Frey 1996; Tannian 1998). For Rydberg states with n in the range of 10 to 60, picosecond or subpicosecond pulses are required, with field strengths in the range of 1 to 10 kV/cm^2.

6.5 HOW TO MAKE A PICOSECOND HALF-CYCLE PULSE

Picosecond HCPs are not quite light and not quite radio waves. Most of their energy lies in the far-infrared part of the electromagnetic spectrum, and techniques for producing and detecting them are borrowed from more established areas of optics and microwaves. Generation mechanisms rely on ultrafast optical pulses with durations of less than 1 ps. When these short light pulses pass through a transparent material or impinge on a suitably prepared surface, they can create an electromagnetic disturbance which will propagate as a half-cycle pulse.

Far-infrared radiation was first observed from a biased GaAs photo-conductive switch illuminated by 35 ps laser pulses (Mourou 1981). Subsequent development of subpicosecond laser sources led to pulsed terahertz transmitters consisting of miniature Hertzian dipole antennas grown directly on an insulating substrate, and excited by ultrafast photoconducting switches (DeFonzo 1987; Auston 1984). A typical geometry is shown in Figure 6.7. Terahertz radiation propagates into the

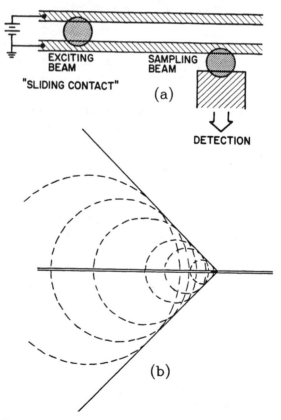

Figure 6.7 (*a*) Electromagnetic shock wave from an ultrafast current pulse excited along a transmission line. (*b*) Experimental geometry. (Reprinted with permission from Grischkowsky 1987.)

dielectric as the photo-induced current travels down a transmission line. This can appear as a coherent electromagnetic shock wave emitted in a Cherenkov cone, whose angle is determined by the mismatch between the group velocity of the current on the transmission line and the slower phase velocity of far infrared propagating in the dielectric (Grischkowsky 1987). It was appreciated early on that the electric field distribution in this shock was that of a single cycle pulse (Fattinger 1989a).

The electromagnetic energy in the shock wave can be coupled into free space through the side of the dielectric, or with a suitable dielectric lens placed on top of the transmission line (van Exter 1989; Fattinger 1989b; Hu 1990a). Several different sources and geometries have been developed, and both the pulses and generating mechanisms have been studied. The freely propagating radiation can be detected using a second antenna in a reciprocal arrangement as shown in Figure 6.8. The laser then acts as a gate to sample the current pulse induced by the terahertz field impinging on the antenna. The field pulse can be detected directly in this way, and the spectroscopy (see Fig. 6.9) carried out by Fourier transform techniques as in ordinary FTIR spectroscopy (Harde 1991).

The highest power ultrafast far infrared pulses have come from large aperture antennas (Zhang 1990; Darrow 1990; Hu 1990b). A typical geometry is shown in Figure 6.10. When a subpicosecond pulse irradiates a large aperture photoconducting antenna, a number of mechanisms can give rise to a pulse of far-infrared radiation. If the semiconductor crystal is noncentrosymmetric with a nonvanishing $\chi^{(2)}$ (GaAs or InP are two examples), then optical rectification is possible. The THz bandwidth in the laser pulse gives rise to THz radiation through this mechanism of

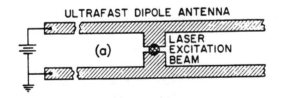

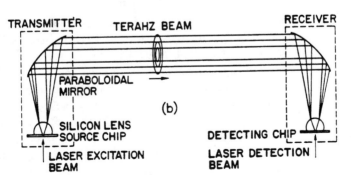

Figure 6.8 Typical arrangement for transmitting and detecting terahertz from photoconducting switches. (Reprinted with permission from Grischkowsky 1987.)

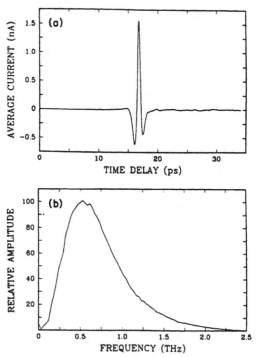

Figure 6.9   Measured terahertz field and the spectrum of the radiation obtained by Fourier transformation. (Reprinted with permission from Grischkowsky 1987.)

difference-frequency generation. When a static electric field is also present, $\chi^{(3)}$ processes can also produce THz. External electrodes need not apply the field, since there is often a static field due to the depletion of bulk carriers near the surface of the semiconductor. Careful studies have shown the presence of significant THz radiation from $\chi^{(2)}$, $\chi^{(3)}$, and $\chi^{(4)}$ processes in externally biased InP (You 1994). FIR has also been generated from electric dipole oscillation in semiconductor quantum wells (Nuss 1994).

Small antennas or unbiased large-aperture devices produce enough terahertz energy for purposes of linear spectroscopy (van Exter 1989; Grischkowsky 1992; Grischkowsky 1990; Chwalek 1991; Greene 1991), but the fields are inadequate for impulsive excitation of Rydberg wave packets. Much higher pulse energies can be produced through a different physical mechanism: the direct radiation of transient current from large-aperture antennas. If the incident ultrafast pulse has a frequency above the band-gap in the material, then conduction electrons are produced which accelerate in the static field. This gives rise to radiation. The field responsible for the acceleration of this transient current decays rapidly due to screening by the displaced electrons and holes. Typically acceleration occurs over a few hundred femtoseconds, while deceleration can take approximately 100 picoseconds. Thus to a good approximation, the electron current is a step function. The fields produced in this

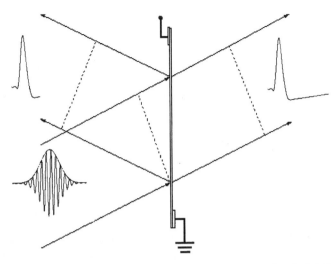

Figure 6.10   Large aperture photoconducting switch biased with an electric field in the surface plane. When subpicosecond pulses illuminate the switch, far-infrared radiation is generated along the transmitted and reflected laser beam direction.

situation have a rather unusual character, as displayed in Figure 6.11, which is described in *Fundamentals of Physical Optics* by Jenkins and White (Jenkins 1937). The transverse propagating component of the field, due to an electron that is accelerated from rest over a short interval $\Delta\tau$ and then left in uniform motion, is a unipolar pulse. Expressed in terms of cycles of the electromagnetic field, this is a *half-cycle pulse*!

The half-cycle pulse emerging from a planar photoconductor actually begins its life as a step function in the applied electostatic field (see Fig. 6.12). That is, if we measure

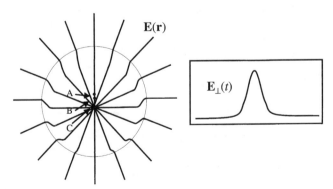

Figure 6.11   Electron accelerated from point A to point B, and then coasting with constant speed to point C. The electric field lines develop a transverse component during the acceleration, which propagates outward at the speed of light as a half-cycle pulse.

the field adjacent to the surface of the semiconductor as the optical pulse arrives, we find that the static field "shorts out" over a few hundred femtoseconds. You might call this step-function field a *quarter-cycle pulse*. This field decay has been measured (see Fig. 6.13) by probing the Pockels effect in an electro-optic crystal placed behind the GaAs photoconducting switch (Budiarto 1998). Budiarto showed that the THz field $u(P_0, t)$ at any point beyond the aperture could be calculated by the Huygens-Fresnel diffraction integral:

$$u(P_0, t) = \int_\sigma \frac{\cos(\hat{n}\mathbf{r}_{01})}{2\pi c|\mathbf{r}_{01}|} \frac{d}{dt} \left[u\left(P_1, t - \frac{|\mathbf{r}_{01}|}{c}\right) \right] d\xi. \qquad (6.4)$$

Here u (P_0, t) is the field at P_0 at time t, and the integral is taken over the field distribution at the surface of the transmitter at the retarded time. Modeling a collecting mirror as a thin lens, Budiarto calculated the appearance of the field distribution at an around a focus, and compared it to a direct measurement performed using a ZnTe electro-optic crystal at the focus. The half-cycle pulses were thus confirmed through direct observation. Feng, Winful, and Hellwarth found analytical solutions as well,

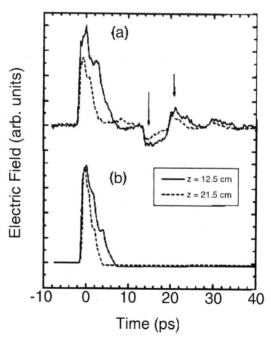

Figure 6.12 (*a*) Experimental and (*b*) simulation results for the THz pulse shape at a distance z directly behind the antenna. Time 0 ps has been set arbitrarily at the peak of the pulse. (Reprinted with permission from Budiarto 1998.)

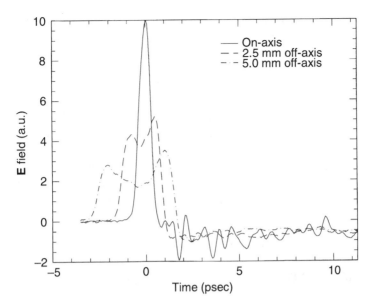

Figure 6.13 Measured half-cycle pulse at the focus of a THz beam produced by a GaAs large area electro-optic switch. (Reprinted with permission from Budiarto 1998.)

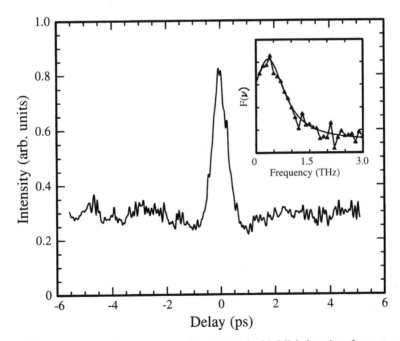

Figure 6.14 Bandwidth of a half-cycle pulse measured with Michelson interferometry. The interferogram, which shows the autocorrelation of the HCP electric field, is shown to the left. The derived spectrum is shown in the inset.

and that the distortion of the "quarter-cycle" edge into a half-cycle pulse can be related to the Gouy phase shift experience by a focused Gaussian beam (Feng 1998).

Since the optical pulse that excites the photoconductor is much shorter that the THz pulse rise time, the latter is evidently constrained by material properties to be in the range of 200 to 400 fs. This then limits the bandwidth of the pulse to about 1 THz (see Fig. 6.14). Far-infrared radiation made from nonlinear interactions in the material such as difference frequency mixing are not constrained in this way, and near-single-cycle pulses of more than 20 THz bandwidth have been produced (Bonvalet 1995).

The peak field in the HCP grows linearly with the applied bias field, but it is a nonmonotonic function of the laser fluence. Sub-single-cycle pulses of electromagnetic radiation with pulse energies of up to 0.8 μJ have been produced from a GaAs transmitter with approximately 12 cm^2 area, biased with more than 10 kV/cm. When focused, this produced a peak field of 150 kV/cm (You 1993).

6.6 IONIZING RYDBERG STATES WITH HALF-CYCLE PULSES

The technique we have just described produces freely propagating electromagnetic pulses with central frequencies around 0.5 THz (16 cm^{-1}), and up to 1 μJ, and peak fields in excess of 100 kV/cm in a nearly unipolar 500 fs electric field pulse. The interactions of these pulses with atomic systems differ dramatically from ordinary laser-atom interactions due to the enormous coherent bandwidth in the pulse. Although most of the energy in the pulse is in the large half-cycle component, of course over a very long time the electric field must integrate to zero. Nonetheless, during the time of interaction with the atoms in a practical experiment, the electric field is essentially unidirectional.

According to the analysis earlier in this chapter, a unipolar electric field could ionize an atom through simple "over-the-barrier" static field ionization. For the ground state of a weakly bound atom like cesium, this requires about 26 MV/cm peak field, more than 100 times more than we can produce at present. Atoms in high Rydberg states, on the other hand, are weakly bound and have transition dipole moments in the THz range, so these are more interesting objects for study with half-cycle pulses. Figure 6.15 shows a part of the Rydberg series for atomic Na. Starting in the 12s state, the coherent bandwidth contains frequencies that are resonant with the 12p state separated by 85 cm^{-1}, but no other dipole-allowed transitions are 1-photon accessible. On the other hand, if we start in $n = 30$, many states are simultaneously resonant, although the ionization limit is still beyond the bandwidth of the pulse. The coherent transfer of population among so many states will convert a single eigenstate into a wave packet! Although this spectral analysis has some use, particularly for weak fields, it is much more informative to consider the problem in the time domain as we did earlier. The 12s state corresponds in energy to a classical orbit with a Kepler period that is quite short compared to the HCP duration: $\tau_{\text{Kepler}} = $ ~180 fs, while $\tau_{\text{pulse}} = $ ~500 fs. So the atomic motion can be averaged over the field

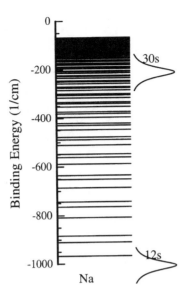

Figure 6.15   Energy levels for Na s-, p-, and d-states, for $n = 10$–40.

pulse, resulting in very little net transfer of momentum over the pulse duration. The 30s state has a much slower orbit time of under 4 ps. To an electron in this orbit, the HCP looks nearly like a delta function impulse.

Early experiments tested these ideas by measuring the ionization of Rydberg states excited by HCPs (Jones 1993a). The geometry is shown in Figure 6.16. A thin (500 μm) GaAs wafer is placed in a vacuum chamber housing an atomic beam and pulsed with a bias field of up to 30 kV/cm². An intense (40 μJ/cm²) 100 fs laser beam incident on the semiconductor creates the HCP. The electric field in the far-infrared pulse is antiparallel to the static bias field, and its magnitude is directly proportional to the bias voltage.

The transmitted radiation is weakly focused with a parabolic mirror. Two pulsed dye lasers excite Rydberg states via two photon excitation through the $3_{p1/2}$ resonance line. The Na beam passes between two parallel metal plates, and then the HCP interacts with these states, exciting or ionizing them. Following the interaction, the plates may be pulsed to extract either ions or electrons. The field plates can also be ramped in voltage to provide a means for state-selective detection of Rydberg states.

The ionization probability is a direct test of strong-field behavior, since in the range of Rydberg states studied, single-photon ionization by the HCP is not energetically possible. Nonetheless, ionization does occur, and the probability can be close to 1.0 for strong HCPs. Figure 6.17 shows the measured probability for ionizing three particular Rydberg states as a function of the peak electric field in the HCP. Note that the ionization probability has a threshold, and then rises slowly, finally saturating near

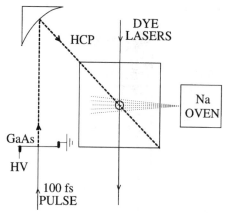

Figure 6.16 Schematic diagram of the interaction region, showing the photoconductive switch which produces the HCP, a parabolic reflector to deflect and focus the HCP into the interaction region, and the Na atomic beam, which is prepared by excitation with dye lasers.

1.0. This behavior is qualitatively different from what is observed during the ionization of Rydberg states with short laser pulses where photoionization by a pulse that is shorter than the Kepler period saturates at a level $\approx \tau_{\text{laser}}/\tau_{\text{Kepler}} \leq 1$, inevitably trapping some population in a wave packet state (Jones 1991, 1993b; Stapelfeldt 1991).

We can also analyze this data by plotting the measured electric field needed to ionize 10% or 50% of the Rydberg population. Recall that ordinary photoionization cross sections for Rydberg states scale as n^{-3}; the critical field for dc field ionization scales like n^{-4}. Figure 6.18 shows that the field required to ionize 10% of the s- or d-states scales as n^{-2}, while for 50% ionization the power law is closer to $n^{-3/2}$.

The n^{-2} scaling of the "threshold" (10% ionization) field can be explained by appealing to the classical dynamics of impulsive excitation of eccentric Kepler orbits. Here is how it works: If the HCP is very short compared to the orbit time, then it imparts a momentum impulse, or kick to the electron:

$$Q = \int e\mathbf{F}(t)dt \tag{6.5}$$

For a classical particle in a Kepler orbit, this changes the electron momentum $\mathbf{P} \rightarrow \mathbf{P}' = \mathbf{P} + \mathbf{Q}$, and therefore also its energy:

$$\Delta E = \frac{\mathbf{P}'^2}{2m} - \frac{\mathbf{P}^2}{2m} = \frac{1}{2m}(Q^2 + 2\mathbf{P} \cdot \mathbf{Q}) \qquad \text{(impulse approximation).} \tag{6.6}$$

At threshold, where $\mathbf{Q}$ is relatively small, the energy gain is greatest for regions of the orbit where the momentum $\mathbf{P}$ is high, namely near the inner turning point. This peak momentum is nearly independent of the principal quantum number n (Goldstein 1980), in line with the fact that the first antinode of the Rydberg–Schrödinger

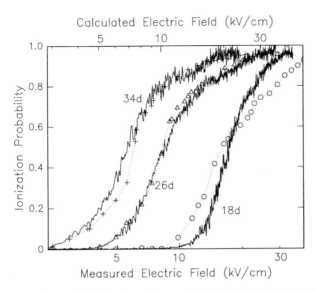

Figure 6.17 Ionization probability plotted in a peak HCP electric field for three different initials states 18*d*, 26*d*, 34*d*. The results of classical simulations are shown as (O), (△), and (+) connected by dotted lines for 18*d*, 26*d*, and 34*d*, respectively. There is a factor of 1.3 between the experimental electric field scale and the scale used in the calculations. (Reprinted with permission from Jones 1993a.)

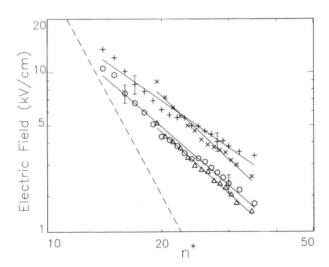

Figure 6.18 Peak field for 10% and 50% ionization of Na and Cs by a half-cycle pulse, plotted as a function of n^*, the effective principal quantum number. Symbols: O, Na 10% threshold; △, Cs 10% threshold; +, Na 50% threshold; and ×, Cs 50% threshold. (Na data are from Jones 1996, and Cs data are from Raman 1997a.) The error bars are for the Na data. The solid lines are power-law fits to the data, and the dashed line provides the $(2n^*)^{-4}$ classical field ionization threshold for comparison. (Reprinted with permission from Raman 1997a.)

wavefunctions is nearly independent of n for fixed l. We can estimate that the inner-turning-point momentum is on the order of the expectation value of the magnitude of the momentum of the lowest n-state in the series, that is, the one with $n = l + 1$. By the Virial theorem,

$$\frac{\langle \mathbf{P}^2 \rangle}{2m} = \frac{Ry}{(l+1)^2}.$$ (6.7)

The momentum transfer for ionization at threshold $\mathbf{Q}_{\text{threshold}}$ is the amount required to achieve $\Delta E \geq Ry/n^2$, which is given by

$$\mathbf{Q}_{\text{threshold}} = \frac{l+1}{\sqrt{2n^2}} mc\alpha.$$ (6.8)

This explains the n^{-2} scaling; furthermore it provides a quantitative, if crude, estimate of the value of the threshold. Note, however, that it also suggests that s- and d-states should differ by a factor of three in their threshold for the same n, and this is certainly not borne out by the data in Figure 6.18! The problem with equation (6.8) is that $\mathbf{Q}_{\text{threshold}}$ is calculated from the peak velocity of the electron in its orbit, whereas we should be averaging over the velocities that occur over the finite (but small compared to τ_{Kepler}) duration of the HCP. The n^2 scaling still holds, however, so long as we replace the principal quantum number n with the *effective principal quantum number* n^*, computed using quantum defect theory (Gallagher 1994). This replacement has been done in Figure 6.18.

One way to achieve a quantitative estimate of the ionization threshold in spite of the problems of a finite pulse is to use instead the integral over the classical motion. The ionization limit is obtained for

$$\frac{1}{2n^{*2}} = \int_{\tau_{HCP}} F(t)\sqrt{2/x(t)}\, dt.$$ (6.9)

The smallest field F that satisfies this equation is found if the electron begins or ends the integration period at the nucleus. If the HCP is modeled as an inverted parabola, this gives a peak value of

$$F_0 = \frac{2}{5n^{*2}} \tau_{HCP}^{2/3}.$$ (6.10)

The 500 fs HCP then implies a threshold peak field of 2.7×10^6 n^{*-2} V/cm, in reasonable agreement the experimental 10% threshold value (Jones 1993a).

A Monte Carlo analysis yields calculated ionization shapes shown in Figure 6.18 as dotted lines. By tracking the classical trajectories of a distribution of Rydberg electrons, and weighting the results according to the quantum probability for finding electrons with the same initial conditions, the ionization probability was estimated.

The threshold laws and general ionization curve shapes were reproduced using this simple model, but the threshold fields were approximately 25% higher.

The most dramatic feature of the impulse limit is the ability to ionize 100% of the state, even though the impulse duration is much less than the Kepler orbit time. This is because the HCP, unlike an ordinary laser pure, is capable of transferring energy to a free electron. To quantify this, we take the limit of large momentum transfer in the equation above. To saturate the state, we require $Q_{sat}^2 \geq E_{binding}$ which for a Rydberg state implies an n^{-1} scaling for 100% ionization. The $n^{-3/2}$ scaling observed for 50% ionization is evidently a transition between these two scaling laws. This transition region was studied by Burgdorfer et al., who concluded that the ionization of all Rydberg states follows a universal curve if one plots the ionization threshold *scaled momentum* $\tilde{Q} = Q/n$ versus the scaled pulse length τ/τ_{Kepler} (Reinhold 1993).

Redistribution of coherent population among different bound Rydberg states occurs following ionization, or even when the field is too weak to observe any ionization of the initial state. This process can be studied by state-selective field

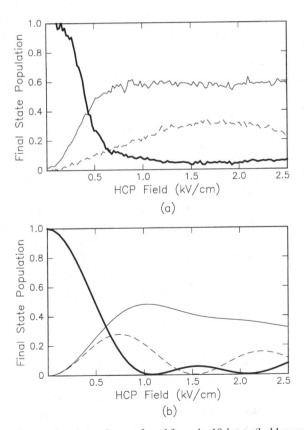

Figure 6.19 Population is coherently transferred from the 19d state (bold curve) in Na to 19p (dashed curve) and 191 > 2 (solid curve). (a) Shows the experimental results and (b) shows the results of a simple theory. (Reprinted with permission from Jones 1996.)

ionization of the Rydberg population following redistribution by the HCP. Tielking and Jones conducted a detailed study of this, and typical results are shown in Figure 6.19 (Tielking 1995). States that field-ionize at different values of the static field can be distinguished and plotted separately. A simple classical model similar to the one already described above was used to analyze the data, and there was reasonable qualitative agreement.

6.7 MAKING AND TRACKING RYDBERG WAVE PACKETS WITH HALF-CYCLE PULSES

The quantum mechanical description of impulsive excitation of a Rydberg state is easy to describe, because it is just momentum translation. For impulse $\mathbf{Q}$,

$$\Psi(\mathbf{x}, t) \rightarrow \Psi'(\mathbf{x}, t) = e^{i\mathbf{Q} \bullet \mathbf{x}} \Psi(\mathbf{x}, t). \tag{6.11}$$

An eigenstate is thus transformed by the HCP into a superposition state. which can be analyzed by projecting it onto the complete set of stationary bound and continuum states of the atom. The projection on the continuum states then represents the ionized fraction, while the projection onto the bound states represents the bound wave packet.

Although this impulsive limit is easy to describe, it is not so easy to measure. The reason is that momentum translation breaks the symmetry of the wavefunction, so the wave packet now contains all angular momentum states as well as different n-states. State-selective detection methods such as ramped-field ionization do not effectively differentiate l-states with the same n. To date, the best technique for measuring these wave packets is to cause them to interact with a second HCP, which acts as an analyzer for the momentum along the direction of the impulse.

Momentum analysis using an HCP can be understood using a simple WKB-type picture of the interaction. In a stationary state the total energy E can be divided into kinetic and potential parts: $E = T + V$. Since V is a function of position $\mathbf{r}$, the kinetic energy must also be $T(\mathbf{r}) = E - V(\mathbf{r})$, and so it is possible to define a local momentum magnitude $p(\mathbf{r}) = |\mathbf{p}(\mathbf{r})| = \sqrt{2mT(\mathbf{r})}$. If the potential is not changing very rapidly compared to $\hbar/p$, the deBroglie wavelength (WKB assumption), then the wavefunction is locally similar to a plane wave, and the effect of the impulse is to simply translate the momentum, and change the energy *locally*:

$$\mathbf{p}(\mathbf{r}) \rightarrow \mathbf{p}'(\mathbf{r}) = \mathbf{p}(\mathbf{r}) + \mathbf{Q},$$

$$E \rightarrow E'(\mathbf{r}) = E + \frac{1}{2m}(\mathbf{p}(\mathbf{r}) \cdot \mathbf{Q} + \mathbf{Q}^2). \tag{6.12}$$

In regions where the initially constant and negative total energy is transformed to a positive number, the local probability density has enough momentum to carry the electron out of the potential. These are the parts of the initial wavefunction that become ionized. The entire wavefunction is transformed into a nonstationary wave packet with an energy spread described by equation (6.12).

This same analysis can be applied to any Rydberg wave packet. The ionization fraction for a given HCP impulse gives information about the integral of the momentum distribution along the HCP polarization axis. This can be demonstrated using a simple Rydberg radial wave packet. Figure 6.20 shows the ramped field ionization spectrum of a radial wave packet in cesium with $l = 1$ and a Kepler orbital period of approximately 2 ps (Raman 1997a). If a half-cycle pulse instead of a ramped field ionizes this wave packet, the ionization fraction oscillates in time as the wave packet moves between its outer and inner turning points. This motion is plotted in Figure 6.2 at the beginning of this chapter. Figure 6.21 shows both the ionization fraction and the remaining bound state fraction for a fixed-amplitude HCP as the time is varied between the launch of the wave packet and arrival of the HCP (Raman 1996). The collapse of the wave packet due to dispersion in the Coulomb potential is clearly evident, as is the later revival.

Changing the amplitude of the HCP for fixed time delay gives information about the integral of the momentum distribution along the HCP polarization axis. The inner and outer turning points have quite different momenta, as seen in Figure 6.22. Equation (6.12) can be used to extract the momentum distribution function from the data. A Rydberg state with energy $E = Ry/n^2$ has a distribution of momenta. Any particular HCP will accelerate to just above threshold energy those electrons with a momentum along $\mathbf{Q}$ of

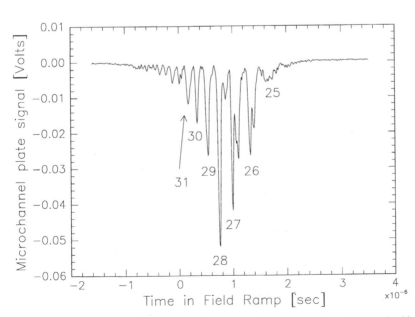

Figure 6.20 Ramped electric field ionization spectrum for l=1 Rydberg states excited by a 100 fs laser pulse in atomic cesium. The principal quantum numbers of each state are given in the wave packet as it is ionized in the ramped electric field.

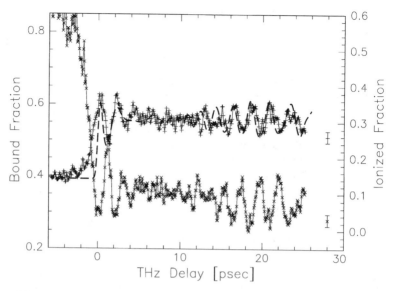

Figure 6.21   HCP ionization of the Rydberg wave packet of Figures 6.2 and 6.16. Fraction of atoms ionized by HCP (+ symbol) as well as the fraction left in bound states with $n > 25$ ($\times$ symbol), as a function of time delay. The HCP had a peak field of 6.3 kV/cm, and duration of approximately 400 fs. The bold curve is the result of the one-dimensional calculation described in the text, for a momentum transfer of 0.02 atomic units. (Reprinted with permission from Raman 1996.)

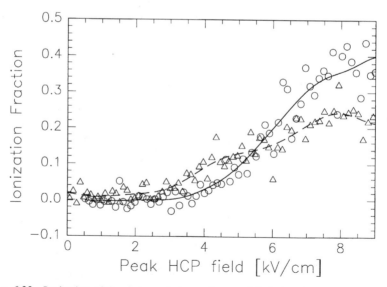

Figure 6.22   Ionization of the wave packet near the core (circles) and near the outer turning point (triangles) as a function of peak field in the HCP. The solid lines are the same data after smoothing. (Reprinted with permission from Raman 1996.)

$$p_Q \approx \frac{m^2 e^4}{\hbar^2 n^2 |\mathbf{Q}|} - |\mathbf{Q}|. \tag{6.13}$$

This means that measurements of the electrons ionized as a function of $\mathbf{Q}$ will yield the momentum distribution function in the $\mathbf{Q}$-direction. Analysis of data curves such as those in Figure 6.23 results in reasonable experimental momentum distributions (Jones 1996).

Equation (6.13) shows that it is not only possible to measure the magnitude but also the direction of the momentum in a quantum state. Thus, if a wave packet is created with an oscillating dipole moment, this could be seen directly by analyzing it with an HCP. The wave packet produced by a half-cycle pulse *itself* can also be momentum-analyzed this way. To do this, one first produces the wave packet by exciting a Rydberg state with an HCP; but instead of analyzing it with a ramped field as was done in Figure 6.20, the analysis is done by probing the wave packet with a second HCP. The result is a family of momentum distributions as a function of time, as shown in Figure 6.24 (Tielking 1995).

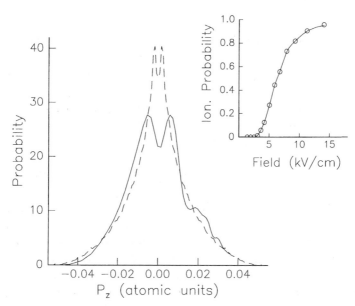

Figure 6.23 Experimental probability distribution for the Na 25d state obtained using impulsive momentum retrieval indicated by equation (6.12). The theoretical momentum distribution is the dashed curve; the inset is the measured ionization probability plotted in the HCP field. (Reprinted with permission from Jones 1996.)

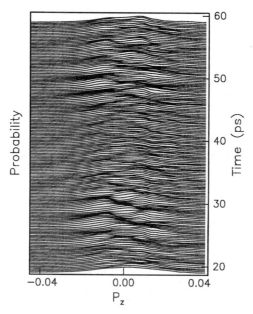

Figure 6.24   Measurement of the oscillating momentum distribution of a wave packet made with an HCP in atomic sodium. To make this wave packet, a 25*d* state was excited with a 500 fs HCP. Note that the oscillations clearly display a +/− asymmetry due to the unipolar nature of the impulse. (Reprinted with permission from Jones 1996.)

6.8   STARK WAVE PACKETS: HCP ANALYSIS OF ANGULAR MOTION

In Section 6.3 we discussed the classical dynamics associated with a Rydberg atom in an electric field, called a *Stark state*. Since HCPs can detect the momentum distribution of a wave packet, the detection of wave packets composed of Stark states should be particularly interesting. A series of experiments has been carried out on this problem, which we now describe.

The semiclassical dynamics described earlier in this chapter displayed three relevant time scales: the Kepler orbital time τ_{Kepler}, the angular momentum precession period τ_{ang}, and the time scale for precession of the Runge-Lenz vector which is associated with the nonhydrogenic nature of the electron trajectory due to interactions with core electrons. This last time scale is more difficult to define precisely. It is related to $\hbar$ divided by the avoided crossing energies seen in the energy level diagram Figure 6.5, but in fact the precession is not gradual or regular. During parts of the classical orbit when the angular momentum of the electron is large, the trajectory does not get near to the core electrons, and so there is essentially no precession. Once per τ_{ang}, the trajectory may dive into the core and emerge in a different direction that is extremely sensitive on the precise angle of approach. This sensitivity to angle gives rise to classical chaos, and its quantum manifestations have been the subject of intensive studies (Courtney 1994, 1995; Naudeau 1997).

Stark quantum wave packets, unlike their classical counterparts, cannot be made with a single energy. Therefore they undergo dispersion, and this can interfere with clear visualization of the trajectory dynamics. To minimize this problem, the wave packet must be sculpted so as to limit the dispersion. This was investigated by Naudeau and coworkers, who found that spectral filtering of the wave packet could reduce the Coulomb dispersion to a level where angular motion could be easily observed (Naudeau 1997). The aim of the spectral filter is to produce a wave packet that is predominately made of only two n-manifolds in the atom. Such a wave packet is very broad in the radial dimension, but it still displays motion that remains periodic with the same visibility over many cycles. Elimination of radial dispersion helps to reveal the angular dynamics. If there is no electric field, a two n-state wave packet will oscillate indefinitely without dispersing.

Naudeau did not use HCPs to probe the Stark wave packet, and therefore was only able to observe the periodic return of the packet to the core. These periodic revivals depend on whether the angular and radial motions are commensurate, as described in equations (6.2) and (6.3) earlier in this chapter. By varying the Stark field, this commensurate condition can be controlled (Fig. 6.25), but the optical probe of wave packet revivals cannot reveal the dynamics far from the core. The wave packet evolves during the angular momentum precession from linear motion towards and away from the core, to circular motion around it.

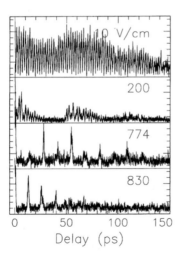

Figure 6.25   Quantum interferograms for Cs Stark states with different values of the electric field **F**. Nearly pure radial orbits occur when **F** = 10 V/cm. The 2.1 ps oscillation is the radial orbit period, and the slow oscillation of approximately 60 ps is a quantum beat due to the spin-orbit interaction. For **F** = 200 V/cm the precession is slow, producing broad angular returns modulated by the radial orbits. At **F** = 830 V/cm, the radial and angular returns are nearly commensurate. Individual angular returns can be completely suppressed (e.g., **F** = 774 V/cm) if the radial wave happens to be at its outer turning point when the angular oscillation is completed. (Reprinted with permission from Bucksbaum 1996.)

Figure 6.26 shows a series of contour plots of the probability density at various times during the precession of a Stark state in Cs with an angular return period of about 35 ps. This relatively small field of 350 V/cm helps to reduce dispersion due to manifold mixing to keep the size of the basis small (94 states) for the simulation. There are 14.5 Kepler orbits per angular momentum precession period at this field. A wave packet with the same parameters was probed by an HCP in a series of experiments. (Raman 1997b). Results are shown in Figure 6.27. The two traces correspond to HCP polarization aligned or anti-aligned with the static field. The wave packet is excited at $t = 0$ and immediately begins to move away from the ion core. The HCP can be used to differentiate between different directions of travel. That is, if the average ionization rate is less than 50%, the HCP only ionizes components of the wave function traveling in a single direction. Note the difference in modulation period for positive and negative HCP polarity near $t = 0$. This causes a phase difference in the two signals to appear immediately. The modulations also disperse rapidly, reviving roughly 16 ps later when the wave packet has acquired high angular momentum. The phase of these modulations changes by π if the unipolar pulse reverses. Figure 6.26 suggests that by this point the radial motion of the wave is replaced by angular motion along a polar orbit. The HCP ionizes the electron only if its own motion is initially in the direction of the impulse imparted by the transient field. Thus the THz can directly detect this angular motion, which persists for several cycles before the torque exerted by the static field renders it unstable. Motion of this type, at far distances from the ion core, has never been observed before.

In addition to the fast Kepler motion, there is a slower modulation of the ionization rate with a phase that depends on the polarity of the HCP. For the first 10 ps the rate of ionization for an HCP anti-aligned with the static field is less than for the HCP aligned, but midway through the angular precession, this asymmetry reverses. These

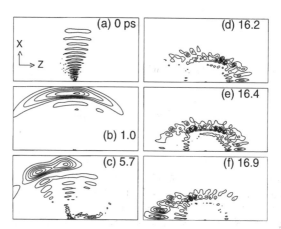

Figure 6.26 Contour plots of calculated wave packet spatial probability distributions in the x, z plane at the 0, 0.5, 2.7, 7.7, 7.8, and 8.0 Kepler orbital periods. The static field is along the horizontal z-axis, which runs from -1000 to 1000 bohr radii; the vertical x-axis runs from 0 to 1200 bohr radii. (Reprinted with permission from Raman 1997b.)

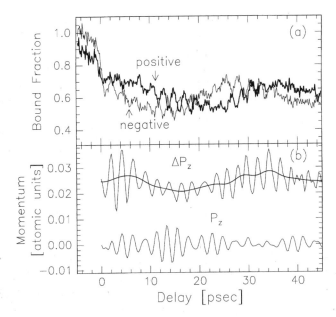

Figure 6.27 (*a*) Evolution of a Stark Rydberg wave packet excited at **F** = 350 V/cm. The heavier and lighter traces are for an HCP anti-aligned and aligned with the static field, respectively. (*b*) Calculated time evolution of wave packet average momentum, p_z (lower trace), and momentum uncertainty Δp_z (upper trace). The wave packet spreads and collapses in momentum space while its angular momentum oscillates. (Reprinted with permission from Raman 1997b.)

features in the data indicate the linear momentum distribution of the wave function along its polarization. Figure 6.27*b* shows that near 31 ps, when the electron angular momentum has returned to its low value, the wave function is more symmetrically distributed in uphill and downhill directions (i.e., p_z is close to zero). However, Δp_z undergoes large oscillations, indicating Kepler radial motion in both directions simultaneously. The ionization curves in Figure 6.23*a* also return to synchronization, at this point, indicating symmetric motion. At times when the wave function is at high angular momentum (between 10 and 20 ps), the expectation value of p_z oscillates about zero with a period of about 2 ps. By contrast, at this point the momentum spread is not varying as rapidly. The HCP ionization curves mirror this oscillation in p_z indicating the periodic asymmetry of the motion in uphill and downhill directions.

6.9   BEYOND THE IMPULSE LIMIT: STARK STATES AND HALF-CYCLE PULSES

Almost all of the theory and analysis described in this chapter makes use of the impulse limit, where the HCP is modeled as a delta function in time. Since we know that the pulses are actually about one-half picosecond in duration, this approximation really is not very good for Rydberg states with $n < 20$. Photoionization of Rydberg states as low as $n = 15$ has been observed, so the transition away from the impulse limit is accessible. The breakdown of the impulse approximation is most dramatic in the case of the ionization of oriented states, such as Stark eigenstates. If the HCP is very short compared to the electron motion, then the average impulse cannot depend on the sense of HCP polarization because the average momentum of all electron trajectories must be zero for any bound state. If the HCP is longer, then it has time to interact with the electron over a significant portion of the electron orbit, so the details of the orbital dynamics can make a difference in the ionization rate. A long HCP will interact differently depending on its sense relative to the Stark field.

The polarization dependent ionization asymmetry can be observed in states below about $n = 20$. Na atoms were placed in a field of 376 V/cm. A tunable dye laser can then excite each of the states in the $n = 15$, $m = 0$ manifold shown in Figure 6.28 (You 1996).

Figure 6.29 shows four different sets of HCP ionization curves, corresponding to four different Stark states. In each view there are two curves: one for the HCP aligned with the static field is labeled B, and one where the HCP is aligned against the static field, labeled A. In the top left figure, the state with the largest positive Stark shift is shown. This is the state excited at 416.4 nm in the spectrum in Figure 6.28. In the top right figure, the state with the most negative Stark shift is shown, which is excited at about 416.7 nm. The lower left figure is for a state at the center of the manifold. The

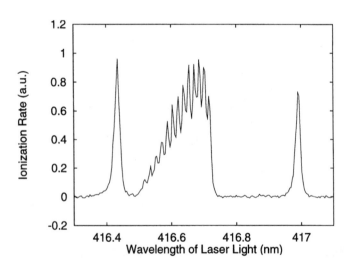

Figure 6.28   Rydberg state excitation spectrum from the $4p$ state in Na, in a field of 375 V/cm, $n = 15$. The manifold consists of 15 clearly resolved states. The relative magnitude shows the different admixture of s- and d-character that controls the dipole cross section from a p-state. The left and right split-off peaks correspond at zero field to the $16p$ and $16s$ states, respectively, which have large quantum defects. (Reprinted with permission from You 1996.)

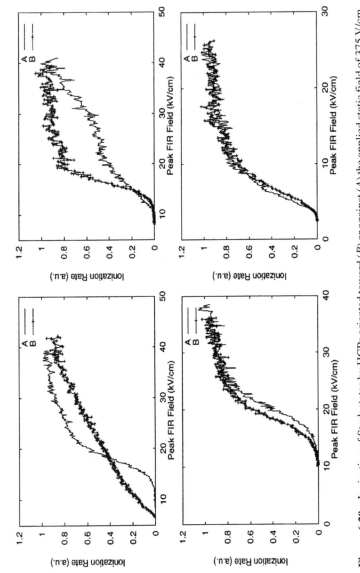

Figure 6.29 Ionization of Stark states by HCPs oriented toward (*B*) or against (*A*) the applied static field of 375 V/cm. *Top left*: *n* = 15 state with the most positive Stark shift. *Top right*: *n* = 15 state with the most negative Stark shift. *Bottom left*: A state in the middle of the *n* = 15 manifold. *Bottom right*: *n* = 25 state with the most positive Stark shift.

lower right figure is for a state with a positive Stark shift for $n = 25$, where the impulse approximation is valid.

The most marked differences occur for the negative "red" and positive "blue" Stark states of the $n = 15$ manifold. The behavior is rather complicated, but one could summarize by noting that Stark states are most easily ionized if the HCP pulls them *away* from the core. In other words, the blue state has a positive dipole moment, which places more electron probability in the direction of the applied field. The HCP can fully ionize this state more easily if it is oriented in the opposite direction, so that it kicks the negatively charged electron away from the core. Likewise for the red states, the full ionization asymptote is reached first for an HCP oriented in the direction of the field, so that it kicks this electron "downhill" in the electrostatic potential. These results have also been compared to analyses using classical models (Jones 1995) and full quantum calculation (Bugacov 1995), which confirm the general results.

6.10 CONCLUSIONS

Half-cycle pulses show great promise as unique probes of momentum structure in Rydberg wave packets. Limitations of space do not permit us to cover some of the recent interesting developments in this subject in detail, but for completeness we will briefly mention them.

Half-cycle pulses not only can ionize an atom; they also can be used to induce recombination. Jones recently introduced a technique whereby full three-dimensional localization of a Rydberg electron may be accomplished by first ionizing a Rydberg atom and then reattaching the electron with a second HCP (Jones 1998). The conditions of reattachment may be made so precise that only electrons in a small region of phase space survive as bound states. A different scheme for using HCPs to localize Rydberg wave packets has also been discussed by Stroud (de Araujo 1998).

Half-cycle pulses can also be shaped using the same techniques that have recently led to a revolution in shaping the amplitude and phase of ultrafast optical pulses. Shaping has been studied by Weling and Auston (Weling 1994). Shaped terahertz radiation has been used to study strong field excitation in Rydberg states, and this work continues (Raman 1997c). These fields may eventually be used to study exotic Rydberg states, such as Trojan wave packets (Kalinski 1996; Delande 1997) (see Chapter 4 of this book; Farrelly, Lee, and Uzer 1995; Kalinski 1996; Delande 1997).

Finally we should note that not all half-cycle pulses need to be in the terahertz regime. Recent work on very high Rydberg states by Dunning and coworkers have reproduced and extended some of the work described here, by making use of Rydberg states with principal quantum numbers above 500! Such states move so slowly, and visit the core so infrequently, that their dynamics may be studied using conventional field pulses at sub-GHz frequencies (Frey 1996; Tannian 1998).

ACKNOWLEDGMENTS

I gratefully acknowledge discussions with Robert Jones, Thomas Weinacht, and Jaewook Ahn. This material is based on work supported by the National Science Foundation under Grant 9414335.

REFERENCES

Auston, D. H., Cheung, K. P., and Smith, P. R. (1984). *App. Phys. Lett.* **45**, 284.

Bensky, T. J., Campbell, M. B., and Jones, R. R. (1998). *Phys. Rev. Lett.* **81**, 3112.

Bonvalet, A., Joffre, M., Martin, J. L., and Migus, A. (1995). *Appl. Phys.* **67**, 2907.

Budiarto, E., Pu., N.-W,. Jeong, S., and Bokor, J. (1998). *Opt. Lett.* **23**, 213–215.

Bugacov, A., Piraux, B., Pont, M., and Shakeshaft, R. (1995). *Phys. Rev.* A**51**, 4877.

Bucksbaum, P. H., Naudeau, M. L., and Sukenik, C. I. (1996). *Laser Spectroscopy: XII International Conference*, eds. M. Inguscio. M. Allegrini, and A. Sasso. World Scientific, Singapore, pp. 299–301.

Casati G., Chirikov, B. V., Shepelyansky, D. L., and Guarneri, I. (1987). *Phys. Rep.* **154**, 77–123.

Chwalek, J. M., Whitaker, J. F., and Mourou, G. A. (1991). *Electron. Lett.* **27**, 447.

Courtney, M., Jiao, H., Spellmeyer, N., and Kleppner, D. (1994). *Phys. Rev. Lett.* **73**, 1340–1343.

Courtney, M., Spellmeyer, N., Jiao, H., and Kleppner, D. (1995). *Phys. Rev.* A**51**, 3604.

Darrow, J. T., Hu, B. B., Zhang, X. C., and Auston, D. (1990). *Optics Lett.* **15**, 323–325.

de Araujo, L., Walmsley, I. A., and Stroud, C.R., Jr. (1998). *Phys. Rev. Lett.* **81**, 955–958.

DeFonzo, A. P., Jarwala, M., and Lutz, C. (1987). *Appl. Phys. Lett.* **50**, 1155.

Delande, D., Zakrzewski, J., and Buchleitner, A. (1997). *Phys. Rev. Lett.* 79, 3541–3544.

Fattinger, Ch., and Grischkowsky, D. (1989a). *Phys. Rev. Lett.* **62**, 2961–2964.

Fattinger, Ch., and Grischkowsky, D. (1989b). *IEEE Journal of Quantum Electronics*, **25**, 2608–10.

Feng, S. M., Winful, H. G., and Hellwarth, R.W. (1998). *Optics Lett.* **23**, 385–387.

Frey M. T., Dunning, F. B., Reinhold, C. O., et al. (1996). *Phys. Rev.* A**53**, R2929–2932.

Gallagher, T. F. (1994). *Rydberg atoms*. Cambridge University Press, New York.

Goldstein, H. (1980). *Classical Mechanics*. Addison-Wesley, Reading, MA.

Greene, B. I., Federici, J. F., Dykaar, D. R., Levi, A. F. J., and Pfeiffer, L. (1991). *Opt. Lett.* **16**, 48.

Grischkowsky, D., Duling III, I. N., Chen, J. C., and Chi, C.-C. (1987). *Phys. Rev. Lett.* **59**, 1663.

Grischkowsky, D., and Keiking, S. (1990). *Appl. Phys. Lett.* **57**, 1055.

Grischkowsky, D., Keiding, S., van Exter, M., and Fattinger, Ch. (1992). *J. Opt. Soc. Am.* B**7**, 2006.

Harde, H., Keiding, S., and Grischkowsky, D. (1991). *Phys. Rev. Lett.* **66**, 1834–1837.

Hu, B. B., Zhang, X.-C., Auston, D. H., and Smith P. R. (1990a). *Appl. Phys. Lett.* **56**, 506–508.

Hu, B. B., Darrow, J. T., Zhang, X.-C., and Auston, D. H. (1990b). *Appl. Phys. Lett.* **56**, 886–888.

Jenkins, F. A., and White, H. E. (1937). *Fundamentals of Physical Optics*. McGraw-Hill, New York.

Jensen, R. V., Susskind, S. M., and Sanders, M. M. (1991). *Phys. Rep.* **201**, 1–56.

Jones, R. R. (1996). *Phys. Rev. Lett.* **76**, 3927–3930.

Jones, R. R., and Bucksbaum, P. H. (1991). *Phys. Rev. Lett.* **67**, 3215–3218.

Jones, R. R., You, D., and Bucksbaum, P. H. (1993a). *Phys. Rev. Lett.* **70**, 1236–1239.

Jones, R. R., Schumacher, D. W., and Bucksbaum, P. H. (1993b). *Phys. Rev.* A**47**, R49–52.

Jones, R. R., Tielking, N.E., You, D., Raman, C., and Bucksbaum, P. H. (1995). *Phys. Rev.* A**51**, R2687–2690.

Kalinski, M., and Eberly, J. H. (1996). *Phys. Rev. Lett.* **77**, 2420–2423.

Koch, P. M., and VanLeeuwen, K. A. H. (1995). *Phys. Rep.* **255**, 290–403.

Mourou, G., Sancampiano, C. V., and Blumenthal, D. (1981). *App. Phys. Lett.* **38**, 470.

Naudeau, M. L., Sukenik, C. I., and Bucksbaum, P. H. (1997). *Phys. Rev.* A**56**, 636–639.

Nuss, M. C., Planken, P. C. M., Brener, I., et al. (1994). *Appl. Phys.* B**58**, 249.

Raman, C., Conover, C. W. S., Sukenik, C. I., and Bucksbaum, P. H. (1996). *Phys. Rev. Lett.* **76**, 2436–2439.

Raman C. S. (1997a). Ph.D. thesis. University of Michigan.

Raman, C. S., Weinacht, T. C., and Bucksbaum, P. H. (1997b). *Phys. Rev.* A**55**, R3995–3998.

Raman, C. S., DeCamp M. F., and Bucksbaum, P. H. (1997c). *Optics Exp.* **1**, 186.

Reinhold, C. O., Melles, M., and Burgdörfer, J. (1993). *Phys. Rev. Lett.* **70**, 4026.

Stapelfeldt, H., Papaioannou, D. G., Noordam, L. D., and Gallagher, T. F. (1991). *Phys. Rev. Lett.* **67**, 3223–3226.

Tannian, B. E., Popple R. A., Dunning F. B., et al. (1998). *J. Phys B* **31**, L455–460.

Tielking, N. E., and Jones, R. R. (1995). *Phys. Rev.* A**52**, 1371.

van Exter, M., Fattinger, Ch., and Grischkowsky, D. (1989). *Optics Lett.* **14**, 1128–1130.

Weinacht, T. C., Ahn, J., and Bucksbaum, P. H. (1998). *Phys. Rev. Lett.* **80**, 5508.

Weling, A. S., Hu, B. B., Froberg, N. M., and Auston, D. H. (1994). *Appl. Phys. Lett.* **64**, 137.

Yeazell, J. A., Mallalieu, M., and Stroud, C. R., Jr. (1990). *Phys. Rev. Lett.* **64**, 2007.

You, D., Jones, R. R., Dykaar, D. R., and Bucksbaum, P. H. (1993). *Optics Lett.* **18**, 290–292.

You, D., Jones, R. R., Bucksbaum, P. H., and Dykaar, D. R. (1994). *J. Opt. Soci. Am.* B**11**, 486.

You, D. (1996). Ph.D. thesis. University of Michigan.

Zhang, X.-C., Hu, B. B., Darrow, J. T., and Auston, D. (1990). *Appl. Phys. Lett.* **56**, 1011–1013.

CHAPTER 7

Exploring Phase Space with Wave Packets

FLORENTINO BORONDO and ROSA M. BENITO

7.1 INTRODUCTION

The last decade has witnessed an extraordinary development of the theoretical methods and experiments based on or related to wave packets. These time dependent methods were forgotten for many years in favor of their counterparts in the energy domain until theoretical techniques such as fast Fourier transform, discrete variable representation, or Chebyshev and suitable experiments became available.

In general terms, propagation of wave packets is a tool to solve the time dependent Schrödinger equation, which has attracted a great deal of attention lately (e.g., see Feit and Fleck 1984; Heather and Metiu 1987; Leforestier et al. 1991; Kulander 1991; Schinke 1993), specially in relation with the study of fundamental problems in atomic and molecular physics. The appearance of new related experimental techniques such as femtosecond chemistry (Zewail and Bernstein 1992), or experiments with atomic Rydberg packets (Yeazell and Stroud 1988; Yeazell et al. 1990; Gaeta et al. 1994) have contributed to trigger this growing interest in the study of wave packets.

Wave packets can also be propagated semiclassically using a variety of strategies (Sepúlveda and Grossmann 1996), some of which are surprisingly successful. These methods have the advantage that can be related to classical trajectories, which carry the necessary information (phases and probability amplitudes) for the calculation. This approach has two advantages. From a fundamental point of view, classical mechanics can provide a useful physical insight into the dynamical aspects of the problem under study that are often hidden in quantum treatments. From a practical point of view, classical and semiclassical mechanics represent the only computational alternatives in situations where the density of states is large, and the quantum methods run into big numerical problems, due to the large size of the matrices that need to be diagonalized. This is true even in the present state of hardware development.

In the recent past there has been a lot of progress in the understanding of nonlinear dynamics of classical Hamiltonian systems (Lichtenberg and Lieberman 1992), which at a high level of excitation can exhibit chaotic behavior. The important role of phase

The Physics and Chemistry of Wave Packets, Edited by John Yeazell and Turgay Uzer
ISBN 0-471-24684-0 © 2000 John Wiley & Sons, Inc.

space structures in low-dimensionality systems, such as periodic orbits (PO), cantori, or broken separatrices is now well understood, due to the appearance of the Kolmogorov-Arnold-Moser theorem and the widespread possibility of extensive numerical simulations. These structures, which constitutes the only remnants of order and regularity existing in the chaotic regions, are also important in the quantum mechanics of chaotic systems. Gutzwiller constructed in the 1960s a semiclassical quantization theory valid for chaotic systems, based solely on POs and their properties. Another striking phenomenon in the field of classical-quantum correspondence is that of "scarring." The term *scar* was coined by Heller (1984) to name the enhancement of probability density that takes place along short unstable POs for certain eigenfunctions of classically chaotic systems. This localization, which implies a departure from the random matrix theory predictions, has also important influence on the long-time quantum transport of this kind of systems, and has been observed experimentally.

In this chapter we will use wave packets to explore the phase space of a small triatomic systems, namely LiNC/LiCN. The vibrational motions of this system will be described with a realistic model. In particular, we are interested in the observation and analysis of the effects that the POs cause at quantum level, in the eigenfunctions and spectra of the molecule. Several examples will be presented and discussed at various levels, and some attention will be paid to the formation of scars and their distribution among the different states of the system.

7.2 SYSTEM AND CALCULATIONS

In this section we present the system used in our study and describe the methodology of the different type of calculations performed. The LiNC/LiCN isomerizing system is considered, as a prototype for a large class of small polyatomic molecules exhibiting chaos due to the existence of a floppy motion along one of the vibrational modes. Theoretical calculations have been performed at classical, quantal, and semiclassical level, paying special attention to the relationship existing between dynamics, exploration of phase space, and spectra specially when low resolution features are considered.

7.2.1 The Model

To illustrate our study, we choose, as an example, the dynamics of the molecular vibrations of the triatomic molecule LiNC/LiCN. This isomerizing system presents two stable isomers at the linear configurations: Li–NC and Li–CN. The absolute minimum corresponds to the first structure, and both are separated by a relatively modest energy barrier.

The motion in the bending coordinate is very floppy, and the Li atom can easily rotate around the CN fragment, sampling extensive regions of the potential energy surface, where anharmonicities and mode couplings are important. Thus, chaos sets in at low values of the excitation energy. Also the C–N vibrational frequency is very high and separates effectively from the remaining modes of the system.

All of these characteristics make the LiNC/LiCN molecule a very interesting system. It constitutes a generic example in molecular physics which can be described very realistically, and it has been often considered in the past, specially in relation with the quantum manifestations of classical chaos.

The vibrational dynamics of this system can be adequately studied by a two degrees of freedom model, where the C–N distance is kept frozen at its equilibrium value of $r_e = 2.186$ a.u. The vibrational (total angular momentum $J = 0$) Hamiltonian in scattering or Jacobi coordinates (Child 1974) is given by

$$H = \frac{P_R^2}{2\mu_1} + \frac{1}{2}\left(\frac{1}{\mu_1 R^2} + \frac{1}{\mu_2 r_e^2}\right)P_\theta^2 + V(R, \theta), \tag{7.1}$$

where R is the distance between the centre of mass of the CN (anionic) fragment to the Li atom, r the N–C distance, θ the angle formed by these two vectors (see Fig. 7.1), and P_R and P_θ the associated conjugate momenta. The reduced masses μ_1 and μ_2 corresponding, respectively, to Li–CN and C–N are given by

$$\mu_1 = \frac{m_{Li}(m_C + m_N)}{m_{Li} + m_C + m_N}, \quad \text{and}$$

$$\mu_2 = \frac{m_C m_N}{m_C + m_N}. \tag{7.2}$$

The potential energy surface is a nine term expansion in Legendre polynomials,

$$V(R, \theta) = \sum_{\lambda=0}^{9} P_\lambda(\cos\theta)v_\lambda(R), \tag{7.3}$$

where the coefficients v_λ, which contain short- and long-range contributions, have been taken from the literature (Essers et al. 1982). They were obtained by fitting expression (7.3) to numerical results obtained within the self-consistent field (SCF) approximation using a large basis set ($11s, 6p, 2d/6s, 3p, 2d$) of Gaussian-type orbitals

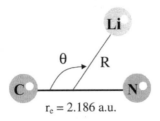

Figure 7.1 Definition of the coordinates for the LiNC/LiCN isomerizing system used in the text.

(Wormer 1981). This potential energy surface is presented in the form of 3D and contours plot in Figure 7.2. The two stable isomers, LiNC and LiCN, appearing at θ = 180° and 0°, respectively, are clearly visible as potential wells separated by 2281 cm^{-1}. Relevant information concerning the stationary points of function (1.3) is summarized in Table 7.1. The minimum energy path,

$$R_e(\theta) = 4.1159 + 0.25510 \cos \theta + 0.49830 \cos 2\theta - 0.053427 \cos 3\theta$$

$$- 0.068124 \cos 4\theta + 0.020578 \cos 5\theta \qquad \text{(in a.u.)}, \qquad (7.4)$$

connecting these two wells has also been plotted superimposed as a dotted line.

Very recently, new improved *ab initio* potential energy surfaces, including electron correlation at MP2 (Makarewitz and Ha 1994) and MP4 (Makarewitz and Ha 1995) level, have been reported in the literature. The geometries of the linear minima predicted by these authors are in close agreement with those of Essers et al., but there are significant differences in their relative energies. Another difference is the existence, in the Makarewitz and Ha surfaces, of a stable T-shaped minimum in the region around $(R, \theta) = (3.65 \text{ a.u.}, 110°)$, where the surface of Essers et al. only shows a small plateau. However, this feature is dynamically not very significant since the motion around this region gets stabilized by an adiabatic separation mechanism for

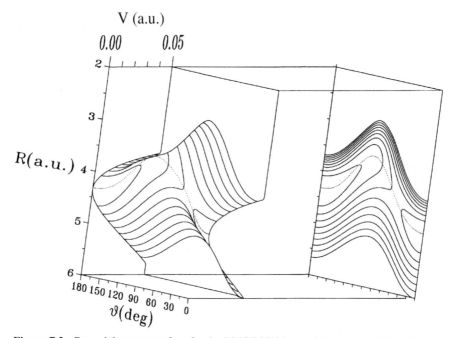

Figure 7.2 Potential energy surface for the LiNC/LiCN isomerizing system. The minimum energy path connecting the two isomer wells, Li–NC and Li–CN, has also been represented as a dotted line.

TABLE 7.1 **Geometries and Energies for the Stationary Points of the Potential Energy Surface Used in the 2D LiNC/LiCN Isomerizing System Model**

Configuration	θ (deg)	R (a.u)	E (cm^{-1})
Li–NC	180°	4.35	0
Li–CN	0°	4.795	2281.0
Saddle	48.41°	4.30	3454.0
"Plateau"[a]	110°	3.65	1207.0

Source: Essers et al. (1982).
[a]See text for details.

high value of the excitation in the R coordinate, as has been demonstrated by us (Borondo et al. 1995). All these surfaces predict LiNC harmonic vibrational frequencies in good agreement with the existing experimental results obtained by microwave spectroscopy (Van Vaals et al. 1983) and infrared spectra in inert gas matrices (Ismail et al. 1972). There is no experimental evidence, however, of the T-shaped minimum.

Finally we would like to point out that LiNC/LiCN is representative of a large class of triatomic molecules that exhibit similar behavior, mainly related to the character of large amplitude motion in one of the vibrational modes. This class includes other cyanides or similar species, like HCN/HNC, RbCN/RbNC, KCN, HCP, and HO$_2$ radical, for which potential energy surfaces are available in the literature. These molecules present very significant differences in their potentials: While HCN and RbCN shows two stable isomers at the linear configuration, very much like LiCN, for HCP the structure corresponding to $\theta = 180°$ is unstable, and for KCN and HO$_2$ the configuration at the minimum is bent. Weakly bound hydrogen bonded or van der Waals complexes, with shallow attractive wells, also fall within this category of floppy molecules.

7.2.2 Classical Calculations

The classical vibrational dynamics of the LiNC/LiCN molecular system has been followed by using the method of classical trajectories. These trajectories have been calculated using a fixed step Gear algorithm (Gear 1964) for the numerical integration of Hamilton equations of motion corresponding to equation (7.1). The isomerization process is best monitored by considering the motion along the θ coordinate, which shows the most interesting dynamics in this molecule. This is most significantly done by means of Poincaré surfaces of section (SOS) in phase space (Lichtenberg and Lieberman 1992). Usually SOS for systems of coupled oscillators are obtained by plotting one coordinate and its conjugate momentum every time that the other coordinate crosses through its equilibrium distance, and the momentum has a predetermined sign. In our case this would correspond to taking the sectioning plane to lie along the minimum energy path. That is, the values of the coordinates (θ, P_θ) should be recorded every time that a trajectory passes through $R = R_e(\theta)$. This presents a problem since R_e depends on θ (Eckhardt 1988). To ensure that the SOS is an area preserving map, it is necessary to make a canonical transformation $(R, \theta, P_R, P_\theta) \rightarrow (\rho, \psi, P_\rho, P_\psi)$ according to the F_2-type (Goldstein 1980) generating function, namely a function of the old coordinates and the new momenta,

$$F_2(R, \theta, P_\rho, P_\psi) = P_\rho[R - R_e(\theta)] + P_\psi\theta. \tag{7.5}$$

It renders the following set of canonical coordinates (Benito et al. 1989)

$$\rho = R - R_e(\theta),$$

$$\psi = \theta,$$

$$P_\rho = P_R, \quad \text{and}$$

$$P_\psi = P_\theta + P_\rho\left(\frac{\partial R_e}{\partial \theta}\right)_{\theta=\psi}. \tag{7.6}$$

The SOS corresponds then to the conjugate pairs (ψ, P_ψ) at the successive intersections with the $\rho = 0$ plane, namely $R = R_e(\psi)$, taking only those points for which P_ρ is in a particular branch of the momentum, that is, in a particular solution of the equation:

$$H(\rho = 0, \psi, P_\rho, P_\psi) = E. \tag{7.7}$$

7.2.3 Quantum Calculations

Quantum vibrational energy levels and the corresponding wave functions for Hamiltonian (7.1) are calculated using the Discrete Variable Representation (DVR) method (Sutcliffe et al. 1988; Bačić and Light 1989). The corresponding volume element is given by

$$d\tau = R^2 \sin\theta \, dR \, d\theta, \qquad 0 \le \theta \le \pi, \quad 0 \le R \le \infty. \tag{7.8}$$

The DVR method is ideally suited for molecules with one or more large amplitude vibrational motions, contrary to traditional variational methods based on finite basis representations (FBR), which are more demanding computationally, since they require larger basis sets. The essence of the DVR is an orthogonal transformation into a representation labeled by a discrete set of angles corresponding to the points of a given quadrature (Dickinson and Certain 1968). This transformation presents the additional advantage of allowing an easier way of handling the involved potential matrix elements. It has also been established that DVR is an efficient method to solve other problems in quantum mechanics (e.g., see Lin and Muckerman 1991; Zhang and Light 1996).

In our case we used the program of Bačić and Light (1986), which consists of a discrete representation in the θ coordinate and a function representation of distributed Gaussian basis (DGB) (Hamilton 1986) in the radial coordinate R. A prediagonalization along each θ ray prepares the final basis set, that consisted of 2016 elements rendering the 900 low-lying eigenvalues converged to within 0.1 cm^{-1}.

Another way to study the quantum dynamics of a system, which can be compared more straightforwardly with the classical results described in the previous subsection, is to consider phase space representations. There is no unique way to define a phase space representation of quantum mechanics (Moyal 1949). Probably the most popular one is that introduced by Wigner in his 1932 paper on the calculation of quantum corrections to statistical thermodynamics (Wigner 1932). In it, the now widespread transform

$$\mathcal{W}(\mathbf{q}, \mathbf{P}) = \frac{1}{(2\pi\hbar)^N} \int \cdots \int d\mathbf{x}\, e^{i\mathbf{x}\cdot\mathbf{P}/\hbar}\, \varphi\left(\mathbf{q} - \frac{\mathbf{x}}{2}\right) \varphi^*\left(\mathbf{q} + \frac{\mathbf{x}}{2}\right) \tag{7.9}$$

was introduced. This function is obtained from $|\varphi(\mathbf{q})\rangle$ by applying standard methods in statistics to obtain joint probabilities (Reif 1965) and choosing the Weyl quantization (Ozorio de Almeida 1998),

$$e^{i(\mu q_k + \nu P_k)} \rightarrow e^{i(\mu \hat{q}_k + \nu \hat{P}_k)} \tag{7.10}$$

or

$$q^n P^m \rightarrow \frac{1}{2^n} \sum_{l=0}^{n} \hat{q}^{n-l}\, \hat{P}^m\, \hat{q}^l, \tag{7.11}$$

to obtain the characteristic function. The Wigner transform has certain properties which suggest that it can be interpreted as a probability density in phase space. Namely it gives the correct marginal probability distributions

$$\int d\mathbf{q}\, \mathcal{W}(\mathbf{q}, \mathbf{P}) = |\langle \mathbf{P}|\varphi\rangle|^2,$$

$$\int d\mathbf{P}\, \mathcal{W}(\mathbf{q}, \mathbf{P}) = |\langle \mathbf{q}|\varphi\rangle|^2. \tag{7.12}$$

Moreover it has been shown (Berry 1977; Ozorio de Almeida and Hannay 1982) that the amplitude peak of $\mathcal{W}$ for pure states of classically integrable systems lies close to the corresponding invariant tori in the semiclassical limit. However, this interpretation is sometimes considered questionable, since W can be negative, and also its dynamical evolution does not follow, in general, the Liouville equation in the semiclassical limit $\hbar \rightarrow 0$.

The first problem is somehow due to the uncertainty principle, which precludes the possibility of defining distribution functions at precise points of phase space. A way to overcome this difficulty was devised by Husimi (1940), who proposed the use of Gaussian convolutions of the Wigner function:

$$\mathcal{H}(\mathbf{q}, \mathbf{P}) = \frac{1}{(\pi\hbar)^N} \int \cdots \int d\mathbf{q}' d\mathbf{P}' \; \mathcal{W}(\mathbf{q}',\mathbf{P}') \; G_{\mathbf{q},\mathbf{P}}(\mathbf{q}', \mathbf{P}'), \qquad (7.13)$$

where the weighting function, G, is given by

$$G_{\mathbf{q},\mathbf{P}}(\mathbf{q}',\mathbf{P}') = \frac{1}{\pi\hbar} \exp\left\{ -\frac{1}{2} \sum_{k=1}^{N} \left[\left(\frac{q_k' - q_k}{\Delta q}\right)^2 + \left(\frac{P_k' - P_k}{\Delta P}\right)^2 \right] \right\} \qquad (7.14)$$

with

$$\Delta q = \sqrt{\hbar/2}\,\sigma, \qquad \Delta P = \sqrt{\hbar/2}\,\frac{1}{\sigma}, \qquad (7.15)$$

so that $\Delta q \Delta P = \hbar/2$ is the minimum allowed by the uncertainty principle. Varying the parameter σ, the relative uncertainties in the coordinates and momenta can be changed. Usually σ is taken equal to the unity, and this is the value used in our calculations. The Husimi function defined in this way has been proven to be everywhere nonnegative (Cartwright 1976), and should be understood, consistently with the uncertainty principle, in the following way: $\mathcal{H}$ does not give the probability density at a point in phase space, but the probability density smoothed over a region of volume $\sim\hbar^N$ around that point. Furthermore it can be shown to be given by the expression.

$$\mathcal{H}(\mathbf{q}, \mathbf{P}) = \frac{1}{(2\pi\hbar)^N} |\langle\chi|\varphi\rangle|^2, \qquad (7.16)$$

where χ is a minimum uncertainty harmonic oscillator coherent state (Cohen-Tannoudji et al. 1977; Zhang et al. 1990)

$$\chi_{\mathbf{q},\mathbf{P}}(\mathbf{q}', \mathbf{P}') = \left[\frac{1}{2\pi(\Delta q)^2}\right]^{N/4} \exp\left[-\frac{(\mathbf{q}' - \mathbf{q})^2}{4(\Delta q)^2} + \frac{i}{\hbar}\mathbf{q}' \cdot \mathbf{P} \right], \qquad (7.17)$$

centered at the phase space point $(\mathbf{q}, \mathbf{P})$. The Husimi function, or coherent state representation (Glauber 1963), can be interpreted in a number of ways: as a smoothed Wigner distribution (Hillery 1984), as a coarse graining of phase space (O'Connell 1983), or as the expectation value of the projection operator on a certain state. Other properties of the Husimi function have been discussed in the literature (Harriman 1988; Takahashi 1989).

Other quasi-probability densities, analogous to (7.9), can be defined by taking quantization conditions alternative to that of Weyl. For example, choosing the Rivier quantization condition,

$$e^{i(\mu q_k + \nu P_k)} \rightarrow \frac{1}{2}\left[e^{i\mu \hat{q}_k}e^{i\nu \hat{P}_k} + e^{i\nu \hat{q}_k}e^{i\mu \hat{P}_k}\right], \tag{7.18}$$

the Margenau-Hill function (Margenau and Hill 1961),

$$\mathcal{M}(\mathbf{q}, \mathbf{P}) = \frac{1}{(2\pi\hbar)^N} \mathcal{R}e\left[\int\cdots\int d\mathbf{x} \; e^{i\mathbf{x}\cdot\mathbf{P}/\hbar}\varphi(\mathbf{q}) \; \varphi^*(\mathbf{q}-\mathbf{x})\right] \tag{7.19}$$

is obtained.

The information contained in the Wigner or Husimi functions can be visualized in a number of ways. From the point of view of the present paper the most interesting one is the construction of quantum analogues to the classical Poincaré surfaces of section, which hereafter will be called QSOS. This can be accomplished, for example, by projection on suitable planes of phase space, or by integrating out the unwanted coordinates and momenta. In our case we have obtained Husimi-based QSOS for the first 900 states of LiNC/LiCN (Borondo and Benito 1995) using the same definition introduced in the previous subsection,

$$\mathcal{H}_{QSOS}(\psi, P_\psi) = \mathcal{H}(\rho = 0, \psi, P_\rho = P_\rho(E), P_\psi), \tag{7.20}$$

calculated by numerical integration of (7.16) using the eigenfunctions obtained with the DVR program of Bačić and Light. For our nonrigid system, the bend and stretch frequencies are quite different, and there is no straightforward way to define a set of coherent states with which to calculate the Husimi function. In the absence of better alternatives, we choose the use of isotropic harmonic oscillator coherent states, with width parameters corresponding to the geometric mean of that for the Li–NC bend and the LiNC/LiCN stretch modes. Another problem that arises when calculating $\mathcal{H}$ for this molecular system is that the wave functions are only defined on a half-plane, which is the volume element in (7.8), whereas the 2D coherent states are defined on the full (x, y) plane. It is therefore necessary to embed the 2D wave functions into a 3D cartesian space, and calculate overlaps between wave functions $\varphi(R, \theta, \phi)$ (independent of angle ϕ) and 3D harmonic oscillator coherent states centered on the (R, θ) plane with zero momentum out of this plane. Moreover, when comparing $J = 0$ classical and quantum mechanics, we have the problem that the classical Hamiltonian is planar, while the quantum is not. Different approximate approaches have been described in the literature for 1D problems (Langer 1937; Adams and Miller 1977) which include an "artificial" barrier into the problem. Extension to multidimensional systems has also been discussed (Watson 1989). In this chapter we will simply fold the SOS into the interval $0 \le \psi \le \pi$ taking into account the invariance under the transformation

$$\psi \rightarrow 2\pi - \psi,$$

$$P_\psi \rightarrow -P_\psi. \tag{7.21}$$

A great deal of attention has been paid in the literature to the study of the maxima of quasi-probability densities like $\mathcal{H}_{QSOS}$ (e.g., see Arranz et al. 1998). Special care has been taken in relating these maxima with fixed points and other relevant classical structures of the corresponding SOS, specially in conditions of widespread chaos. Recently it has been proposed (Leboeuf and Voros 1990; and Leboeuf 1991) to use the distribution of the zeros of the Husimi function to assess the chaotic character of quantum states: In the case of regular states (Stratt 1979) the zeros of the Husimi function all lie on a line; while for the more stochastic states the zeros appear uniformly distributed over the whole available phase space. The original article is very mathematical, and the results were subject to strong restricting conditions. In later publications (Cibils et al. 1992; Arranz et al. 1996), it was numerically shown that these conditions can be relaxed. Moreover it was shown by us (Arranz et al. 1996) that the distribution of zeros is an efficient method to detect scarred states. In such cases some zeros separate from the distribution curve and locate on points whose positions are determined by the scarring PO.

7.2.4 Wave Packet Propagation and Spectra

Another very popular and convenient way to compare quantum dynamical results with classical trajectory calculations is through the use of wave packet propagation (Heller 1975), especially when it is done in a semiclassical fashion. According to the ideas of Ehrenfest (Ehrenfest 1927; Schiff 1968), the center of these nonstationary functions follows during a certain time classical paths, thus representing a natural way of establishing the quantum-classical correspondence (Heller 1991; Tomsovic and Heller 1991; Littlejohn 1986).

In our case the dynamics of wave packets can be calculated quite easily by projection of the initial function into the basis set formed by the stationary eigenstates of the system, followed by the application of the corresponding evolution operator

$$|\Psi(t)\rangle = e^{-i\hat{H}t/\hbar} |\Psi(0)\rangle = \sum_n |\varphi_n\rangle\langle\varphi_n|\Psi(0)\rangle \, e^{-iE_n t/\hbar}. \tag{7.22}$$

This evolution can be studied either in time domain, by following the recurrences of the correlation function:

$$C(t) = \langle\Psi(0)|\Psi(t)\rangle, \tag{7.23}$$

or in energy domain through the corresponding (infinite resolution) spectrum:

$$I(E) = \sum_n |\langle\Psi(0)|\varphi_n\rangle|^2 \, \delta(E - E_n), \tag{7.24}$$

where the coefficients are known as Franck-Condon factors. It is well known that these two quantities are related by Fourier transform,

$$I(E) = \frac{1}{2\pi\hbar} \int\limits_{-\infty}^{+\infty} dt \; e^{iEt/\hbar} \langle \Psi(0)|\Psi(t)\rangle. \tag{7.25}$$

This equation constitutes the basis of the time dependent formulation of spectroscopy (Heller 1981), which has been widely used in other fields such as in examining the properties of condensed matter (McQuarrie 1976).

As stated at the beginning of this subsection, wave packets follow for some time close to classical paths, so they can be used to explore phase space and reveal which features are important quantum mechanically. For example, if a packet is placed at the turning point of an unstable PO of the system, it will move initially away from the starting position causing recurrences on this time scale in the correlation function. This fall takes place in a characteristic time of T_1. At a later time, T_2, of the order of the period of the PO, the wave packet will come back to the neighborhood of the initial point, causing new recurrences on this new time scale. Finally the wave packet will disperse, and the maxima of the correlation function will progressively decrease, since it is the corresponding envelope characterized by a much longer time, T_3, associated to the Lyapunov exponent or instability parameter of the PO. A schematic illustration of this argument (Heller 1981) is presented in Figure 7.3. Moreover this dynamical argument has a translation in the energy domain where, due to the uncertainty principle, bigger energy ranges into the corresponding spectrum are associated to shorter time scales, and vice versa. Accordingly times T_1, T_2, and T_3 determine, respectively, the envelope, the spacing, and the peak widths in $I(E)$ (see Fig. 7.3).

This influence of the periodic motions of the system, or pieces of longer trajectories mimicking them, in the low resolution features of spectra can be revealed by using appropriate techniques of spectral analysis. See, for example, Taylor (1989), Johnson and Kinsey (1989), or Gomez-Llorente and Pollak (1992) and the references they contain for an interpretation of different molecular spectra of classically chaotic systems. Essentially the method consists in a convolution of the infinite resolution or stick spectrum, (7.24), with Lorentzian or Gaussian function of increasing width until a simple, regular pattern of broad bands can be identified. Each of these bands includes

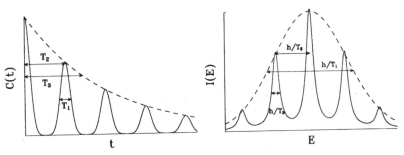

Figure 7.3 Schematic illustration of the relationship existing between the correlation function of a wave packet and its associated spectrum.

a clump of eigenstates, affected by different Franck-Condon factors, which coherently contribute to its shape.

The method can be used the other way around, that is, using classical mechanics to simulate the characteristics of quantum spectra. This can be accomplished through the survival function, $S_q(t)$, which is defined as the squared magnitude of the correlation function

$$S_q(t) = |C(t)|^2 = \text{Tr}\{\hat{\rho}(0)\hat{\rho}(t)\},$$

(7.26)

where $\hat{\rho}(t)$ is the Heisenberg density operator,

$$\hat{\rho} = e^{-i\hat{H}t/\hbar}|\Psi(0)\rangle\langle\Psi(0)|e^{i\hat{H}t/\hbar}.$$

(7.27)

A classical analogue of this function can be easily constructed by changing "Tr" by the integral over all positions and momenta and $\hat{\rho}$ by a suitable density distribution in phase space (e.g., Wigner or Husimi),

$$S_c(t) = \int \cdots \int d\mathbf{q}\, d\mathbf{P}\, \langle\rho(0)\rho(t)\rangle.$$

(7.28)

An example of this technique can be found in Borondo et al. (1994) and Iken et al. 1994 where an application to the hydrogen atom in strong parallel magnetic and electric fields was presented. The integral in (7.28) is usually calculated by a Monte Carlo method using a swarm of trajectories representative of the initial distribution. From the classical point of view, analyzing the magnitude of the different terms allows also to detect which classical trajectories or PO contribute more importantly to any given (low-resolution) structure that may appear in the spectra.

Moreover the wavefunctions associated with these features can also be calculated easily. Since they correspond to structures that are not well defined in energy, they must have some sort of a resonant character. This character can be unveiled by appropriate calculation techniques that come from resonance theory (Fano 1961; Moiseyev 1998). For example, one can obtain the corresponding wave functions by Fourier transforming the time dependent wave packet $|\Psi(t)\rangle$ using a finite time span

$$|\Psi^k\rangle = \frac{1}{2\pi\hbar}\int_{-\tau}^{\tau} dt\, |\Psi(t)\rangle\, e^{iE_k t/\hbar},$$

(7.29)

where E_k represents the energy of the center of the kth band, and τ is of the order of the period of the scarring PO. In the way that we calculate $|\Psi(t)\rangle$, (7.22), this expression reduces to

$$|\Psi^k\rangle = \sum_n |\varphi_n\rangle\langle\varphi_n|\Psi(0)\rangle \frac{\sin[(E_k - E_n)\tau/\hbar]}{\pi(E_k - E_n)}$$

(7.30)

which can be approximated to

$$|\Psi^k\rangle = \sum_n' |\varphi_n\rangle \langle\varphi_n|\Psi(0)\rangle, \tag{7.31}$$

where the prime indicates that the sum extend only to the states under the kth band. Equation (7.31) clearly indicates the projection character implied in the procedure used to calculate $|\Psi^k\rangle$ (Gomez-Llorente et al. 1992). Moreover this method provides information about the localization or scarring in the probability density, either in phase or configuration space, produced by POs (De Polavieja et al. 1994).

7.3 RESULTS

In this section some examples of the method described above are presented. Gaussian wave packets are launched from different initial positions in the chaotic region of phase space of LiNC/LiCN so that they can sample different aspects of the vibrational dynamics of this molecular system. As was mentioned earlier, the advantage of this method is twofold. For one thing, wave packets explore phase space (for some time) in a fashion similar to classical trajectories, but on the other hand, they have superimposed to this dynamical behavior the wavelike character and interference effects inherent to the quantum description.

As a typical energy we have chosen $E = 6100$ cm^{-1}, which is close to that of the 188th quantum state. This energy is well above the isomerization barrier, so that LiNC/LiCN can exhibit most of its dynamical complexity. The wave packets that will be considered in this work are shown in Figure 7.4 as contours plots representing the initial shape of the function. Some relevant POs (see Section 7.2) are included in the figure.

To give an idea of the structure of the phase space that the wave packets will explore, we present in Figure 7.5 a classical composite Poincaré SOS for our system

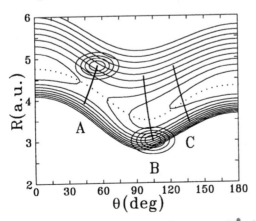

Figure 7.4 Gaussian wave packets used to generate the spectra of Section 7.3, represented over the contours of the potential energy surface. Some periodic orbits relevant for the discussion presented in the text are also included.

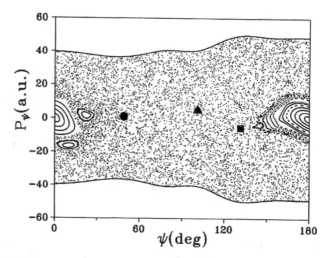

Figure 7.5 Classical composite Poincaré surface of section for the LiNC/LiCN system at a vibrational energy of 6100 cm^{-1}. It includes the results for 50 trajectories that follow each one until 300 crossings with the minimum energy path take place. The points $(\psi, P_\psi) = (0°, 0)$ and $(180°, 0)$ represent, respectively, the fixed points corresponding to the Li–CN and Li–NC stretching motions, and are surrounded by (relatively) large islands of regularity. The fixed points corresponding to periodic orbits A, B, and C of Figure 7.4 have also been represented.

at the vibrational energy under study. The motions around the two isomer wells appear as big islands of regularity centered at the points $(\psi, P_\psi) = (0°, 0)$ and $(180°, 0)$, respectively. The stretching motion corresponding to the Li–CN is more regular, and this region of phase space appears dominated by a prominent 1 : 4 resonant chain of islands. On the other side, in the Li–NC well, larger portions of regular motion have been destroyed, and several chains of islands, with their corresponding bands of stochasticity, appear. The rest of phase space is filled with a chaotic sea corresponding to irregular isomerizing trajectories connecting the two isomers. One interesting area exists inside this chaotic region, which is the result of the evolution of the two main POs originated in the saddle-node bifurcation which are described in detail in a later subsection. These orbits are B and C of Figure 7.4, and their corresponding fixed points, together with that of A, are also plotted.

7.3.1 Transition States

Saddle points of potential energy surfaces are the origin of many interesting phenomena in atomic and molecular systems. Principal among them are saddle-point ionization in atomic collisions (Olson et al. 1987; Rost et al. 1989) and separatrix crossing in time dependent Hamiltonians (e.g., see Hänggi et al. 1990). Transition states are of special interest for chemical reactivity (Levine and Bernstein 1987; Wyatt and Zhang 1996). They separate different reaction channels for example in bimolecular reactions (Miller 1995), or in unimolecular reactions provide energy barriers for the inverse reaction (Uzer 1991; Benito and Santamaria 1988). Associated with such regions are the so-called Siegert complex eigenvalues (Siegert 1939) which are known to be related to the scattering resonances observed in the cross sections of many chemical reactions. Reaction rates are also connected to these saddle resonant states (Seideman 1991). These transition states have been observed experimentally (Arrowsmith et al. 1983; Weaver et al. 1990; Scherer et al. 1990).

In our system the saddle point existing in the potential surface at $\theta = 48.41°$ separates the two nonequivalent isomer wells, thus constituting the bottleneck for the corresponding isomerization conversion. According to Moser's theorem (Moser 1976) a single family of unstable POs is generated from this point. This orbit (marked with an A in Fig. 7.4) runs approximately perpendicular to the minimum energy path, representing the semiclassical equivalent to the transition state, and leaving an unstable hyperbolic fixed point at the SOS (full circle in Fig. 7.5). The manifolds emanating from it separate the librational motions of species LiNC and LiCN from the rotational isomerizing trajectories (Lichtenberg and Lieberman 1992), and the homoclinic tangle (Wiggins 1990) formed by their crossings organizes the chaotic region around it in a broken separatrix. The transport across it determines the interconversion between the different isomers.

To study the dynamics in this region, we propagate a Gaussian wave packet initially located approximately at the outer turning point of the PO labeled with an A in Figure 7.4 at $(R, \theta)_0 = (4.830 \text{ a.u.}, 55.41°)$. The resulting stick spectrum is presented in Figure 7.6. When examined, it looks quite complicated, showing all characteristics of chaotic spectra: irregular distribution of intensities, chaotic spacing between lines, and so forth (Brody et al. 1981). However, dynamical information can still be extracted by calculating the survival probability, $S(t)$, corresponding to this spectrum and its classical analog, by means of equations (7.26) and (7.28). There were 1200 trajectories required to converge the classical survival probability function. The

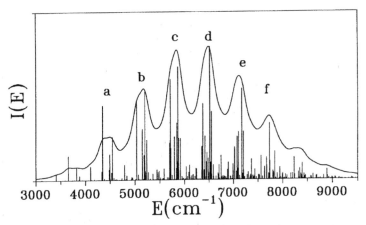

Figure 7.6 Stick spectrum and low-resolution version of it for a Gaussian wave packet initially centered on the outer turning point of the transition state periodic orbit A of Figure 7.4 in the LiNC/LiCN system. The low-resolution spectrum consists of a series of bands equally spaced.

results are shown in Figure 7.7. Up to a time of 150 fs the agreement between quantum and classical results is quite satisfactory, which entitles us to use classical mechanics to explain the dynamics of the wave packet. At longer times the classical function decays faster than its quantum analogue, which also shows additional structure. The relatively fast decay of the classical results has its origin in the chaotic nature of the majority of the trajectories involved in the calculation of $S_c(t)$. More dynamical information can be obtained in frequency domain, by calculating the spectral densities, $D_q(\omega)$ and $D_c(\omega)$, by Fourier transforming the corresponding survival probabilities. These spectral densities are presented in the right part of Figure 7.7 after convolution with a Lorentzian function, whose width was increased until classical-quantum correspondence appeared. This is done with the purpose of smoothing out structure coming from longer times, in which we are not interested in the present work. Both spectral densities can be explained in terms of two fundamental frequencies, approximately equal to $\omega_1 = 650 \text{ cm}^{-1}$ and $\omega_2 = 160 \text{ cm}^{-1}$. The origin of these peaks can now be elucidated by examination of the classical trajectories contributing most to each one of them. The biggest frequency, ω_1, corresponds roughly to a Li–CN stretch motion at the fixed θ angle, which is associated with an unstable PO at the saddle or transition state. This picture is typical of resonance spectra, the width of the bands, centered at multiples of ω_1, are related to the corresponding lifetimes. The resonances correspond to different excitations of the transition state for the isomerization reaction, with the bending motion playing the role of a quasi-continuum. Moreover the bands show a triplet structure with peaks separated by ω_2; this is an indication that the decay mechanism is complex. Several families of unstable POs exist, connecting the transition state region with the LiNC

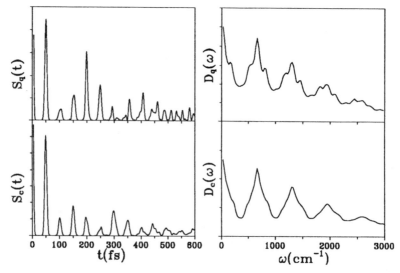

Figure 7.7 Quantal and classical survival probability functions, S_q and S_c, and spectral densities, $D_q(\omega)$ and $D_c(\omega)$, corresponding to the stick spectrum presented in Figure 7.6.

and LiCN wells, and even some of them are of rotating (isomerizing) type. Which one appears in the spectrum depends very sensitively on the precise initial position of the packet, but this is not particularly relevant for the discussion presented here. For example, for energies close to the isomerization barrier (Gomez-Llorente 1992), the result can change completely if we move the initial packet from a position closer to a well or to the other. In any case the mechanism for the energy relaxation takes place along the following lines: Energy put into the stretch mode of the transition state is first transferred to the motion along the isomerization coordinate through coupling to the bending mode, and then uniformly distributed from here to all over the chaotic phase space. The behavior of the system is far from statistical.

Finally let us discuss the consequences that these results have on the complicated stick spectrum of Figure 7.6. Coherent with the resonance picture, when a low-resolution version of it is considered, more regular pattern emerges. The result obtained by convoluting the stick spectrum with a Lorentzian of width 180 cm^{-1} is presented as a continuous line superimposed in Figure 7.6. The smoothed spectrum shows nine equally spaced bands (six are labeled) of approximately the same width. Moreover our hypothesis of the resonance origin of the spectrum can be further checked by calculating the wavefunctions associated to each of these bands as in expression (7.31). The results are plotted in Figure 7.8. The resonance functions correspond to the transition state, with increasing excitation in a mode (approximately R motion) and no excitation in the other (approximately θ motion). Each stationary eigenstate contributes to the resonance functions with a coefficient $\langle \Psi(0)|\varphi_n \rangle$ so that the coherent combination of them eliminates contributions from highly delocalized states. This point is clearly illustrated in Figure 7.9 where we have plotted the four most important components to the wavefunction of band c in the spectrum presented in Fig. 7.6; all of them appear very delocalized over the whole range of the angular coordinate.

7.3.2 Saddle-Node States

Saddle-node or tangent bifurcations are the origin in 2D systems of pairs of POs, one stable and one unstable, that appear "out of nowhere" from no previous PO. The corresponding fixed point in the SOS is called parabolic, and it is characterized for having the two eigenvalues of the monodromy matrix equal to one. As energy increases, and we move away from the bifurcation point, the two newly created POs separate in the phase space. This type of bifurcation is important for molecular systems, since they are the origin of stability islands inside chaotic regions. Moreover, as will be described below, they are able to produce important quantum manifestations (Borondo et al. 1995, 1996; Zembekov et al. 1997). Also massive tangent bifurcations can produce marginally stable sets (Zembekov 1994), which result in regular recurrences that are readily observable in molecular spectra.

The saddle-node bifurcation that will be considered here occurs at $E_{SN} = 3440.64$ cm^{-1} (approximately the 50th quantum state), which is the parabolic point located at $(\psi, P_\psi)_{SN} = (113.24°, 0.057617$ a.u.$)$. The associated POs correspond to a 1 : 1 resonant motion, mainly along the R coordinate. They were represented in Figure 7.4

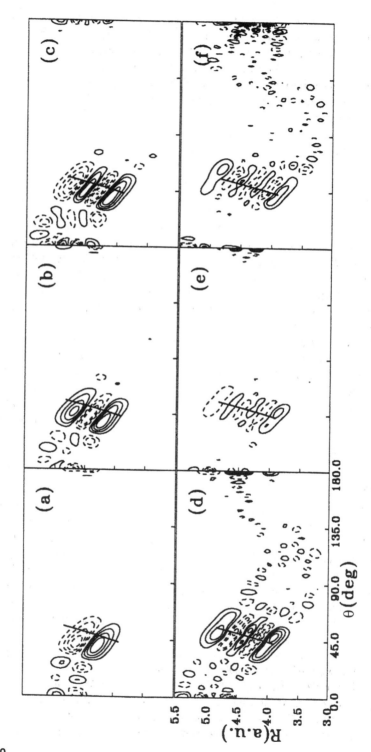

Figure 7.8 Wavefunctions associated to the different bands appearing in the smoothed spectrum of Figure 7.6. They correspond to the different excitation of the transition state along a mode approximately equal to the R motion.

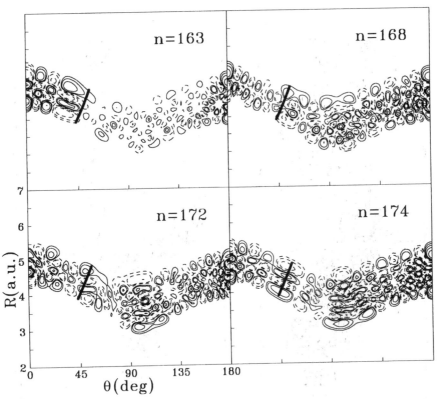

Figure 7.9 The four most important contributions to band *c* wave function of Figure 7.8. None of them shows a specially big degree of localization.

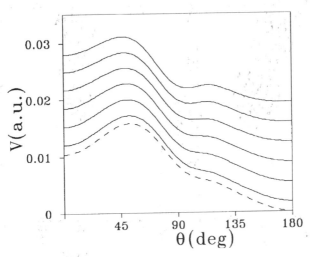

Figure 7.10 Adiabatic potential energy curves for the bending motion of LiNC/LiCN. The minimum energy profile is also shown in dashed line.

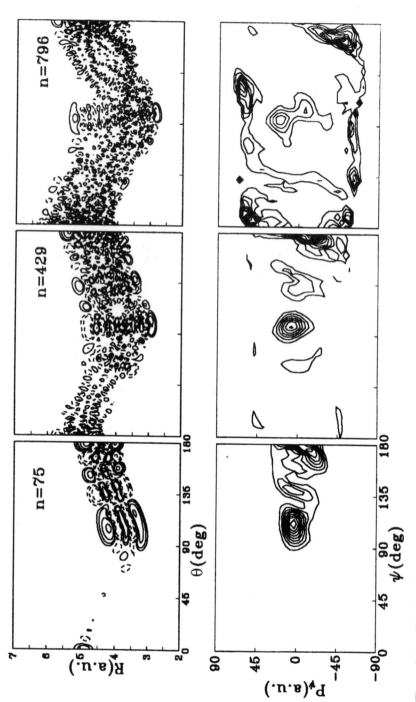

Figure 7.11 Wavefunctions (upper part) and QSOS (lower part) of some vibrational states of LiNC/LiCN scarred by saddle-node POs above the bifurcation energy.

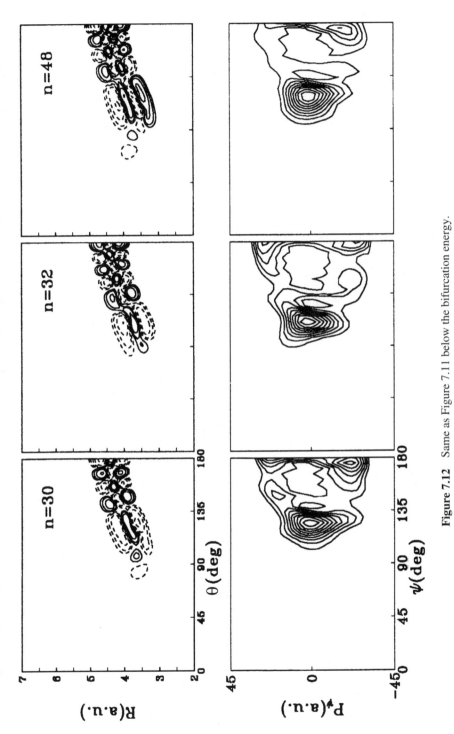

Figure 7.12 Same as Figure 7.11 below the bifurcation energy.

213

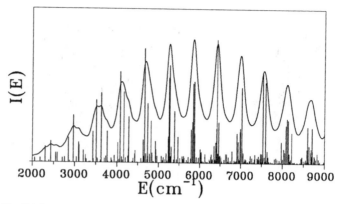

Figure 7.13   Stick spectrum and low-resolution version of it for a Gaussian wave packet initially centered on the inner turning point of the saddle-node periodic orbit *B* of Figure 7.4 in the LiNC/LiCN system. The low-resolution spectrum consists of 12 bands equally spaced.

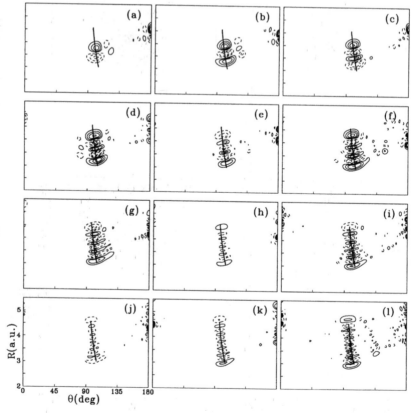

Figure 7.14   Wavefunctions associated to the different bands appearing in the smoothed spectrum of Figure 7.13. They correspond to the different excitation along the *R* coordinate in the saddle-node region.

(orbits B and C); the associated fixed points have also been marked (full triangle and square) in Figure 7.5. It is precisely this excitation in R that is the origin of the bifurcation. Since the time scales of the R and θ motions are very different, an adiabatic effective potential along the angular coordinate can be constructed by averaging the fast radial coordinate

$$V_{\text{eff}}(\theta) = V_{\text{MEP}}(\theta) + \hbar\omega(\theta) \left(n_R + \frac{1}{2}\right), \tag{7.32}$$

where V_{MEP} is the potential along the minimum energy path, and the motion along the perpendicular direction has been fitted at every value of the angle θ to a harmonic oscillator. In Figure 7.10 we have represented V_{eff} for various values of the quantum number. It can be observed that for $n_R > 2$, a small potential well appears; this explains the appearance of the tangent bifurcation: The two new singular points turn into a pair of POs when two dimensions are considered.

The local validity of this adiabatic separation (Hose and Taylor 1983) has a profound influence in the eigenfunctions of LiNC/LiCN which are highly excited in R. When the 900 calculated functions are examined, many are found that are scarred by the stable, the unstable, or both POs originated at the bifurcation. Some representative examples are shown in Figure 7.11. The accumulation of quantum density along the POs is obvious, and it is even more clear in the phase space pictures (QSOS) that are shown in the lower part of the figure. Notice that this effect is quite strong, and persists at high energies where many more bifurcations take place.

More interestingly, this localization also happens *below* the bifurcation point, as is clearly seen in the plots of Figure 7.12. This constitutes a manifestation in the configuration and phase space of "ghost" orbits (non-real trajectories with imaginary momentum), whose influence in the spectra has been discussed (Kuś et al. 1993). The effect can be understood in an alternative way by using real orbits. In finite resolution spectra only short time dynamics are revealed. Thus it is not surprising that pieces of longer trajectories or short closed orbits, which mimic the behavior of the prebifurcated PO, can give a significant contribution to the spectrum. Below the bifurcation there is a dense set of such *precursor* orbits from which the bifurcated one can be generated by pruning. This has been explained in detail in a previous work (Borondo et al. 1996), and an application to the semiclassical quantization of saddle-node states of LiNC/LiCN embedded in a chaotic region was later presented (Zembekov et al. 1997) in which the Einstein-Brillouin-Keller condition was applied to surrogates of invariant curves extracted from the islands around islands structure originated around the saddle-node POs.

The localization effect induced by the saddle-node POs and their surroundings can be further explored by propagating the wave packet B of Figure 7.4 corresponding to the initial conditions $(R, \theta)_0 = (3.009 \text{ a.u.}, 104.38°)$. The corresponding stick and smoothed spectra are presented in Figure 7.13. Again the lines of the stick spectrum are chaotic but appear grouped in clumps. This is more obvious in the low-resolution version, which is obtained by convolution with Lorentzian functions 250 cm^{-1} wide and consists of a series of equally spaced bands. The separation between bands (~ 560 cm^{-1}) coincides approximately with the saddle-node POs frequency in this range of energies, and it is in reasonable agreement with the Li–NC harmonic frequency of 785 cm^{-1} (Losada et al. 1998). The wavefunctions, calculated using (7.31), are presented in Figure 7.14. Each corresponds to a consecutive higher excitation in the R coordinate.

7.4 CONCLUDING REMARKS

In this chapter we have presented a study of the correspondence between classical and quantum mechanics applied to the vibrational dynamics of the floppy LiNC/LiCN molecular system.

The system is described by a realistic Hamiltonian model of just two degrees of freedom, with an *ab initio* interaction potential presenting two wells and the possibility of isomerization. The anharmonicity in the bending motion is high, so the Li atom can explore ample portions of the potential energy surface, and chaos sets in at (relatively) low excitation energies. From the classical point of view, this model constitutes a prototype of a generic Hamiltonian system exhibiting soft chaos. The phase space shows a mixture of embedded regions of regularity and chaos, as dictated by the Kolmogorov-Arnold-Moser theorem (Lichtenberg and Lieberman 1992), where the dynamics in the Li–CN well are more regular than in the other well. At energies above the isomerization barrier the regular isomer regions are surrounded by a chaotic sea of isomerizing trajectories with dynamical bottlenecks inside.

Among all possible regions with interesting dynamics, we have focused in this work on two of them. The first one is the saddle point of the potential, which is related to the transition state for the corresponding isomerization reaction. As a second example we have considered the region around $\theta \simeq 113°$ in which the motion along the R coordinate gets stabilized by an adiabatic separation of frequencies, originating a saddle-node bifurcation that has a profound influence in the quantum dynamics of the system.

Following an approach strongly relying on Heller's work, we have studied the dynamics in these areas by propagating wave packets for short periods of time, and Fourier transforming them. The results in the form of low-resolution spectra and associated band wavefunctions reveal which periodic motions on a certain phase space region have strong influence in the dynamics of the system from a quantum mechanical point of view. The scarring effect of the POs is clearly revealed in the band wavefunctions, which in turn show its distribution in the different eigenstates. Further analysis of the spectral densities provides additional information on the time scales and mechanisms of divergence away from the unstable PO paths and/or interactions with other coexisting orbits.

The formation of scars has been recently studied by our group (Arranz et al. 1998) at its most elementary level. Instead of doing it, as usual, in the high state density regime (to attain the semiclassical limit), we have constructed a correlation energy level diagram by changing the value of the constant $\hbar$. This allows us to go to the edge of the chaotic region, which appears marked by a series of broad avoided crossings, where the scars are formed as the result of mixing of pairs of interacting wavefunctions.

Further developments will include the analysis of intraband structure, band shoulders, and so forth. We are especially interested in the decaying mechanisms of localized wavefunctions, and its variations with energy and associated phase space evolution, such as due to the appearance of new bifurcations or the destruction of invariant curves.

ACKNOWLEDGMENTS

This work has been supported in part by DGES (Spain) under Projects PB95-425 and PB96-76.

REFERENCES

Adams, J. E., and Miller, W. H. (1977). *J. Chem. Phys.* **67**, 5775.

Arranz, F. J., Borondo, F., and Benito, R. M. (1996). *Phys. Rev.* E**54**, 2458.

Arranz, F. J., Borondo, F., and Benito R. M. (1998). *J. Mol. Struct.* **426**, 87.

Arrowsmith, P., Bly, S. H. P., Charters, P. E. and Polanyi, J. C. (1983). *J. Chem. Phys.* **79**, 283.

Bačić, Z., and Light, J. C. (1986). *J. Chem. Phys.* **85**, 4594.

Bačić, Z., and Light, J. C. (1989). *An. Rev. Phys. Chem.* **40**, 469.

Benito, R. M., Borondo, F., Kim, J.-H., Sumpter, B. G., and Ezra, G. S. (1989). *Chem. Phys. Lett.* **161**, 60.

Benito, R. M., and Santamaria, J. (1988). *J. Phys. Chem.* **92**, 5028.

Berry, M. V. (1977). *Phil. Trans. R. Soc.* **287**, 237.

Borondo, F., and Benito, R. M. (1995). In *Frontiers of Chemical Dynamics*, ed. E. Yurtsever. NATO ASI Series C. Kluwer Academic, Dordrecht, p. 371.

Borondo, F., Benito, R. M., Iken, M. A., and Uzer, T. (1994). *An. Fis.* **90**, 230.

Borondo, F., Zembekov, A. A., and Benito, R. M. (1995). *Chem. Phys. Lett.* **246**, 421.

Borondo, F., Zembekov, A. A., and Benito, R. M. (1996). *J. Chem. Phys.* **105**, 5068.

Brody, T. A., Flores, J., French, J. B., Mello, P. A., Pandey, A., and Wong, S. S. M. (1981). *Rev. Mod. Phys.* **53**, 385.

Cartwright, N. D. (1976). *Physica* A**83**, 210.

Child, M. S. (1974). *Molecular Collision Theory*. Academic Press, London.

Cibils, M. B., Cuche, Y., Leboeuf, P., and Wreszinki, W. F. (1992). *Phys. Rev.* A**46**, 4560.

Cohen-Tannoudji, C., Din, B., and Laloë, F. (1977). *Quantum Mechanics*. Wiley, New York.

De Polavieja, G. G., Borondo, F., and Benito, R. M. (1994). *Phys. Rev. Lett.* **73**, 1613.

Dickinson, A. S., and Certain, P. R. (1968). *J. Chem. Phys.* **49**, 4209.

Eckhardt, B. (1988). *Phys. Rep.* **163**, 205.

Ehrenfest, P. (1927). *Z. Physik* **45**, 455.

Essers, R., Tennyson, J., and Wormer, P. E. S. (1982). *Chem. Phys. Lett.* **89**, 223.

Fano, U. (1961). *Phys. Rev.* **124**, 1866.

Feit, M. D., and Fleck, J. A. Jr. (1984). *J. Chem. Phys.* **80**, 2578.

Gaeta, Z. D., Noel, M. W., and Stroud, C. R. Jr. (1994). *Phys. Rev. Lett.* **73**, 636.

Gear, C. W. (1964). *SIAM J. Num. Anal.* B2, 69.

Glauber, R. J. (1963). *Phys. Rev. Lett.* **10**, 84.

Goldstein, H. (1980). *Classical Mechanics.* Addison-Wesley, Reading, MA.

Gomez-Llorente, J. M., Berenger, N., Borondo, F., and Benito, R. M. (1992). *Chem. Phys. Lett.* **192**, 430.

Gomez-Llorente, J. M., and Pollak, E. (1992). *An. Rev. Phys. Chem.* **43**, 91.

Gutzwiller, M. C. (1990). *Chaos in Classical and Quantum Mechanics.* Springer, New York.

Hamilton, I. P., and Light, J. C. (1986). *J. Chem. Phys.* **84**, 306.

Hänggi, P., Talkner, P., and Borkovec, M. (1990). *Rev. Mod. Phys.* **62**, 251.

Harriman, J. E. (1988). *J. Chem. Phys.* **88**, 6399.

Heather, R., and Metiu, H. (1987). *J. Chem. Phys.* **86**, 5009.

Heller, E. J. (1975). *J. Chem. Phys.* **62**, 1544.

Heller, E. J. (1981). *Acc. Chem. Res.* **14**, 368.

Heller, E. J. (1984). *Phys. Rev. Lett.* **53**, 1515.

Heller, E. J. (1991). In *Chaos and Quantum Physics*, eds. M. J. Gianonni, A. Voros, and J. Zinn-Justin. NATO Les Houches Lecture Notes. Elsevier Science, Amsterdam.

Hillery, M., O'Connell, R. F., Scully, M. O., and Wigner, E. P. (1984). *Phys. Rep.* **106**, 121.

Hose, G., and Taylor, H. S. (1983). *Phys. Rev. Lett.* **51**, 947.

Husimi, K. (1940). *Proc. Phys. Math. Soc. Jpn.* **22**, 264.

Iken, M. A., Borondo, F., Benito, R. M., and Uzer, T. (1994). *Phys. Rev.* A**49**, 2734.

Ismail, Z. K., Hauge, R. K., and Margrave, J. L. (1972). *J. Chem. Phys.* **57**, 5137.

Johnson, B., and Kinsey, J. L. (1989). *Phys. Rev. Lett.* **62**, 1607.

Kulander, K. C., ed. (1991). *Comp. Phys. Comm.* **63** (thematic issue).

Kuś, M., Haake, F., and Delande, D. (1993). *Phys. Rev. Lett.* **71**, 2167.

Langer, R. E. (1937). *Phys. Rev.* **51**, 669.

Leboeuf, P. (1991). *J. Phys.* A**24**, 4575.

Leboeuf, P., and Voros, A. (1990). *J. Phys.* A**23**, 1765.

Leforestier, C., Bisseling, R., Cerjan, C., Feit, M. D., Friesner, R., Guldberg, A., Hammerich, A., Jolicard, G., Karrlein, W., Meyer, H. D., Lipkin, N., Roncero, O., and Kossloff, R. (1991). *J. Comp. Phys.* **94**, 59.

Levine, R. D., and Bernstein, R. B. (1987). *Molecular Reaction Dynamics and Chemical Reactivity.* Oxford University Press, New York.

Lichtenberg, A. J., and Lieberman, M. A. (1992). *Regular and Chaotic Dynamics*, 2nd ed. Springer, Berlin.

Littlejohn, R. G. (1986). *Phys. Rep.* **138**, 193.

Losada, J. C., Estebaranz, J. M., Benito, R. M., and Borondo, F. (1998). *J. Chem. Phys.* **108**, 63.

Makarewitz, J., and Ha, T. (1994). *J. Mol. Struct.* **315**, 149.

Makarewitz, J., and Ha, T. (1995). *Chem. Phys. Lett.* **232**, 497.

Margenau, H., and Hill, R. N. (1961). *Prog. Theor. Phys. (Kyoto)* **26**, 722.

McQuarrie, D. A. (1976). *Statistical Mechanics*. Harper and Row, New York.

Miller, W. H. (1995). In *Frontiers of Chemical Dynamics*, ed. E. Yurtsever. NATO ASI Series C. Kluwer Academic, Dordrecht, p. 1.

Moiseyev, N. (1998). *Phys. Rep.* **302**, 211.

Moser, J. (1976). *Commun. Pure Appl. Math.* **29**, 727.

Moyal, J. E. (1949). *Proc. Camb. Philos. Soc.* **45**, 99.

Lin, F. J., and Muckerman, J. T. (1991). *Comp. Phys. Comm.* **63**, 538.

O'Connell, R. F. (1983). *Found. Phys.* **13**, 83.

Olson, R. E., Gay, T. J., Berry, H. G., Hale, E. B., and Irby, V. D. (1987). *Phys. Rev. Lett.* **59**, 36.

Ozorio de Almeida, A. M. (1998). *Phys. Rep.* **295**, 265.

Ozorio de Almeida, A. M., and Hannay, J. H. (1982). *An. Phys. (NY)* **138**, 115.

Reif, F. (1965). *Fundamentals of Statistical and Thermal Physics*. McGraw-Hill, New York.

Rost, J. M., Briggs, J. S., Greeland, P. T. (1989). *J. Phys.* B**22**, L353.

Scherer, N. F., Sipes, C., Bernstein, R. B., and Zewail, A. H. (1990). *J. Chem. Phys.* **92**, 5239.

Schiff, L. I. (1968). *Quantum Mechanics*. McGraw-Hill, New York.

Schinke, R. (1993). *Photodissociation Dynamics*. Cambridge University Press, Cambridge.

Seideman, T., and Miller, W. H. (1991). *J. Chem. Phys.* **95**, 1768.

Sepúlveda, M. A., and Grossmann, F. (1996). *Adv. Chem. Phys.* **96**, 191.

Siegert, A. J. F. (1939). *Phys. Rev.* **56**, 750.

Stratt, R. M., Handy, N. C., and Miller, W. H. (1979). *J. Chem. Phys.* **71**, 3311.

Sutcliffe, B. T., Miller, S., and Tennyson, J. (1988). *Comp. Phys. Comm.* **51**, 73.

Takahashi, K. (1989). *Prog. Theor. Phys. (suppl.)* **98**, 109.

Taylor, H. S. (1989). *Acc. Chem. Res.* **22**, 263.

Tomsovic, S., and Heller, E. J. (1991). *Phys. Rev. Lett.* **67**, 664.

Uzer, T. (1991). *Phys. Rep.* **199**, 75.

Van Vaals, J. J., Leo Meerts, W., and Dymanus, A. (1983). *Chem. Phys.* **82**, 385.

Watson, J. K. G. (1989). *J. Chem. Phys.* **90**, 6443.

Weaver, A., Metz, R. B., Bratforth, S. E., and Neumark, D. M. (1990). *J. Chem. Phys.* **93**, 5352.

Wiggins, S. (1990). *Introduction to Applied Non Linear Dynamical Systems and Chaos*. Springer, New York.

Wigner, E. (1932). *Phys. Rev.* **40**, 749.

Wormer, P. E. S. (1981). *J. Chem. Phys.* **75**, 1245.

Wyatt, R. E., and Zhang, J. Z. H., eds. (1996). *Dynamics of Molecules and Chemical Reactions*. Dekker, New York.

Yeazell, J. A., Mallalieu, M., and Stroud, C. R. Jr. (1990). *Phys. Rev. Lett.* **64**, 2007.

Yeazell, J. A., and Stroud, C. R. Jr. (1988). *Phys. Rev. Lett.* **60**, 1494.

Zembekov, A. A. (1994). *J. Chem. Phys.* **101**, 8842.

Zembekov, A. A., Borondo, F., and Benito, R. M. (1997). *J. Chem. Phys.* **107**, 7934.

Zewail, A., and Bernstein, R. (1992). In *The Chemical Bond: Structure and Reactivity*, ed. A. Zewail. Academic Press, San Diego, CA, p. 223.

Zhang, W.-M., Feng, D. H., and Gilmore, R. (1990). *Rev. Mod. Phys.* **62**, 867.

Zhang, D. H., and Light, J. C. (1996). *J. Chem. Phys.* **104**, 6184.

CHAPTER 8

Wave Packet Dynamics in Small Molecules

MICHAEL BRAUN, HOLGER DIETZ, STEFAN MEYER, OLIVER RUBNER, and VOLKER ENGEL

8.1 INTRODUCTION

This chapter treats the dynamics of molecules—or *Mollycules*—as Mr. O'Brien tells us in *The Dalkey Archive*:

> "Did you ever study the Mollycule Theory when you were a lad?" he asked. Mick said no, not in any detail. "That is a very serious defalcation and an abstruse exacerbation," he said severely, "but I'll tell you the size of it. Everything is composed of small mollycules of itself and they are flying around in concentric circles and arcs and segments and innumerable various other routes too numerous to mention collectively, never standing still or resting but spinning away and darting hither and thither and back again, all the time on the go. Do you follow me intelligently? Mollycules?"

Classically we could imagine these particles as very alive, moving and bouncing *as lively as twenty punky leprechauns doing a jig on the top of a flat tombstone*. This highlights the dynamics of molecular systems which is the central topic in what follows.

We learn about molecules using the laws of classical mechanics: Atoms that are bound together in a compound perform oscillatory motions around their equilibrium positions; the molecule as a whole moves through space, rotates, and changes orientation with respect to a fixed spatial coordinate system.

The dynamic behavior of systems described quantum mechanically is fixed by the time dependent Schrödinger equation. The time evolution of a state that fulfills the time independent Schrödinger equation is given by a phase factor which, for the majority of applications, can be ignored. In general, in the equations that describe the properties of molecules (or other quantum systems) the time variable has vanished altogether. This may seem strange to the human mind which is accustomed to follow events in time, but it is not so strange in quantum mechanics. Let us take a look at a specific example.

The Physics and Chemistry of Wave Packets, Edited by John Yeazell and Turgay Uzer
ISBN 0-471-24684-0 © 2000 John Wiley & Sons, Inc.

In a spectroscopic experiment, an electromagnetic field interacts with a sample of molecules thus inducing a transition from an initial to a final state. The transition rate is, in most cases, sufficiently described by Fermi's golden rule formula, which relates the rate to the matrix element of a transition operator, taken with respect to the initial and final state wavefunction (Loudon 1983). The resulting expression is time independent, and one might wonder where the temporal variation of the field and the dynamics of the transition is hidden. The golden rule formula is derived from time dependent perturbation theory. A closer look at the derivation shows that time vanishes, since the limit $t \rightarrow \infty$ is taken. This implies that the interaction time between the field and the molecule is in fact much longer than any time scale that characterizes the dynamics of the molecular system. Some numbers are of help here. The system that is best characterized in traditional molecular spectroscopy is the I_2 molecule, and in particular, its transition from the electronic ground state $|X\rangle$ to the excited $|B\rangle$ state. The energy difference $\Delta E = 1.95$ eV between the states (the difference between the minima of the corresponding potential curves) may be converted to a time Δt by

$$\Delta t = \frac{h}{\Delta E}, \tag{8.1}$$

where h is Planck's constant. We then find $\Delta t \sim 2$ fs, which gives the period for electronic motion.

The energy difference between two vibrational states, say, those with quantum numbers $v' = 20, 21$ in the $|B\rangle$ state, is about 0.0116 eV. Using the relation (8.1), we find a time scale for vibrational motion that is about 350 fs. Now, regarding rotations in the B state for a rotational quantum number of $J = 60$ which corresponds to the maximum of the rotational intensity distribution at a temperature of $T = 373$ K, a rotational period of 4700 fs is calculated.

The numbers given above characterize the internal time scales for the atomic or electronic motion within a molecule. These time scales have to be compared to the interaction time between the electromagnetic field and the sample of molecules being investigated. Standard laser sources produce powerful pulses of several nanosecond lengths which thus are much longer than typical periods of electronic, vibrational, or rotational motion. This is of course a relative statement: Highly excited electronic (and likewise vibrational) states are very close in energy, so their corresponding time scales become longer. This is not the topic of this chapter, and relevant information can be found somewhere else in this book.

The fact that the molecule-field interaction time is much longer than the characteristic times for the internal molecular motion justifies the limit of time approaching infinity to derive the equations describing spectroscopic transitions. A different situation arises if pulses with a temporal width of about 100 fs interact with molecular systems. In this case the interaction time between field and system is comparable to (vibrational motion) or less than (rotational motion) the internal dynamics. As a consequence the notion of stationary quantum mechanical states has to be abandoned as will be discussed extensively in what follows. We will see below that the interaction of a molecular sample with a femtosecond laser pulse results in the preparation of a wave packet which is not a stationary state but a superposition of such states.

The structure of this chapter follows the structure of the experiment. First, we describe how molecular wave packets are prepared. After some discussion of nuclear dynamics, we show how the wave packet motion can be detected. Several experimental techniques are summarized as we discuss pump/probe experiments.

The discussion is restricted to systems under gas-phase conditions so that intermolecular interactions can be ignored. Experimentally these conditions are fulfilled in measurements on a sample with low density and, ideally, in a molecular beam (Ramsey 1986). This consideration will keep the discussion simple: It is sufficient to describe the experiments using the time dependent Schrödinger equation, so we do not need to use the more general formulation of density matrix theory (Blum 1996; Mukamel 1995).

8.2 PREPARATION OF MOLECULAR WAVE PACKETS

In this section we describe different experimental methods to prepare molecular wave packets. These approaches are illustrated in Figure 8.1.

8.2.1 Short-Pulse Electronic Excitation

Many gas-phase experiments with ultrashort laser pulses have been performed where wave packets in electronically excited states are prepared. Some examples that will be discussed below are I_2 (Gruebele and Zewail 1993), Na_2 (Baumert and Gerber 1995), and K_2 (Berry et al. 1997; Schwoerer et al. 1997). In these three cases the excitation takes place from the electronic ground state to one or more electronically excited

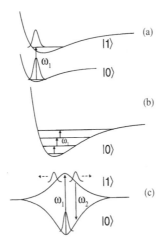

Figure 8.1 Preparation of vibrational wave packets. Several experimental approaches are illustrated, as discussed in the text.

states. In what follows we want to describe the preparation process using time dependent perturbation theory, namely we assume weak electromagnetic fields.

Theoretical Considerations Let us regard a molecule in the initial state $|\alpha\rangle|0\rangle$. Here $|\alpha\rangle$ is the state vector of the nuclei corresponding to the energy E_α and $|0\rangle$ designates the electronic state. The Born-Oppenheimer approximation is assumed so that nuclear and electronic degrees of freedom separate. The nuclear state in the excited electronic state $|1\rangle$, which is created through the interaction with an external electric field, is calculated within first-order perturbation theory as (atomic units are used throughout)

$$|\psi_1\rangle|1\rangle = \frac{1}{i} \int_{-\infty}^{t} dt'\, U_1(t-t')\, W_{10}(t')\, U_0(t')\, |\alpha\rangle|0\rangle. \tag{8.2}$$

Here U_0, U_1 are the propagators for the nuclear motion in the two electronic states and thus contain the respective Hamiltonians H_1 and H_0. Within the dipole approximation the interaction energy $W_{10}(t)$ is

$$W_{10}(t) = -|1\rangle\, \boldsymbol{\mu}_{10}\, \mathbf{E}(t)\, \langle 0|. \tag{8.3}$$

In this expression the transition dipole moment $\boldsymbol{\mu}_{10}$ appears together with the electric field

$$\mathbf{E}(t) = \frac{1}{2}\, \boldsymbol{\varepsilon}\, f_0 f(t)\, e^{-i\omega_1 t}, \tag{8.4}$$

where $\boldsymbol{\varepsilon}$ is the polarization vector of the field, ω_1 its frequency, f_0 the field strength and $f(t)$ an envelope function. In the equations above we included absorption processes only. Defining $\tilde{\mu}_{10} = -(\boldsymbol{\mu}_{10}\boldsymbol{\varepsilon})/2$ and projecting the state (2) on $\langle 1|$ yields

$$|\psi_1(t)\rangle = \frac{1}{i} \int_{-\infty}^{t} dt'\, U_1(t-t')\, f_0 f(t')\, \tilde{\mu}_{10}\, e^{-i\omega_1 t'}\, U_0(t')\, |\alpha\rangle. \tag{8.5}$$

Let us now expand this state in the set of nuclear eigenstates defined by

$$H_1|\beta\rangle = E_\beta|\beta\rangle. \tag{8.6}$$

This results in the expression

$$|\psi_1(t)\rangle = \frac{1}{i} \sum_\beta e^{-iE_\beta t}\, |\beta\rangle\, \langle\beta|\tilde{\mu}_{10}|\alpha\rangle \int_{-\infty}^{t} dt'\, f_0 f(t')\, e^{i(E_\beta - E_\alpha - \omega_1)t'}, \tag{8.7}$$

where we employed that $U_0(t)|\alpha\rangle = e^{-iE_\alpha t}|\alpha\rangle$ and the completeness of the expansion is assumed. To be more specific, we use a Gaussian envelope function

$$f(t) = e^{-at^2} \tag{8.8}$$

and regard a time t_f where $f(t)$ has almost decreased to zero so that the upper limit in the time-integral (8.7) can be extended to infinity. We then obtain

$$|\psi_1(t)\rangle = \frac{1}{i} \sum_\beta e^{-iE_\beta t} |\beta\rangle \, c_{\beta\alpha} \, F_{\beta\alpha}(\omega_1), \tag{8.9}$$

with the constants (which for $\tilde{\mu}_{10} = 1$ are the Franck-Condon factors)

$$c_{\beta\alpha} = \langle\beta|\tilde{\mu}_{10}|\alpha\rangle, \tag{8.10}$$

and the Fourier-transform

$$F_{\beta\alpha}(\omega_1) = f_0 \sqrt{\frac{\pi}{a}} \, e^{-(E_\beta - E_\alpha - \omega_1)^2/(4a)}. \tag{8.11}$$

For a very long Gaussian pulse, namely in the limit $a \to 0$, $F_{\beta\alpha}(\omega_1)$ becomes a representation of the δ-function:

$$\lim_{a \to 0} F_{\beta\alpha}(\omega_1) = c \, \delta(E_\beta - E_\alpha - \omega_1), \tag{8.12}$$

where c collects all constants that are of no interest in our perturbation. For resonant excitation to a final state β' we then obtain

$$|\psi_1(t)\rangle \sim e^{-iE_{\beta'}t}|\beta'\rangle \tag{8.13}$$

which is a stationary state.

If, on the other hand, a short pulse is used (large value of a), the spectral width $F_{\beta\alpha}(\omega_1)$ is broad so that several terms in the sum (8.9) are nonzero and the prepared state $|\psi_1(t)\rangle$ is a coherent superposition of stationary states $|\beta\rangle$, which is a wave packet.

We note that the initial nuclear state $|\alpha\rangle$ for a molecule is a ro-vibrational state so that pulse excitation in general results in a coherent superposition of rotational and vibrational states. The rotational degree of freedom is neglected in what follows, since rotational motion takes place on a much longer time scale.

Example: The $|B\rangle \leftarrow |X\rangle$ Transition in K_2

Experiments on the potassium dimer have been performed (Schwoerer et al. 1997) using laser pulses of approximately 100 fs widths and a wavelength of $\lambda = 625$ nm, which is resonant with the $|B\rangle \leftarrow |X\rangle$ electronic transition. To keep the notation, these electronic states will be denoted as $|0\rangle$ ($|X\rangle$) and $|1\rangle$ ($|B\rangle$). Figure 8.1a shows the typical excitation scheme for the kind of preparation process we describe here. The potential energy curves V_0, V_1 of the interaction energy between the nuclei as a function of the bond length R are shown schematically for the two electronic states. The initial nuclear wavefunction in $|0\rangle$ is sketched as the vibrational ground state and the arrow indicates

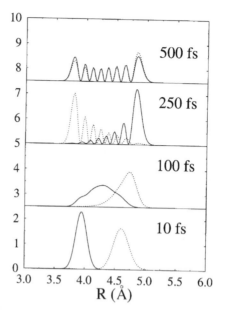

Figure 8.2   Excited state vibrational functions $\langle R|\psi_1(t)|$ in K_2, calculated for different pulse widths τ, as indicated. The functions are shown at the end of the molecule-field interaction $t = 2\tau$ (solid line) and after 200 fs of free propagation (dotted line).

the center wavelength of the pulse. Only bound states in $|1\rangle$ are accessed so that all of the equations (8.1) to (8.13) apply.

Let us calculate the wavefunction $\langle R|\psi_1(t)\rangle$ for different widths T of the Gaussian shape function (8.8). Figure 8.2 shows the modulus $|\langle R|\psi_1(t)\rangle|$ obtained for several pulse widths (full widths at half maximum, FWHM), as indicated. The functions are plotted as solid lines at times $t = T$, where the Gaussians are almost zero, so effectively the molecule-field interaction stopped. Also shown are the functions 200 fs later. During this time the time-evolution of $\langle R|\psi_1(t)\rangle$ proceeds unperturbed by the external field. At $T = 10$ fs we find the excited state wavefunction to be very similar to the initial vibrational state in $|0\rangle$. This wave packet then moves toward larger distances. An interaction with a 100 fs pulse prepares a broad probability density distribution, and so produces the vibrational motion of the wave packet seen in the figure. For the case of a 250 fs pulse, one finds a nodal structure in the prepared function, but still the nonstationary dynamics can be observed. Finally interaction with a 500 fs pulse prepares a state that changes very little in time; thus we approach the case of a stationary state which by definition does not exhibit any dynamics. For the present parameters it is the vibrational state with quantum number $\beta' = 7$.

8.2.2   Intense Infrared Excitation

Pulses with wavelengths in the infrared (IR) regime are able to excite transitions between ro-vibrational states within a single electronic state. In this section we consider femtosecond excitation processes that take place in the ground electronic state $|0\rangle$ of a molecule, as sketched in Figure 8.1b.

Theoretical Considerations Let us first discuss weak field excitation so that perturbation theory applies. The theoretical description which is given in Section 8.3 must be modified slightly in the present case. The first-order nuclear state in $|0\rangle$ is given within perturbation theory as

$$|\psi_0(t)\rangle = e^{-iE_i t}|\psi_i\rangle + i \int\limits_{-\infty}^{t} dt'\, U_0(t-t')\, \tilde{\mu}_{00} f_0 f(t')\, e^{-i\omega_1 t'} e^{-iE_{\alpha}' t}|\alpha\rangle, \qquad (8.14)$$

where $|\alpha\rangle$ is the initial vibrational state. $\tilde{\mu}_{00}$ is the projection of the permanent dipole-moment in the ground state on the laser polarization vector, and all other quantities are the same as defined before.

It is interesting, that we do not need a short pulse (in the sense that its spectral width must be broader than the energy differences between vibrational levels) to prepare a ground state wave packet. This is obvious in the case of a long-pulse resonant excitation into the state $|\alpha'\rangle$. Using our earlier arguments in this section, we find that

$$|\psi_0(t)\rangle = e^{-iE_\alpha t}|\alpha\rangle + A\, e^{-iE_\alpha' t}|\alpha'\rangle, \qquad (8.15)$$

where A collects all constants. Thus a wave packet consisting of a coherent superposition of two states is prepared. Of course in the perturbative regime, where $|A| \ll 1$, the packet will not exhibit any observable dynamics. Furthermore within the perturbative treatment the norm of the wavefunction is not conserved.

In intense field excitation, perturbation theory no longer applies, so one has to solve the time dependent Schrödinger equation

$$i\frac{\partial}{\partial t}|\psi_0(t)\rangle = \{H_0 - \mu_{00}\varepsilon\, f_0 f(t)\cos(\omega_1 t)\}\, |\psi_0(t)\rangle \qquad (8.16)$$

with the appropriate initial condition. Here H_0 is the nuclear Hamiltonian in state $|0\rangle$.

Example: Ground State Dynamics in an Organometallic Fragment

As a numerical example we use a linear model of the complex R–Fe–Si–H_3 where R($=C_5H_5$) and H_3 are treated as single particles. This metallosilane fragment is described within a model of three coupled Morse-oscillators (Markert et al. 1997).

Because of the large mass differences one expects classically that the R–Fe and Fe–Si vibrational motion decouple from the Si–H_3 mode. This expectation is confirmed by quantum calculations. In Figure 8.3a we display the dynamics of the vibrational wave packet prepared by an intense 50 fs pulse. The center frequency was chosen to be equal to the energy difference between the ground and first excited vibrational state of the system. The calculation was performed within the three-dimensional model, and we show plots along the Jacobi coordinates S_2 (distance

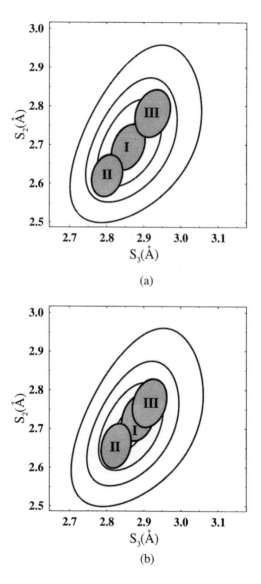

(a)

(b)

Figure 8.3   Wave packet dynamics in a metallosilane fragment. Contours of the potential are indicated as solid lines. (*a*) Preparation of a wave packet via an intense 50 fs pulse. The full width at half maximum of the modulus squared is shown at different times. I: pulse maximum; II: 100 fs later; III: 710 fs later. (*b*) Preparation of a wave packet via a weak pulse of 800 fs. The times correspond to I: pulse maximum; II: 170 fs later; III: 970 fs later.

Fe to the center of mass of SiH_3) and S_3 (distance of R to the center of mass of $FeSiH_3$). The distance $Si-H_3$ was fixed at its equilibrium position. Also shown are contours of the potential surface. At the center of the pulse (I), the wave packet resembles very much the initial state, which is the vibrational ground state. Evidence that a wave packet is being prepared is confirmed at later times, where we find essential displacements from the initial position (II, III).

As stated above, it is also possible to create a wave packet with a long pulse of low intensity. This is illustrated in Figure 8.3b. Here we chose a width of 800 fs with an intensity of a factor of 100 less than in the former case (see Fig. 8.3a). Again the vibrational dynamics can clearly be observed in the plot.

8.2.3 Multiple-Pulse Interaction

Another way to prepare molecular wave packets is to use a sequence of two or more pulses of different frequency (see Fig. 8.1c). Let us illustrate this scheme for the generation of wave packets in the electronic ground state of a molecule. We will consider the case of a two-pulse interaction which proceeds resonantly via an electronic excited state.

The photodissociation process of the water molecule in its first absorption band is probably the best understood fragmentation process of a triatomic molecule (Schinke 1993). Many sophisticated experiments have been performed and full *ab initio* studies, including quantum chemical and dynamical calculations contributed to a complete characterization of the absorption spectrum and fully state-to-state resolved photofragment distributions (Engel et al. 1992).

Here we regard the simultaneous interaction of the HOD-molecule with two femtosecond laser pulses. The excitation scheme is displayed in Figure 8.1c, which shows cuts through the potential energy surfaces in the ground and excited state along the reaction coordinate in the antibonding excited state. The reaction coordinate is defined as the minimum energy path on the potential surface V_1 connecting the asymptotic channels H + OD and D + OH. The frequency of the first pulse (ω_1) is tuned to resonance, whereas the second frequency (ω_2) is taken to be smaller, as indicated. The first pulse prepares a wave packet in the excited state which in the present case describes a dissociation process. Accordingly the packet, while splitting into two parts, will move into the two arrangement channels. Through the interaction with the second pulse, however, parts of the wave packet are stimulated down to the ground state and a vibrational wave packet is built.

Since in the dissociation process the angular motion can be separated from the vibrational degrees of freedom, it is sufficient to treat only the two bond lengths as internal coordinates in the theoretical description. Figure 8.4 illustrates the wave packet preparation process for pulses of 12 fs width and peak energies of 7.6 and 3.3 eV. To achieve a visible effect, strong pulses of intensities in the order of 10^{14} W/cm^2 were employed in the simulation. At $t = 0$ the wave function is the vibrational ground state of HOD. Already at t_1, which corresponds to the maxima of the pulse envelopes, deviations from the initial state can be seen. This is even more dramatic at the end of the interaction (t_2) where a bifurcation of the vibrational wave packet has occurred.

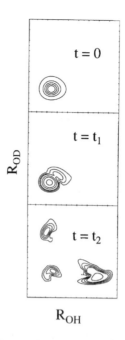

Figure 8.4 HOD: Wave packet preparation by a two-photon process.

8.3 VIBRATIONAL WAVE PACKET DYNAMICS

In the preceding section we showed how vibrational wave packets can be prepared by the interaction of an ensemble of molecules with one or several laser pulses. In this section we will describe the vibrational dynamics of such packets. Several examples will be presented to illustrate the general features of the dynamical behavior.

8.3.1 Bound-State Motion in Diatomic Molecules

The interaction potential between two atoms in a diatomic molecule depends only on the bond distance R. In this subsection we treat the simple one-dimensional dynamics within a bound electronic state using the ground state $|0\rangle$ and an excited electronic state $|1\rangle$ of Na_2 as an example. A femtosecond excitation (50 fs pulse) from the vibrational ground state in $|0\rangle$ with a peak energy of 2 eV prepares a vibrational wave packet in the excited state (Meier and Engel 1995). Figure 8.5 illustrates the vibrational dynamics in coordinate and momentum space. For short times ($t_1 = 90$ fs, $t_2 = 270$ fs) the coordinate-space wave packet moves back and forth between the outer and inner potential wall. This means that in the molecular ensemble the bond lengths change periodically in time, just as one might imagine the classical vibrational case of two mass points connected by a spring. So far the dynamics of wave packets is very close to our classical intuition. The dynamics may also be observed in momentum space. At the times t_1 and t_2 we find a positive and negative momentum distribution, respectively.

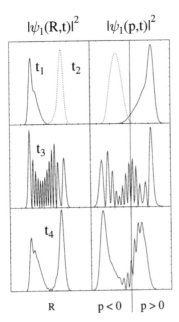

Figure 8.5 Wave packet dynamics in Na_2. Shown are vibrational wave packets in coordinate and momentum space at different times. t_1, t_2: short-time behavior of localized packet; t_3: dispersion; t_4: fractional revival.

The motion proceeds similarly to the one in a harmonic potential where after one period any initial wave packet is exactly the same as in the beginning. However, molecular potentials are not harmonic, and the anharmonicities cause dispersion of the packet, as can be seen at time $t_3 = 10$ ps. The localized packet has spread and fills the whole classically accessible region between the potential walls. For even longer times one finds that the packet has split into two localized parts which move out of phase (see time t_4 in Fig. 8.5). This effect is called a *fractional revival*. The implication of this phenomenon is rather strange because it means that at certain times one group of molecules is likely to have a short bond length and another group a large bond length. This is counterintuitive to our perception of the distribution of bond lengths in a cell containing many molecules.

On a very long time the wave packet reconstructs itself and resembles very much the initial wave packet. This *revival* effect has been studied not only in molecular systems (Baumert et al. 1992; Fischer et al. 1995) but earlier in atomic Rydberg wave packets (Alber and Zoller 1991) and quantum optical systems (Yoo and Eberly 1985).

The same effect occurs in a system of classical oscillators with different discrete frequencies. These oscillators initially move in phase and then dephase, and at one point they move against each other. Some time later a rephasing will take place, and the initial motion is recovered. However, for our systems the effect is purely quantum mechanical, since a quantization of the energy (appearing in the different phase factors) is necessary: Revivals are not found in systems with a continuous spectrum.

8.3.2 Bound-State Motion in Several Degrees of Freedom

A molecule consisting of N atoms possesses $3N - 6$ internal degrees of freedom ($3N - 5$ for a linear molecule). Accordingly the vibrational wave packet dynamics is much more complex than that which we encountered in the case of the diatomic molecule. In what follows we will take a look at three- and two-dimensional problems in order to generalize our description of molecular wave packets.

Three-Dimensional Vibrational Dynamics We use the R–Fe–Si–H_3 fragment from our earlier discussion as a first example of multidimensional vibrational dynamics. As seen in Figure 8.3, an intense infrared excitation with a pulse of 50 fs width results in the preparation of a vibrational wave packet. In our three-dimensional model the internal coordinates are the distances x (R–Fe), y (Fe–Si), and z (Si–H_3). Instead of plotting cuts through the three-dimensional time dependent wavefunction, we calculate the expectation values

$$\langle r \rangle(t) = \frac{\langle \psi(x, y, z, t) | r - r_0 | \psi(x, y, z, t) \rangle}{\langle \psi(x, y, z, t) | \psi(x, y, z, t) \rangle}, \tag{8.17}$$

where $r = x$, y, or z and r_0 is the corresponding equilibrium distance. The expectation values are shown in Figure 8.6. The preparation pulse has its maximum at 500 fs, so before this time the system is in a stationary state and the expectation values are constant. The bond lengths x and y change periodically in time after the interaction is finished. This reflects the vibrational quantum dynamics. On the other hand, the z coordinate is only slightly excited which indicates the decoupling of this mode from the others.

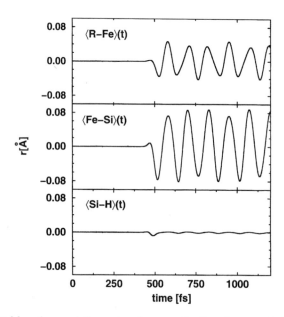

Figure 8.6 Bond-length expectation values in a metallosilane fragment before, during, and after intense infrared excitation.

Revivals in Two Dimensions Revivals can also be found in more dimensional systems. As a simple example we regard the vibrational dynamics in the HOD molecule (Meyer and Engel 1997a). A wave packet can be prepared in the electronic ground state, as has been described in Section 8.2. Within a two-dimensional model that includes the OH and OD distances as variables and keeps the bending angle fixed, the vibrational motion will proceed along both degrees of freedom. If the vibrational modes are not coupled strongly, nearly independent motion should occur in the two coordinates. This is the case in our numerical example which is presented in Figure 8.7. The wavefunctions are shown at different times, where $t = 0$ fs corresponds to the revival time of the separated OH vibrational motion. A localized wave packet moving back and forth in the OH coordinate can be observed. On the other hand, the packet is delocalized in the OD vibrational mode.

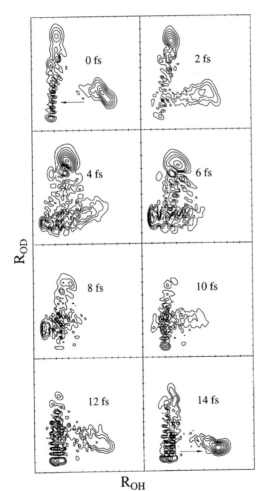

Figure 8.7 Vibrational revival in the HOD molecule. Shown are contours of the two-dimensional wave packet. The zero of the time scale corresponds to the revival time of the OH mode. Snapshots during one vibrational period are shown.

8.3.3 Quasi-bound Vibrational Motion

Theoretical Treatment We consider a diatomic molecule with two electronic states $|0\rangle$ and $|1\rangle$. These states are decoupled within the Born-Oppenheimer approximation. In this section we discuss the case where a nonadiabatic coupling occurs, so the electronic and nuclear degrees of freedom cannot be treated independently. The time dependent Schrödinger equation for the nuclear wavefunctions $\psi_0(t)$ and $\psi_1(t)$, which belong to the different electronic states, now takes the form

$$
i\frac{\partial}{\partial t}\begin{pmatrix}\psi_0(R,t)\\\psi_1(R,t)\end{pmatrix}=\begin{pmatrix}H_0 & W^a(R)\\W^a(R) & H_1\end{pmatrix}\begin{pmatrix}\psi_0(R,t)\\\psi_1(R,t)\end{pmatrix}. \tag{8.18}
$$

The kinetic coupling elements $W^a(R)$ contain derivatives of the electronic wavefunctions (which are solutions of the time independent Schrödinger equation for fixed nuclear distance R) with respect to R. As a consequence of the coupling, the potential curves can show an avoided crossing. One also may employ the so-called diabatic representation in which the potential curves for state $|0\rangle$ and $|1\rangle$ cross each other and the coupling elements are potential couplings $W^d(R)$ that are smooth functions of the internuclear distance (Köppel et al. 1984). The coupling (adiabatic or diabatic) induces transitions between the different states, and we will now regard an example for such nonadiabatic transitions that has been studied extensively in the connection with wave packet dynamics induced by femtosecond excitation.

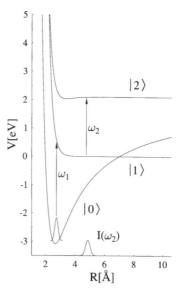

Figure 8.8 Diabatic potential energy curves for the NaI molecule. Indicated are the pump (ω_1) and probe (ω_2) processes, and also the Franck-Condon region $I(\omega_2)$ for resonant excitation $|2\rangle \leftarrow |1\rangle$.

NaI-Predissociation The electronic predissociation of NaI has served as a model system in many studies of femtosecond excitation of small molecules (Rose et al. 1989). Figure 8.8 displays two diabatic potential curves for the NaI molecule. They belong to the ground and a first excited state of equal symmetry. The curves intersect each other at a distance of about 7 Å. In this region the Born-Oppenheimer approximation fails, and we find a coupling of electronic and nuclear motion. This has consequences for the wave packet dynamics in the system. Figure 8.9 shows the dynamics of the coupled nuclear wave functions $\psi_0(R, t)$ and $\psi_1(R, t)$. Initially a wave

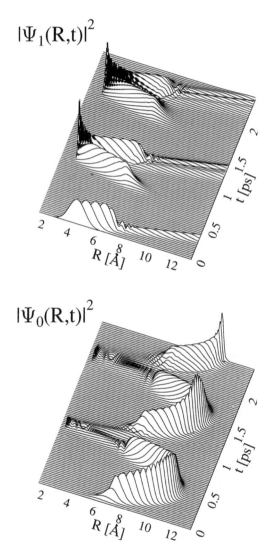

Figure 8.9 Wave packet dynamics in the NaI molecule. The two nuclear components are coupled to each other for several periods of the quasi-bound vibrational motion.

packet is prepared in the adiabatic state $|1\rangle$ by femtosecond excitation from the ground state. It can be seen that when the coupling region is reached, the packet splits: The component in $|0\rangle$ is populated, and parts of $\psi_1(R, t)$ move into the asymptotic region. This part of the wavefunction represents the molecules that have dissociated into Na and I. The remaining part of the wavefunction performs a quasi-bound vibrational motion, showing splittings each time the coupling region is passed. We note that even in the present case revivals can be found for the component ψ_1 (Meier et al. 1991). Nevertheless, the situation is more complicated because the wave function is a mixture of pure bound and continuums states that are coupled to each other.

8.4 DETECTION OF VIBRATIONAL WAVE PACKETS

8.4.1 Pump/Probe Experiment

It is now appropriate to ask how the time dependent quantum motion can be detected in a spectroscopic experiment. Very clever schemes have been invented, and they can be summarized as *pump/probe experiments*. In general such an experiment proceeds as follows:

1. *Pump process.* The interaction with a short laser pulse prepares the system under investigation in a nonstationary state. This preparation process sets the zero point of the time scale.

2. *Probe process.* A second ultrashort pulse interacts with the ensemble of molecules at a delay time τ. The interaction prepares the system in a state that then enters into an experimental signal, as will be described below.

3. *Delay time.* The delay-time τ is varied so that the signal can be detected as a function of time. Since the intermediate state of the system is nonstationary one expects a temporal variation of the signal, which is characteristic for the quantum mechanical wave packet motion.

We will now give examples of different pump/probe experiments as sketched in Figure 8.10.

8.4.2 Pump/Probe Fluorescence

If the probe pulse excites the molecule in a state that decays via fluorescence, the total fluorescence signal may be detected as a function of delay time. This scheme was used in many pioneering experiments on *femtochemistry* (Zewail 1994; Manz and Wöste 1995).

Let us look at the signal from a theoretical point of view. The probe process populates the electronic state $|2\rangle$, as is indicated in Figure 8.10a. Since the potentials in both electronic states depend on the nuclear coordinates a resonant excitation takes place only in certain spatial regions. The spatial window (*transient Franck-Condon region*) where excitation is most probable can be estimated within time dependent perturbation theory as follows: The wave function prepared by the probe pulse is

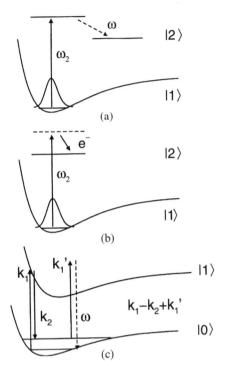

Figure 8.10   Wave packet detection schemes. (*a*) Pump/probe fluorescence experiment; (*b*) pump/probe ionization; (*c*) time-resolved four-wave mixing measurement.

$$\psi_2(t, \tau) \sim \int_{-\infty}^{t} dt' \, e^{-iH_2(t-t')} \tilde{\mu}_{21} f_0 f(t') e^{-i\omega_2 t'} \psi_1(t'). \tag{8.19}$$

Here the notation is the same as was used in Section 8.2. Let us neglect the kinetic energy operators in the Hamiltonians H_1 and H_2. This is a good approximation for short enough excitation pulses (Braun et al. 1995). Furthermore we fix the moving wave packet at its position at the delay time τ. Within these approximations the wavefunction ψ_2 takes the form (within unimportant phase factors)

$$\psi_2(\tau) \sim \tilde{\mu}_{21} \psi_1(\tau) I(\omega_2), \tag{8.20}$$

where we extended the limits of the time integration to infinity to obtain the Fourier-transform

$$I(\omega_2) = \int_{-\infty}^{+\infty} dt' \, e^{i(V_2 - V_1 - \omega_2)t'} f_0 f(t'). \tag{8.21}$$

The latter integral defines the Franck-Condon window, which has its maximum at positions where the difference potential $V_2 - V_1$ equals the laser frequency. At times

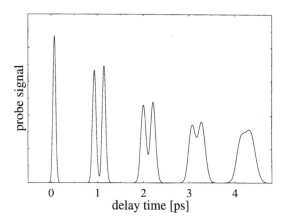

Figure 8.11 Pump/probe fluorescence signal in the NaI system. The peaks reflect the quasi-bound motion and the decay of the instable complex.

when the wave packet ψ_1 is located in this spatial window, there is an enhanced excitation probability. This gives rise to a time dependence of the pump/probe signal.

The Franck-Condon window for electronic excitation $|2\rangle \leftarrow |1\rangle$ with a pulse of peak energy $\omega_2 = 2.5$ eV is noted in Figure 8.8 which gives the potential curves for the NaI molecule. Whenever a probability density falls within this window, the excitation probability is large, and correspondingly a large fluorescence signal is obtained. The population in the upper state $|2\rangle$, which is calculated as a function of the pump/probe delay, is plotted in Figure 8.11. A periodic signal is found that reflects the quasi-bound motion in the coupled electronic states. The doublet in the peaks occurs because the Franck-Condon window is passed during the inward and outward motion of the packet ψ_2. The decrease of the peaks tells us that the complex decays in time, so one can monitor the kinetics of a chemical reaction on the femtosecond time scale.

8.4.3 Time-Resolved Photoelectron Spectroscopy

If the probe laser induces a transition to an ionic state, the pump/probe signal can consist of molecular ions or photoelectrons. Here we discuss the energy- and time-resolved measurement of electrons that are ejected during the ionization process. Under the assumption that perturbation theory applies and that the photoelectron does not couple to the rest of the system, one can derive a first-order expression for the nuclear ionic wavefunction (Seel and Domcke 1991). The wavefunction which corresponds to the ejection of a photoelectron with kinetic energy E is

$$\psi_2(E, t) \sim \int_{-\infty}^{t} dt'\, e^{-iH_2(t-t')}\, \tilde{\mu}_{E1} f_0 f(t')\, e^{-i(\omega_2 - E)t'} \psi_1(t'). \tag{8.22}$$

Here H_2 corresponds to the nuclear Hamiltonian in the ionic state and $\tilde{\mu}_{E1}$ denotes the dipole moment for the neutral-to-ionic transition. Note that the photoelectron energy appears in the exponent containing the laser frequency. With the help of the wavefunction $\psi_2(E, t)$, we can calculate the photoelectron spectrum

$$P(E, \tau) = \lim_{t \to \infty} \langle \psi_2(E, t) | \psi_2(E, t) \rangle. \tag{8.23}$$

As a numerical example we regard the pump/probe ionization spectroscopy of Na_2. The intermediate state here has a double minimum potential, as displayed in Figure 8.12. From what we said above, it is clear that for different positions of the wave packet one has to expect a rather different photoelectron distribution. For a fixed position the Franck-Condon window can be calculated. The window now depends not only on the nuclear coordinate R but also on the energy E, and we find the resonance condition

$$\omega_2 - E = V_2 - V_1. \tag{8.24}$$

This equation gives us a reflection principle: The probability density distribution of the wave packet $|\psi_1|^2$ (multiplied by the transition dipole moment) is reflected in the photoelectron spectrum. The arrows in Figure 8.12 show the maxima of the photoelectron spectrum which are expected for two different positions of the moving wave packet. Figure 8.13a shows the vibrational wave packet motion in the double minimum potential (Meier and Engel 1994). The photoelectron spectrum which is obtained at different delay times of a 50 fs probe pulse is displayed in Figure 8.13b. Clearly, the temporal changes reflect the oscillatory motion of the wave packet. In more detail, the dynamics of the probability density distribution is seen in the spectra. Thus a measurement of the time-resolved photoelectron kinetic energy distribution

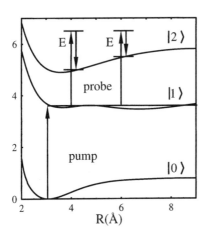

Figure 8.12 Schematic illustration of the mapping of vibrational wave packets through time-resolved photoelectron measurements. Photoionization at different times yields different kinetic energy distributions that can directly be connected to the probability density distribution.

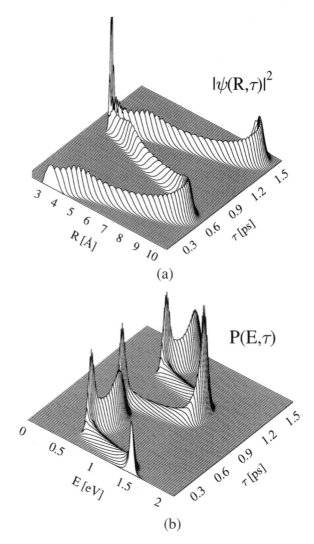

Figure 8.13 (*a*) Wave packet dynamics in the double minimum state of Na_2. (*b*) Photoelectron distributions that result from probe-pulse ionization at different delay times.

allows for the observation of the wave packet itself (the modulus squared). This fact has been verified in recent experiments (Assion et al. 1996).

8.4.4 Nonlinear Techniques

In our last example of a sophisticated detection scheme for vibrational wave packet dynamics, we describe the interaction of a molecule with three ultrashort pulses (Stock and Domcke 1997). Let us discuss the case where two pulses of different frequencies ω_1, ω_2 and wave vectors k_1, k_2 interact simultaneously with a molecular sample and a third pulse (ω_1, k'_1) is time delayed with time τ. The fields induce a nonlinear polarization in the molecules, which can be calculated as

$$P(t) = \langle \psi(t) | \mu | \psi(t) \rangle, \tag{8.25}$$

namely the time dependent polarization is the expectation value of the dipole moment with respect to the wavefunction of the system. In a four-wave mixing experiment the radiation emitted by the molecule is detected in the direction $k_1 - k_2 + k'_1$ (see Fig. 8.10c), so we need to calculate the corresponding third-order polarization. Under the assumptions that the time-delayed pulse does not overlap with the initial ones, a perturbative treatment yields three contributions:

$$P^{(3)}(t) = 2Re\{\langle \psi_0 | \mu | \psi_1(k_1 - k_2 + k'_1, t)\rangle + \langle \psi_0(k_2 - k_1, t) | \mu | \psi_1(k'_1, t)\rangle$$

$$+ \langle \psi_0(k_2 - k'_1, t) | \mu | \psi_1(k_1, t)\rangle\}, \tag{8.26}$$

where Re denotes the real part of the complex function. We used the notation that k_i appearing in the argument of the ket-state means that an absorption process initiated by the pulse of wave vector k_i takes place. Emission processes are indicated by negative wave vectors. In the bra-states emission and absorption processes are denoted by positive and negative wave vectors.

A time-resolved four-wave mixing signal may now be calculated by integration of $P^{(3)}(t)$ over time for a time interval that is longer than the interaction time between the field and the sample. Let us illustrate this with a numerical example. Recently the $|B\rangle \leftarrow |X\rangle$ transition in gas phase I_2 was investigated using the time-resolved four-wave mixing technique (Schmitt et al. 1997). Both electronic states have bound-state potentials, so we can expect to observe a pure bound-state vibrational motion. To see which vibrational dynamics to expect, let us look at two terms in equation (8.26). The first term contains the overlap integral between the initial state and the third-order wavefunction $\psi_1(k_1 - k_2 + k'_1, t)$. The first two pulses prepare a vibrational wave packet in the ground state $|0\rangle$. This packet moves and serves as initial state for another excitation with the time-delayed third pulse. Naturally the first term will show the dynamics in the ground state. The same reasoning shows that the third term reflects the dynamics in the excited state. Figure 8.14 displays the time-integrated polarization which was calculated for wavelengths of 559 nm and 579 nm. Also two terms that contribute to the overall signal are shown. The oscillatory behavior occurs with the vibrational period of the ground (middle panel) and of the excited electronic state (lower panel). Due to the intensity of the first contribution, the total signal reflects only the ground state dynamics. Note also that around zero delay time other terms exist which we do not consider here. A comparison to experiment shows that the rotational degree of freedom has to be taken into account. The signal then shows a superimposed oscillation due to the rotational motion of the diatomic molecule (Meyer et al. 1997b).

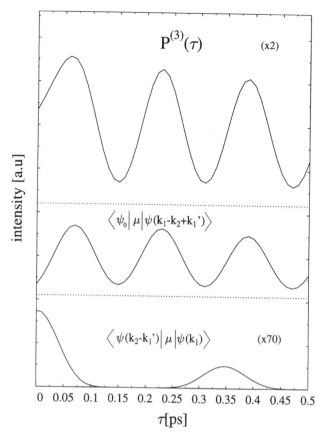

Figure 8.14 Third-order polarization calculated as a function of the time-delay between two pulses that act simultaneously with the sample (k_1, k_2) and a time-delayed third pulse (k_1') (upper panel). Two terms that contribute to the total signal are shown separately.

8.5 SUMMARY

In this chapter we discussed the vibrational wave packet motion in small molecules. The interaction with one or several short pulses results in the preparation of a linear combination of vibrational states in bound electronic states of a molecular sample. The following quantum dynamics proceeds with the characteristic vibrational periods. Anharmonicities in the potential energy result in the dispersion and, in certain cases, revival of the wave packets. The dynamic behavior that can be obtained from the solution of the time dependent Schrödinger equation can be detected experimentally with pump/probe experiments. We gave several examples that illustrate experimental approaches based on, for example, fluorescence or ion detection and nonlinear techniques.

We are now in the position to follow quantum dynamics in real time with a time resolution on the femtosecond time scale.

ACKNOWLEDGMENTS

We gratefully acknowledge financial support by the Deutsche Forschungs-gemeinschaft for several projects. Also we thank our experimentalist colleagues T. Baumert, G. Gerber, W. Kiefer, A. Materny, and their numerous coworkers for a fruitful collaboration.

REFERENCES

Alber, G., and Zoller, P. (1991). *Phys. Rep.* **199**, 231.

Assion, A., Geisler, M., Helbing, J., et al. (1996). *Phys. Rev.* A**54**, R4605.

Baumert, T., Engel, V., Röttgermann, C., et al. (1992). *Chem. Phys. Lett.* **191**, 639.

Baumert, T., and Gerber, G. (1995). *Adv. At. Mol. Opt. Phys.* **35**, 163.

Berry, R. S., Bonačić-Koutecký, V., Gaus, J., et al. (1997). *Adv. Chem. Phys.* **101**, 101.

Blum, K. (1996). *Density Matrix Theory and Applications*, 2nd ed. Plenum Press, New York.

Braun, M., Meier, C., and Engel, V. (1995). *J. Chem. Phys.* **103**, 7907.

Domcke, W., and Stock, G. (1997). *Adv. Chem. Phys.* **100**, 1.

Engel, V., Staemmler, V., Vander Wal, R. L., et al. (1992). *J. Phys. Chem.* **96**, 3201.

Fischer, I., Villeneuve, D. M., Vrakking, M. J. J., et al. (1995). *J. Chem. Phys.* **102**, 5566.

Gruebele, M., and Zewail, A. H. (1993). *J. Chem. Phys.* **98**, 883.

Loudon, L. (1983). *The Quantum Theory of Light*. Clarendon Press, Oxford.

Köppel, H., Domcke, W., and Cederbaum, L. S. (1984). *Adv. Chem. Phys.* **57**, 59.

Manz, J., and Wöste, L. (1995). *Femtosecond Chemistry*. VCH, Weinheim, Germany.

Markert, T., Malisch, W., and Engel, V. (1997). *Chem. Phys. Lett.* **270**, 222.

Meier, C., and Engel, V. (1994). *J. Chem. Phys.* **101**, 2673.

Meier, C., and Engel, V. (1995). In *Femtochemistry*, eds. J. Manz and L. Wöste. VCH, Weinheim, Germany, Ch. 11.

Meier, C., Engel, V., and Briggs, J. S. (1991). *J. Chem. Phys.* **95**, 7337.

Meyer, S., and Engel, V. (1997a). *J. Phys. Chem.* A**101**, 7749.

Meyer, S., Schmitt, M., Materny, A., et al. (1997b). *Chem. Phys. Lett.* **281**, 332.

Mukamel, S. (1995). *Principles of Nonlinear Optical Spectroscopy*. Oxford University Press, New York.

Ramsey, N. F. (1986). *Molecular Beams*. Oxford University Press, New York.

Rose, T. S., Rosker, M. J., and Zewail, A. H. (1989). *J. Chem. Phys.* **91**, 7415.

Schinke, R. (1993). *Photodissociation Dynamics*. Cambridge University Press, Cambridge.

Schmitt, M., Knopp, G., Materny, A., et al. (1997). *Chem. Phys. Lett.* **270**, 9.

Schwoerer, H., Pausch, R., Heid, M., et al. (1997). *J. Chem. Phys.* **107**, 9749.

Seel, M., and Domcke, W. (1991). *J. Chem. Phys.* **95**, 7806.

Yoo, H. I., and Eberly, J. H. (1985). *Phys. Rep.* **118**, 239.

Zewail, A. H. (1994). *Femtochemistry*, Vols. 1, 2. World Scientific, Singapore.

CHAPTER 9

Time Dependent and Time Independent Wave Packet Approaches to Reactive Scattering

STUART C. ALTHORPE, DONALD J. KOURI, and DAVID K. HOFFMAN

9.1 INTRODUCTION

It has long been recognized since the earliest days of quantum mechanics that wave packets are the most general form of solution of the time dependent Schrödinger equation (TDSE), corresponding to a superposition of states of various total energies (Kemble 1958). Until recently the most detailed discussion of wave packets was that of Goldberger and Watson (1964). However, typical computations made essentially no use of the possibilities of obtaining scattering information over a range of energies contained in wave packets (e.g., see Clary 1985). This is probably due in part to the fact that, in references like Goldberger and Watson, the analysis of the scattering information was carried out by assuming that the packet contained a relatively narrow range of energies, thus requiring it initially to be very far from the target (in order that the "experiment" being described corresponded to a noninteracting target and projectile). This tended to prejudice all theoretical discussions of the time dependent wave packet description of scattering in favor of beginning the experiment in the far distant past (under conditions that the initial composition in energy of the packet was unknown) and ending it in the far distant future. With very few exceptions (Gray et al. 1980), this resulted in very high inefficiency of computational methods (although many of these studies yielded fascinating results; e.g., see McCullough and Wyatt 1986).

This was changed in papers by Mowrey and Kouri (1986), and Mowrey, Bowen, and Kouri (1987), which derived, for the collisions of structured molecules off corrugated "soft" surfaces, detailed expressions for the S-matrix elements at *any* energy sufficiently contained in the initial packet. Furthermore they not only obtained results over a wide range of energies in a single wave packet propagation, they also were able to localize the initial packet in a smaller region that could be located closer to the target. Making use of the "initial value" nature of the TDSE (so that the scaling

The Physics and Chemistry of Wave Packets, Edited by John Yeazell and Turgay Uzer
ISBN 0-471-24684-0 © 2000 John Wiley & Sons, Inc.

with problem size is *at worst* quadratic), Mowrey and Kouri (1986) showed that enormous scattering problems could be solved (the largest being converged calculations for N_2 scattering off a corrugated, soft surface, involving 18,711 quantum states, over a wide range of energies; Bowen et al. 1993).

Two approaches to solving the TDSE for wave packet evolution have dominated calculations done since 1983–84, these being the symmetric split operator technique of Feit and Fleck (1984) and the Chebyshev expansion of the time-evolution operator due to Tal-Ezer and Kosloff (1984). The adaptation of the close coupled wave packet approach to treat atom-diatom scattering was done by Sun, Mowrey, and Kouri (1987), and Sun, Judson, and Kouri (1989). Another major contribution making direct solution of the TDSE efficient and practical for the most difficult collisions (i.e., reactive scattering) has been the use of negative imaginary absorbing potentials (NIPs).[1] Although such potentials had been used earlier (e.g., by Leforestier and Wyatt 1983; Kosloff and Cerjan 1984; Kosloff and Kosloff 1986), it was the work of Neuhauser and Baer (1989) and Neuhauser et al. (1989, 1991, 1992) that first showed the enormous power of NIPs to yield quantitatively accurate results, even at the detailed level of *S*-matrix elements and transition probabilities. The original development (Neuhauser and Baer 1989) was restricted to wave packets that were extremely broad in coordinate space (so only probabilities at a single energy were obtained), while the subsequent generalization showed how to obtain results for a wide range of energies in a single propagation, using packets that were highly localized in space. In addition the work of Neuhauser et al. (1991) showed that *total* reaction probabilities into specific final arrangements could be obtained by placing the NIPs just outside an appropriate "surface of no return." They pointed out that this corresponded to an essentially exact quantum transition state approach (Neuhauser, Baer, and Kouri 1990). Subsequently Seideman and Miller (1992) showed that NIPs could be used to develop beautiful and powerful expressions for the cumulative reaction probability without employing initial packets that correspond to specific initial states of the reactants or final states of the products. Their treatment employed the flux–flux autocorrelation formalism of Miller, Schwarz, and Tromp (1983), and it provides a complete, formally exact quantum transition state analysis of reactive scattering.

Independently Kouri, Hoffman, and coworkers (see Kouri, Arnold, and Hoffman 1993; Hoffman et al. 1994; Huang et al. 1993; Kouri and Hoffman 1995) showed that the standard time independent Schrödinger equation (TISE),

$$(E - H)\Psi = 0, \tag{9.1}$$

is *not* the most general equation possible. Rather, this equation is the result of assuming complete ignorance of the initial wave packet's makeup, and evolving it from a time in the remote past to a time in the remote future; namely

---

[1]Early uses of such potentials to model channels not explicitly included in calculations can be found in the nuclear physics literature. See, for example, Brown (1969).

$$\Psi(E) = \frac{1}{2\pi} \int\limits_{-\infty}^{\infty} dt \, e^{iEt/\hbar} \chi(t), \tag{9.2}$$

where $\chi(t)$ is the solution of the TDSE. As a result the formal solution of the TISE satisfying scattering boundary conditions is that first obtained by Lippmann and Schwinger (1950),

$$\psi_k^+(E) = \phi_k(E) + \frac{1}{E - H + i\varepsilon} V\phi_k(E), \tag{9.3}$$

where

$$H_0\phi_k(E) = E\phi_k(E) \tag{9.4}$$

and

$$V = H - H_0. \tag{9.5}$$

The Lippmann-Schwinger equation (LSE) contains absolutely *no* information about the initial packet and provides scattering information at energy E regardless of the details of the experimental conditions. However, in the work of Kouri, Hoffman, and coworkers, it was shown that if the experiment starts the initial wave packet from a finite distance from the target so that its distribution among different energy and momentum states is known, then one should define a time independent "scattering state" $\xi^+(E)$ as

$$\xi^+(E) = \frac{1}{2\pi} \int\limits_{0}^{\infty} dt \, e^{iEt/\hbar} \chi(t). \tag{9.6}$$

Then one finds

$$(E - H)\xi^+(E) = \frac{i}{2\pi} \chi(0), \tag{9.7}$$

referred to as the time independent wave packet Schrödinger equation (TIWSE), with scattering solution

$$\xi^+(E) = \frac{i}{2\pi} \frac{1}{(E - H + i\varepsilon)} \chi(0). \tag{9.8}$$

This is a fundamentally new result for scattering problems, and it has been proved that one can obtain the correct scattering information from (9.8) so long as energy E is sufficiently contained in the initial packet $\chi(0)$. Some confusion in the literature (Mandelshtam and Taylor 1995a) has arisen because of not properly recognizing that inhomogeneous equations obtained from (9.3) for $\Psi_k^+(E)$ differ fundamentally from that for $\xi^+(E)$. The corresponding inhomogeneous equation resulting from the LSE is for the *scattered wave*, defined as

$$\psi_{SW}^+(E) = \frac{1}{(E - H + i\varepsilon)} \, V\phi_k(E). \tag{9.9}$$

Clearly, in the limit $\varepsilon \to 0^+$,

$$(E - H)\psi_{SW}^+(E) = V\phi_k(E). \tag{9.10}$$

It has been tempting to some to assume that $V\phi_k(E)$ is simply a special case of the $i\chi(0)/2\pi$ in equation (9.7). This violates the well-defined physical meaning of $\chi(0)$, as well as the fact that the inhomogeneity $V\phi_k(E)$ results from assuming one has *no* information about $\chi(0)$.

However, the most important difference between equations (9.8) and (9.9) (or (9.7) and (9.10)) comes when one realizes that a Chebyshev expansion of $1/(E - H + i\varepsilon)$ substituted into each yields the results

$$\psi_{SW}^+(E) = \sum_n a_n^+(E) \, T_n(H) V\phi_k(E) \tag{9.11}$$

and

$$\xi^+(E) = \frac{i}{2\pi} \sum_n a_n^+(E) \, T_n(H)\chi(0). \tag{9.12}$$

Essentially this expresses the scattering information at energy E in terms of the "basis vectors"

$$\bar{\eta}_n \equiv T_n(H)V\phi_k(E) \tag{9.13}$$

and

$$\eta_n \equiv T_n(H)\chi(0). \tag{9.14}$$

The point is that the $\bar{\eta}_n$ *change with energy*, whereas the η_n are *energy independent*. As a result one must do *much* more work to solve (9.11) than (9.12) for a range of energies. The *only* energy dependence in (9.12) is in the coefficients $a_n^+(E)$, and this is *known analytically*. As a consequence scattering calculations for systems requiring lots of energies are most efficiently done using the TIWSE approach.

To provide background for the detailed discussion of recent advances in time dependent and time independent wave packet methods (TDW and TIW), we now discuss the incorporation of NIPs into the Chebyshev-type expansion of Green's functions. It is found that the usual recursion formula for the Chebyshev polynomials is unstable when H contains NIPs. In fact, even without NIPs, the eigenspectrum of H

in equation (9.12) must be between ± 1, or the recursion generating the η_n basis becomes unstable. It is necessary to estimate the maximum and minimum eigenvalues H_{max} and H_{min} of the truncated hamiltonian matrix and introduce (Tal-Ezer and Kosloff 1984)

$$H_{norm} = \frac{H - \overline{H}}{\Delta H} \qquad (9.15)$$

with $\overline{H} = (H_{max} + H_{min})/2$ and $\Delta H = (H_{max} - H_{min})/2$. This simple procedure, however, breaks down when H is not Hermitian. Huang, Kouri, and Hoffman (1994) showed how to correct this using the so-called Faber polynomials, which essentially generalize the normalization technique of equation (9.15) to complex eigenvalues. Subsequently a simple procedure was introduced by Mandelshtam and Taylor (1995a,b), and variants of this approach are now used.

Another major modification of both the TDWSE and TIWSE approaches is due to the work of Peng and Zhang (1996), Kouri et al. (1996), and Althorpe et al. (1997). Essentially this is the partitioning of state-resolved quantum reactive scattering calculations into subregions, in which the TD wave packet or TIW state is subdivided into pieces satisfying completely uncoupled equations in the various subregions (essentially a divide-and-conquer strategy). Peng and Zhang have pursued this in the TDW context, whereas Althorpe and coworkers have stressed the TIW context. The latter approach is based fundamentally on the TIW S-matrix Kohn variational expression (Kouri et al. 1994)

$$S(\beta \leftarrow \alpha) = \frac{i\hbar^2 \sqrt{k_\alpha k_\beta}}{\mu(2\pi)^2 A_\alpha(k_\alpha) A_\beta^*(k_\beta)} \langle \chi(\beta) \mid G^+(E) \mid \chi(\alpha) \rangle, \qquad (9.16)$$

where $G^+(E)$ is the full Green's function, $\chi(\alpha)$ and $\chi(\beta)$ are appropriately positioned initial and final *wave packets* (*not* definite energy states), k_α and k_β are wave numbers for relative motion in initial channel α and final channel β, and $A_\alpha(k_\alpha)$ ($A_\beta(k_\beta)$) the amplitude of wave number k_α (k_β). (We note that the techniques being pursued by Mandelshtam and Taylor 1995a,b, for reactive scattering are also based on equation (9.16), and that the time dependent form of equation (9.16) was derived by Tannor and Weeks 1993.)

In the remainder of this chapter we describe a wave packet method recently developed by the authors for calculating state-to-state reaction probabilities. The method is the original work of the authors, but it draws on much of the previous work mentioned above. This includes the time dependent wave packet method, the use of negative imaginary absorbing potentials (NIPs), the method of Neuhauser et al. (1991) for calculating total reaction probabilities and individual S-matrix elements, the method of Seideman and Miller (1992) for calculating cumulative reaction probabilities, the time independent wave packet (TIW) method, the damped version of the TIW method of Mandelshtam and Taylor (1995a,b), the time dependent reactant-product decoupling (RPD) method of Peng and Zhang (1996), and the TIW and extended versions of the RPD method recently introduced by Kouri et al. (1996) and Althorpe et al. (1997a,b). We will discuss each of these topics in turn as we explain the wave packet method, presenting in this way a detailed review of much of the work done in the past decade or so on wave packets in reactive scattering.

9.2   WAVE PACKET THEORY OF REACTIVE SCATTERING

9.2.1   Simple Theory of Reactive Scattering

In this section we present a general theory of state-to-state reactive scattering which is applicable to all chemical reactions. For simplicity, we shall illustrate the theory with the reaction

$$A(\alpha) + BC(\beta) \rightarrow A(\alpha') + BC(\beta')$$

$$\rightarrow AC(\gamma) + B(\delta) \tag{9.17}$$

which is assumed to take place on a two-dimensional potential energy surface, as shown in Figure 9.1. The letters α, β denote the internal quantum states in which A and BC are prepared before they react; α', β' denote the internal quantum states of the products $A + BC$ of inelastic scattering; γ, δ denote the internal quantum states of the products $AC + B$ of reactive scattering. Everything that we say in this chapter about reaction (9.17) is immediately applicable to more complicated reactions, including those with more than two dimensions, more than three atoms, more than one set of

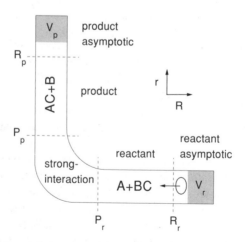

Figure 9.1 Diagram showing how the coordinate space of the two-dimensional $A + BC \rightarrow AC + B$ reaction is partitioned into the reactant asymptotic, reactant, strong-interaction, product, and product asymptotic regions. The partitioning is explained in the text and in Table 9.1. The shaded areas V_r and V_p represent absorbing potentials. The oval to the right of the line R_r represents the initial wave packet $\chi(0)$.

products, or that take place on more than one potential energy surface. We note that the theory we are to present is essentially a summary of previous work (including that of Neuhauser et al. 1989 and Seideman and Miller 1992), but that our application of it to state-to-state reactive scattering (to be described later in this chapter) is original.

The theory is based on the fact that the coordinate space of the $A + BC$ reaction can be partitioned into five regions as shown in Figure 9.1. Here the energetically accessible "tube" of the potential energy surface (connecting $A + BC$ with $AC + B$) is partitioned into the reactant asymptotic region, the reactant region, the strong-interaction region, the product region, and the product asymptotic region. In each region along the tube, the potential energy surface has a different property that gives rise to a different type of nuclear dynamics as summarized in Table 9.1. It may seem somewhat arbitrary to partition the coordinate space into different "regions" in this way: The potential energy surface changes gradually on passing from $A + BC$ to $AC + B$, as does the type of nuclear dynamics. The partitioning is rigorous, however, provided that we specify a numerical accuracy to which we want to calculate the wavefunction. We can then say, referring to Figure 9.1 and Table 9.1, that *to our specified accuracy*, there is a point R_r, along the $A + BC$ coordinate R, beyond which AB and C behave as though they were isolated molecules. We can make similar statements about the points R_p, P_p, and P_r. When discussing the partitioning of Figure 9.1 later in this chapter, we will often leave out specific reference to the "specified accuracy"; this is, however, always to be understood. In a practical calculation, the locations of R_r, R_p, P_p, and P_r are found by numerical convergence tests.

TABLE 9.1 Regions into which the coordinate space of the $A + BC \rightarrow AC + B$ reaction is partitioned in Figure 9.1.

Region	States Mixed by Potential	Type of Dynamics
Reactant asymptotic	None	Free motion of A away from or toward BC
Reactant	α, β	Motion of A away from or toward BC due to intermolecular forces
Strong interaction	$\alpha, \beta, \gamma, \delta$, arrangement ($A + BC$ or $AC + B$)	Formation of ABC transition state(s) and reactive intermediates; AB and AC bond breaking and formation
Product	γ, δ	Motion of AC away from B due to intermolecular forces
Product asymptotic	None	Free motion of AC away from B

Note: Reading the table from top to bottom corresponds to following the reaction of Figure 9.1 from the reactant asymptotic region (bottom right) to the product asymptotic region (top left).

We now consider what happens to a wave packet that is prepared in the given quantum states α of A, and β of BC, and that is given translational energies sufficient to initiate reaction (9.17). The wave packet is initially located in the reactant asymptotic region (to the right of R_r, as shown in Fig. 9.1). It then travels through the reactant region, and into the strong-interaction region, where it divides into two pieces. One piece travels into the reactant region and then back out into the reactant asymptotic region, to yield the inelastic products $A + BC$, in a distribution of internal quantum states α' and β'. The other piece travels into the product region and then into the product asymptotic region, to yield the reacted products $AC + B$, in a distribution of internal quantum states γ and δ. If we consider, say, the piece of the wave packet that travels into the product region, we can see that after passing P_p, the wave packet is traveling in the outward direction *only*, because the potential here is (by definition) too weak to mix $AC + B$ states with $A + BC$ states. Thus we can refer to P_p as the $AC + B$ or product "surface of no return," since after passing P_p, the wave packet cannot turn round and re-enter the strong-interaction region. We can similarly refer to P_r (at the start of the reactant region) as the $A + BC$ or reactant "surface of no return."

This completes our simple theory of state-to-state reactive scattering. For clarity, we have illustrated the theory on the two-dimensional $A + BC$ reaction of Figure 9.1, although, as we said earlier, it is applicable to more general reactions. The theory is in fact applicable to any chemical reaction for which it is possible to isolate the reactants and products. We will show in the following sections how the theory enables us to develop a very efficient wave packet method for calculating the state-to-state reactive scattering wavefunction.

9.2.2 The Time Dependent Wave Packet Method

Before describing the method just mentioned, we need to give a quick overview of how to solve the nuclear dynamics Schrödinger equation using wave packet propagation methods. We discuss the time dependent wave packet method in this section, and then the related time independent wave packet method in Section 9.2.3.

The time dependent close coupled wave packet method (Mowrey and Kouri 1986; Mowrey, Bowen, and Kouri 1987) was developed as a method of solving the time dependent Schrödinger equation

$$i\hbar \frac{\partial}{\partial t} \chi(t) = H\chi(t) \qquad (9.18)$$

by propagating $\chi(t)$ according to

$$\chi(t) = U(t)\,\chi(0). \qquad (9.19)$$

The action of the time-evolution operator $U(t) = \exp(-iHt/\hbar)$ is usually evaluated numerically, using one of a variety of propagation schemes (Feit and Fleck 1984; Tal-Ezer and Kosloff 1984) to generate $\chi(t)$ at a series of time steps.

When applying the time dependent wave packet method to reaction (9.17), $\chi(0)$ is constructed from a superposition of time independent wavefunctions, each of which describes the free motion of A toward BC at a given energy, and for a given α and β. The superposition is made in such a way that $\chi(0)$ is concentrated in a small region of space in the reactant-asymptotic region (as shown in Fig. 9.1). Thus $\chi(0)$ takes the form

$$\chi(0) = f(k_0, R)e^{-ik_0R} \qquad (9.20)$$

where the envelope function $f(k_0, R)$ (usually a Gaussian) is centered about a point to the right of R_r, and k_0 is the average momentum of $\chi(0)$. Expressed as a superposition of plane-waves, $\chi(0)$ can be written as

$$\chi(0) = \int_{-\infty}^{\infty} g(k_0, k)e^{ikR} \, dk, \qquad (9.21)$$

where $g(k_0, k)$ is the momentum-space analogue to $f(k_0, R)$. Typically $g(k_0, k)$ is chosen such that effectively all the free waves are propagating towards the strong-interaction region (i.e., $\underline{k} \cdot \underline{R} \leq 0$), and both $f(k_0, R)$ and $g(k_0, k)$ are taken as "window functions" of finite width. Although this condition is impossible to impose *exactly*, it can always be imposed at a specified accuracy. In a typical calculation (e.g., Althorpe, Kouri, and Hoffman 1998) the initial wave packet is constructed such that $f(k_0, R)$ extends just a few bohr along R, and $g(k_0, k)$ contains a finite range of momenta, centered about k_0, which satisfy $\underline{k} \cdot \underline{R} \leq 0$.

We have already described in Section 9.2.2 how the wave packet $\chi(t)$ propagates throughout the coordinate space of the reaction (shown in Fig. 9.1). When solving $\chi(t)$ numerically, then, the initial wave packet $\chi(0)$ will propagate into the strong-interaction region where it will split into two pieces; one piece will propagate out into the product region, and the other piece back into the reactant region. Since we only want to calculate $\chi(t)$ to our required accuracy, there exists a time t_{max} at which (to our required accuracy) all of the wave packet has passed beyond either R_r or R_p. The propagation of (9.19) is thus terminated at $t = t_{max}$. The value of t_{max} is of course different for each reaction; it is related to how long the wave packet spends in the strong-interaction region. Typically t_{max} is of the order of picoseconds for a direct reaction, but it can be of the order of nanoseconds for reactions with long-lived reactive intermediates (resonances). After propagating $\chi(t)$ up to $t = t_{max}$, one takes the time-to-energy transform of $\chi(t)$ to obtain the state-to-state reaction data as a function of total energy.

This almost completes our overview of the time dependent wave packet method as applied to reactive scattering. For a full introduction to the method, the reader is referred to the papers of Mowrey and Kouri (1986) and Mowrey, Bowen, and Kouri (1987). As recent examples of the application of the method to reactive scattering, we mention the calculations of Zhu et al. (1996) on the $OH + H_2$ reaction, and Zhang and Light (1996b) on the $H + H_2O$ reaction. We now discuss the advantages of the time dependent wave packet method over standard time independent methods.

Clearly, one advantage of the wave packet method is that it yields results over the entire range of energies contained in the initial wave packet (i.e., those energies at which $g(k_0, k)$ is nonzero), albeit at the cost of including the extra variable t in the calculation. This is in contrast to most time independent methods, in which the results must be re-calculated at each energy. Another advantage of the wave packet method is that each step of the calculation yields a physically meaningful result (i.e., $\chi(t)$). Thus it is easier to understand what is happening in a wave packet calculation than in a time independent calculation (for which the only physically meaningful results are those obtained right at the end of the calculation).

There is, however, another much more important advantage of the wave packet method which is not immediately apparent. This is that as the basis size N used to represent the Hamiltonian H increases, the effort required to calculate $\chi(t)$ by the wave packet method scales as better than N^2. This is in contrast to the standard time independent methods, which typically scale as N^3. Thus, even though the wave packet method includes the extra variable t, there will always be a certain value of N above which the wave packet method is *much more efficient* than the standard time independent methods! This rather surprising result was first predicted by Mowrey and Kouri (1986), and it has since been confirmed by recent wave packet calculations, such as the ones on $H + H_2O$ and $OH + H_2$ mentioned above. These calculations would have been very difficult, if not impossible, using standard time independent methods.

9.2.3 The Time Independent Wave Packet Method

We now describe a more direct (but equivalent) way of formulating the time dependent wave packet calculation just described, which is called the *time independent wave packet* (TIW) equation (Kouri, Arnold, and Hoffman 1993; Huang et al. 1993; Hoffman et al. 1994; Kouri and Hoffman 1995). The TIW equation is based on recognizing that in a reactive scattering calculation one usually wants to calculate not $\chi(t)$ itself but its time-to-energy transform

$$\xi^+(E) = \frac{1}{2\pi} \int_0^\infty dt \, \chi(t) e^{-iEt/\hbar}, \tag{9.22}$$

since it is from $\xi^+(E)$ that one obtains the energy dependent data (such as S-matrix elements and reaction probabilities) that can be experimentally compared. The function $\xi^+(E)$ is a time independent wavefunction but *not* a solution of the time independent Schrödinger (TIS) equation, as usually defined. $\xi^+(E)$ is in fact the solution of the TIW equation, which is obtained by applying the time-to-energy transform of equation (9.22) to equation (9.18) to yield,

$$(E - H)\xi^+(E) = \frac{i}{2\pi} \chi(0). \tag{9.23}$$

A formal expression for $\xi^+(E)$ is obtained by applying the same time-to-energy transform to (9.19) to yield

$$\xi^+(E) = \frac{i\hbar}{2\pi} G^+(E)\chi(0) \tag{9.24}$$

in which the time-to-energy transform of $U(t)$ appears as the causal Green's function $G^+(E)$. The reason the TIW equation differs from the TIS equation is that $\xi^+(E)$ is the time-to-energy transform of a time dependent wavefunction $\chi(t)$ which is "prepared" at an initial time $t = 0$ (i.e., $\chi(t) = 0$ for $t < 0$ and $\chi(t) = \chi(0)$ at $t = 0$). The (Lippmann-Schwinger) solution $\Psi(E)$ of the TIS equation, on the other hand, is the time-to-energy transform of a $\chi(t)$ which is prepared at time $t = -T$ in the limit as $T \rightarrow \infty$. The TIW equation thus "remembers" the initial wave packet $\chi(0)$ (which appears as an inhomogeneous source term on the right of (9.23)), whereas the TIS equation does not. It can be shown that the TIW solution $\xi^+(E)$ is proportional to $\Psi(E)$ throughout the entire coordinate space of the reaction, except for the small region to the right of $\chi(0)$ in Figure 9.1; the proportionality constant is a function of $g(k_0, k)$.

We can thus regard the time dependent wave packet calculation of $\chi(t)$ (equations (9.18) and (9.19)) and the subsequent time-to-energy transformation of $\chi(t)$ to $\xi^+(E)$ as an indirect method of solving the TIW equation (9.23). We now consider a *direct* method of solving the TIW equation, which we will refer to as the *TIW method* (Huang, Kouri, and Hoffman 1994; Huang et al. 1996). This method is based on the well-known Chebyshev propagator, which was introduced some years ago (Tal-Ezer and Kosloff 1984) as a propagator for evaluating equation (9.19). In the TIW method the solution $\xi^+(E)$ of (9.24) is expanded as

$$\xi^+(E) = \frac{1}{2\pi\Delta H \sin \phi} \sum_{n=0}^{N} e^{-in\phi}\eta_n, \tag{9.25}$$

where the energy independent basis functions η_n are generated from the Chebyshev recursion relations

$$\eta_0 = \chi(0),$$

$$\eta_1 = 2H_{norm}\eta_0,$$

$$\eta_2 = 2H_{norm}\eta_1 - 2\eta_0$$

$$\eta_n = 2H_{norm}\eta_{n-1} - \eta_{n-2} \quad \text{for } n > 2. \tag{9.26}$$

The Hamiltonian appears in the recursion in the scaled form $H_{norm} = (H - \overline{H})/\Delta H$, in which the scaling factors $\overline{H}$ and ΔH are chosen such that the spectrum of H_{norm} is confined to the interval $[-1{:}1]$. The phase ϕ in (9.25) is defined by $\cos \phi = (E - \overline{H})/\Delta H$.

Thus, to calculate $\xi^+(E)$ by the TIW method, one generates the set of functions η_n by repeatedly acting on $\chi(0)$ with the (scaled) Hamiltonian H_{norm}. The index n correlates strongly with the time t, such that everything we said above about the evolution of $\chi(t)$ (with respect to t) also holds for the evolution of η_n with respect to n. In particular, there is a maximum value of n (N in (9.25)) analogous to the $t_{\max}$ defined above such that, for all $n > N$, the η_n will have passed beyond either R_r or R_p, (having split in two in the strong-interaction region, and propagated outward through either the reactant or product region).

We shall often refer back to the TIW method in the rest of this chapter, and we will present results in Section 9.3 that were calculated using the method. We prefer the TIW method to the (more commonly used) time dependent wave packet method because it is a simple, direct way of calculating a numerically exact $\xi^+(E)$ (once we have specified our accuracy and hence N). There is a similar time-dependent method (Neuhauser et al. 1991) based on the Chebyshev propagator, but this is not as efficient and direct as the TIW method.

We will shortly return to the partitioning of coordinate space described in Section 9.2.1 and Figure 9.1, showing how to use it to make the TIW method (or time dependent wave packet method) efficient for reactive scattering. In doing this, we will make use of imaginary absorbing potentials, which we now discuss briefly in the following section.

9.2.4 Absorbing Potentials

Absorbing potentials were introduced into the wave packet propagation method by several authors (e.g., see Leforestier and Wyatt 1983), in order to absorb the wave packet once it entered the asymptotic regions. The first absorbing potentials to do this efficiently and practically were introduced by Neuhauser and Baer (1989) and Neuhauser et al. (1989). These absorbing potentials are now used in all wave packet calculations of reactive scattering, since without them the calculations are much less efficient. We give here a brief explanation of the absorbing potentials in order to explain the use we will make of them in Sections 9.2.5 to 9.2.9.

Let us begin by considering a standard calculation of the time independent wavefunction $\Psi_{\alpha\beta}(E)$ for the state-to-state reactive scattering of $A(\alpha) + BC(\beta)$. Whichever method one uses, one calculates a set of solutions to the TIS which are confined to what we will call the *reaction coordinate space*. This is the coordinate space made up of the reactant, strong-interaction, and product regions (but not the asymptotic regions). The solutions are then matched at R_r and R_p such that $\Psi(E)$ satisfies the outgoing boundary conditions

$$\lim_{R \to \infty} \langle\, \alpha'\beta' \mid \Psi_{\alpha\beta}(E)\rangle = \delta_{\alpha'\alpha}\delta_{\beta'\beta}f^-(R) + I^{\alpha'\beta'}_{\alpha\beta}(E)f^+(R),$$

$$\lim_{r \to \infty} \langle\gamma\delta \mid \Psi_{\alpha\beta}(E)\rangle = R^{\gamma\delta}_{\alpha\beta}(E)g^+(r), \qquad (9.27)$$

or an equivalent set. These conditions are a more formal way of stating our definitions of the reactant and product asymptotic regions given in Table 9.1. They say that in the reactant or product asymptotic regions, the potential is too weak to mix the internal quantum states of $A + BC$ or $AC + B$, so the wavefunction is a distribution of free waves, each describing either the free motion of A toward BC ($f^-(R)$), A away from BC ($f^+(R)$), or AC away from B ($g^+(r)$). The square moduli of the S-matrix elements $I^{\alpha'\beta'}_{\alpha\beta}(E)$ and $R^{\gamma\delta}_{\alpha\beta}(E)$ are proportional to the probabilities that A (α) + BC (β) scatter inelastically to give $A(\alpha')$ + $BC(\beta')$ or react to give $AC(\gamma)$ + $B(\delta)$.

In a standard time independent calculation, then, one is able to take advantage of the boundary conditions of (9.27) so as to restrict the calculation to the reaction coordinate space. Unfortunately, this cannot be done in a wave packet calculation, as may be seen by taking the time-to-energy transform of (9.27) (to obtain the equivalent condition in the time domain). The outgoing free wave $g^+(r)$ transforms to a wave packet traveling out into the product asymptotic region such that, in the limit as $t \to \infty$, AC and B can be considered isolated molecules. The outgoing free wave $f^+(R)$ transforms to a similar wave packet traveling out into the reactant asymptotic region. Thus the equivalent condition to (9.27) in a wave packet calculation is that the wave packet should be allowed to propagate out into the reactant or product region until the time $t_{\max}$ (or, equivalently, until $n = N$ in a TIW calculation). We do not have to worry about the incoming free wave $f^-(R)$, since, as we mentioned in Section 9.2.3, we prepare the wave packet at a finite separation of A and BC at time $t = 0$; $f^-(R)$, on the other hand, transforms to a wave packet prepared in the limit as $t \to -\infty$.

For most reactions it is inefficient to allow the wave packet to travel out into the asymptotic regions because, after entering the strong-interaction region, the wave packet spreads out, and some components of it stay much longer in this region than others. By the time the whole packet has passed R_r and R_p (i.e., at $t = t_{\max}$ or at $n = N$, with N defined as above), the end of the packet will extend out into the asymptotic regions by typically hundreds of bohr. Fortunately this distance can be reduced to just a few bohr by placing artificial absorbing potentials just after R_r and R_p. Instead of propagating solutions subject to equation (9.18) (or equivalently (9.23)), one propagates solutions subject to the equation

$$i\hbar \frac{\partial \chi(t)}{\partial t} = H\chi(t) - iV_r\,\chi(t) - iV_p\,\chi(t), \tag{9.28}$$

where V_r and V_p are usually (Neuhauser and Baer 1989; Neuhauser et al. 1989) taken of the form

$$V_\lambda(R) = a\left(\frac{R - R_\lambda}{\Delta R}\right)^b, \qquad \lambda = r, p, \tag{9.29}$$

and are placed a short distance after R_r and R_p (as shown in Fig. 9.1). In the limit as $a \to 0$ and $\Delta R \to \infty$, $-iV_r$ and $-iV_p$ act as perfect absorbers, gently damping the wave packet to zero without reflecting any of it. In practice, a and ΔR are chosen such that V_r and V_p absorb all of the wave packet within our specified accuracy, which usually requires ΔR to be only a few bohr. There is a small region of space just after R_r and R_p in which the wave packet behaves as though it were about to continue propagating into the asymptotic region. The subsequent absorption of the packet by

V_r or V_p does not affect the packet in this small region (which needs to be just big enough for one to extract $I_{\alpha\beta}^{\alpha'\beta'}(E)$ and $R_{\alpha\beta}^{\gamma\delta}(E)$ from the packet), nor does it affect the packet in the reaction coordinate space.

Mandelshtam and Taylor (1995a,b) have recently developed a very efficient way of including absorbing potentials into the TIW method. They modify the Chebyshev expansion (9.25) to

$$\xi^+(E) = \frac{1}{2\pi\Delta H \sin\phi} \sum_{n=0}^{N} e^{-in\phi}\,\overline{\eta}_n, \tag{9.30}$$

where

$$\overline{\eta}_n = 2e^{-\gamma}H_{\text{norm}}\overline{\eta}_{n-1} - e^{-2\gamma}\overline{\eta}_{n-2} \tag{9.31}$$

and γ is a (coordinate-dependent) damping factor. It may be shown (Mandelshtam and Taylor 1995b) that including γ into (9.31) is equivalent to including an energy dependent, complex absorbing potential

$$\Gamma(E) = \Delta H[\cos\phi(1 - \cosh\gamma) - i\sin\phi\sinh\gamma] \tag{9.32}$$

into the Hamiltonian. The energy dependence and the real part of $\Gamma(E)$ both turn out to be unimportant, so $\Gamma(E)$ absorbs the wave packet in the asymptotic region just as an energy independent, imaginary absorbing potential does. We will be making use of another form of damped Chebyshev propagator below, which is closely related to this one.

9.2.5 Calculating Total and Cumulative Reaction Probabilities

We now return to the partitioning of coordinate space described in Section 9.2.1 and shown in Figure 9.1 for the $A + BC$ reaction (9.17). In this section we discuss two well-known methods in which the partitioning is used to calculate efficiently the total reaction probabilities and the cumulative reaction probabilities. What we say about these methods will be relevant to Sections 9.2.6 to 9.2.9, when we show how to use the partitioning to calculate efficiently the state-to-state reaction probabilities.

The total reaction probability is defined as the sum of state-to-state probabilities $P_{\alpha\beta}^{\gamma\delta}(E)$,

$$\overline{P}_{\alpha\beta}(E) = \sum_{\gamma\delta} P_{\alpha\beta}^{\gamma\delta}(E) \tag{9.33}$$

and is thus the total probability that $A(\alpha) + BC(\beta)$ reacts to give $AC + B$ at the energy E. One can of course calculate $\overline{P}_{\alpha\beta}(E)$ by calculating all the state-to-state probabilities $P_{\alpha\beta}^{\gamma\delta}(E)$ and then summing over γ, δ, but this is inefficient if one does not want the explicit values of $P_{\alpha\beta}^{\gamma\delta}(E)$. A more efficient way to calculate $\overline{P}_{\alpha\beta}(E)$ is to truncate the coordinate space a short distance into the product region, placing the absorbing potential V_{pq} just beyond P_p, as shown in Figure 9.2. This method was introduced by Neuhauser et al. (1989, 1991) and has since become one of the standard methods of calculating $\overline{P}_{\alpha\beta}(E)$. It is obviously more efficient than obtaining $\overline{P}_{\alpha\beta}(E)$ from the state-to-state calculation of $P_{\alpha\beta}^{\gamma\delta}(E)$, since the calculation is restricted to the *reactant-interaction* region of coordinate space (which we define to be the reactant region plus the strong-interaction region).

To calculate $\overline{P}_{\alpha\beta}(E)$ using the method of Neuhauser et al., the initial wave packet is prepared in exactly the same way as in a wave packet calculation of the state-to-state reaction probabilities. The packet is propagated into the strong-interaction region, where it splits into two pieces (just as in a state-to-state calculation), each piece eventually moving out of the strong-interaction region. The piece that travels into the reactant region continues through to the reactant asymptotic region, exactly as it would have done in a state-to-state calculation. The piece that travels into the product region, on the other hand, is absorbed by V_{pq}. This absorption has no effect on the wave packet up to P_p, for the same reason that the absorption by V_p in the state-to-state calculation (see Fig. 9.1) has no effect on the wave packet up to R_p: Beyond P_p the wave packet is propagating outward only (i.e., P_p is a surface of no return, as we explained in Section 9.2.1), so it makes no difference to the wave packet up to P_p whether the wave packet after P_p is absorbed or allowed to continue. At the end of the calculation, $\overline{P}_{\alpha\beta}(E)$ is obtained by calculating the flux of the wave packet into V_{pq}.

The cumulative reaction probability $N(E)$ is defined as the sum

$$N(E) = \sum_{\alpha\beta} \overline{P}_{\alpha\beta}(E) = \sum_{\alpha\beta\gamma\delta} P_{\alpha\beta}^{\gamma\delta}(E), \tag{9.34}$$

and it is thus the total probability that $A + BC$ reacts to give $AC + B$, at the energy E. $N(E)$ is an important quantity because it yields the rate constant $k(T)$ of the reaction from the expression

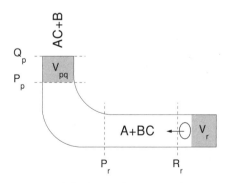

Figure 9.2 Coordinate space used in the method of Neuhauser et al. (1989, 1991) for calculating the total reaction probabilities $\overline{P}_{\alpha\beta}(E)$ of the $A + BC$ reaction. The lines P_r and P_p are the same as in Figure 9.1.

$$k(T) = \frac{1}{2\pi\hbar Q_r(T)} \int_{-\infty}^{\infty} dE \, e^{-E/kT} N(E). \tag{9.35}$$

It is possible to calculate $N(E)$ by calculating all the $P_{\alpha\beta}^{\gamma\delta}(E)$ (or all the $\overline{P}_{\alpha\beta}(E)$), but again, such a calculation is inefficient if one does not want the explicit values of $P_{\alpha\beta}^{\gamma\delta}(E)$ (or $\overline{P}_{\alpha\beta}(E)$). An efficient method for calculating $N(E)$ was developed by Seideman and Miller (1992), in which the coordinate space of Figure 9.2 is further truncated a short distance into the reactant region, by placing the absorbing potential V_{rq} just beyond P_r, as shown in Figure 9.3. In this method a set of wavefunctions is calculated in the strong-interaction region, using either a time independent method (Seideman and Miller 1992) or a wave packet method (Zhang and Light 1996a). In the latter, a set of initial wave packets are placed in the middle of the strong-interaction region and are propagated throughout the coordinate space of Figure 9.3; the packets eventually split into two pieces, each of which enters either the reactant or the product region, where it is absorbed by V_{rq} or V_{pq}. Since V_{rq} and V_{pq} are placed after the surfaces-of-no-return P_r and P_p, they do not affect the wave packets in the strong-interaction region, and they are thus propagated exactly (within the required accuracy). The cumulative reaction probabilities $N(E)$ are obtained by calculating the relative flux of the wave packets into V_{rq} and V_{pq}.

The method just discussed is one of the most important recent developments in chemical reaction theory; It is the most efficient method of calculating the rate constant of a reaction using rigorous quantum mechanics. Having specified the accuracy to which one wants to calculate the rate constant, the method allows one to make the calculation as small as possible, enclosing the coordinate space of the $A + BC$ reaction in the "potential-energy box" of Figure 9.3 (which is made up of V_{rq}, V_{pq}, and the repulsive walls of the $A + BC$ potential energy surface). In Sections 9.2.8 and 9.2.9, we will show how to use the reduced coordinate space of Figure 9.3 to calculate the wavefunction very efficiently, not just in the strong-interaction region

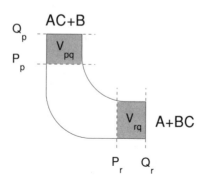

Figure 9.3 Coordinate space used in the method of Seideman and Miller (1992) for calculating the cumulative reaction probabilities $N(E)$ of the $A + BC$ reaction. The lines P_r and P_p are the same as in Figure 9.1.

but also in the reactant and product regions. We lead up to this in the next two sections by showing how to calculate the wavefunction in the product region, after having calculated it in the reactant-interaction region of Figure 9.2.

9.2.6 The Reactant-Product Decoupling Method

In the method shown in Figure 9.2 (which was described above) the wavefunction is calculated exactly in the reactant-interaction region in order to yield the total reaction probabilities $\overline{P}_{\alpha\beta}(E)$. Such a calculation cannot of course yield the state-to-state reaction probabilities, since these require a knowledge of the wavefunction in the product region (and also in a small portion of the product asymptotic region). Very recently, however, Peng and Zhang (1996) have developed a way of extending the method of Figure 9.2 so that, after calculating the wavefunction in the reactant-interaction region, one goes on to calculate the wavefunction in the product region. This method is called the *reactant-product decoupling* (or RPD) method, and it is illustrated schematically in Figure 9.4 for the $A + BC$ reaction.

The basis of the RPD method is simply that we can write down the Schrödinger equation that is solved in the method of Figure 9.2 as

$$i\hbar\frac{\partial\chi_r(t)}{\partial t} = H\chi_r(t) - iV_{pq}\chi_r(t) \tag{9.36}$$

and can then add this to another equation (Peng and Zhang 1996)

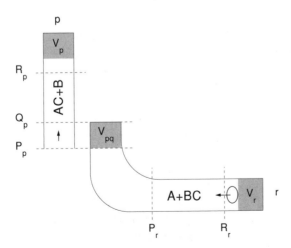

Figure 9.4 Coordinate space of the $A + BC$ reaction partitioned into regions r and p in the reactant-product decoupling (RPD) method of Peng and Zhang (1996). Regions r and p overlap between P_p and Q_p, the overlap containing the absorbing potential V_{pq} in region r, and the source term $V_{pq}\chi_r(t)$ (represented by the vertical arrow) in region p. The area of coordinate space in region r between R_r and P_p is the reactant-interaction region.

$$ i\hbar \frac{\partial \chi_p(t)}{\partial t} = H\chi_p(t) + iV_{pq}\chi_r(t) \tag{9.37} $$

to obtain

$$ i\hbar \frac{\partial [\chi_r(t) + \chi_p(t)]}{\partial t} = H[\chi_r(t) + \chi_p(t)]. \tag{9.38} $$

This last equation is clearly the time dependent Schrödinger equation (9.18) for the A + BC reaction (whose solution $\chi(t)$ extends throughout the entire reaction coordinate space shown in Fig. 9.1). Thus we can write

$$ \chi(t) = \chi_r(t) + \chi_p(t) \tag{9.39} $$

and calculate $\chi(t)$ by solving the *reactant-product decoupling equations* (9.36) and (9.37), to yield the components $\chi_r(t)$ and $\chi_p(t)$. We know that $\chi_r(t)$ is confined to region r of Figure 9.4, and that in the reactant-interaction region (made up of the reactant and strong interaction regions) $\chi_r(t)$ is equal to $\chi(t)$. Thus we also know from equation (9.39) that $\chi_p(t)$ is zero in the strong-interaction region, and hence that $\chi_p(t)$ is confined to region p of Figure 9.4, where it is equal to $\chi(t)$ for $r > Q_p$.

In the RPD method (Peng and Zhang 1996; Zhu, Peng, and Zhang 1997) one first solves equation (9.36) to yield $\chi_r(t)$ by propagating an initial wave packet $\chi(0)$ exactly as if one were calculating the total reaction probabilities using the method of Figure 9.2. One then solves equation (9.37) to yield $\chi_p(t)$ by propagating the time dependent source term $iV_{pq}\chi_r(t)$ throughout region p. We know that $iV_{pq}\chi_r(t)$ must propagate in one direction only, toward the product asymptotic region (as shown by the vertical arrow in Fig. 9.4), since otherwise $\chi_p(t)$ would be able to re-enter the strong-interaction region (where $\chi_p(t)$ is zero). After calculating $\chi_p(t)$, one extracts from it the S-matrix elements of (9.27) and hence the state-to-state reaction probabilities.

Thus the RPD method enables the calculation of $\chi(t)$ for the $A + BC$ reaction to be split into the calculation of $\chi_r(t)$, which is confined to (mostly) the reactant-interaction region, and the calculation of $\chi_p(t)$, which is confined to the product region. These two calculations may be carried out completely separately because equations (9.36) and (9.37) are completely decoupled. To our required accuracy, the RPD method is *exact*, since it splits up the calculation, not by making an approximation but by making use of the partitioning of coordinate space introduced in Figure 9.1. In other words, the RPD method is a way of simplifying the calculation of $\chi(t)$ by recognizing that the dynamics in the strong-interaction region is unaffected by the dynamics in the product region. As we said above, this is because P_p is a surface of no return, beyond which $\chi(t)$ cannot re-enter the strong-interaction region.

We mentioned at the start of Section 9.2.1 that the partitioning of coordinate space shown in Figure 9.1 for the $A + BC$ reaction is also applicable to more complicated reactions. Hence the RPD method is also applicable to reactions in which there are N

different sets of products: One has simply to replace V_{pq} by a set of N absorbing potentials, each of which is placed at the start of one of the product regions, analogously to V_{pq} in Figure 9.4; one then obtains a set of N equations in place of equation (9.37), each of which yields a solution that is confined to one of the product regions.

9.2.7 The TIW Reactant-Product Decoupling Method

It should be apparent from Section 9.2.6 that the RPD method is potentially a much more efficient method of calculating $\chi(t)$ than is the wave packet propagation method described in Section 9.2.2. Instead of propagating a wave packet that spreads throughout the entire reaction coordinate space, one propagates, first, a packet that spreads through the reactant-interaction region, and, second, a packet that spreads through the product region. There is, however, also a potential disadvantage of the RPD method: The first RPD equation, (9.36), can be solved by propagating $\chi(0)$ exactly as was described in Section 9.2.2; the second equation, (9.37), on the other hand, cannot be, since instead of propagating a time independent source (e.g., $\chi(0)$), one must propagate the time dependent source $iV_{pq}\chi_r(t)$. To propagate $iV_{pq}\chi_r(t)$ using the time dependent wave packet method (or to propagate the equivalent n-dependent source using the TIW method) requires a separate propagation of $iV_{pq}\chi_r(t)$ to be evaluated at each value of t (or at each value of n in the TIW method).

We now describe an efficient method of solving the RPD equations, whereby the second equation, (9.37), is solved with just one wave packet propagation. This method solves the TIW version of equations (9.36) and (9.37), which we will give shortly, and requires the damped Chebyshev recursion to be modified so that it differs from that of Mandelshtam and Taylor (1995a) in equation (9.31). We mention that Peng and Zhang (1996) and Zhu, Peng, and Zhang (1997) have developed a method of solving the original time dependent equations (equations (9.36) and (9.37)), which also solves equation (9.37) with just one wave packet propagation.

The TIW version of the RPD equations may be derived by taking the time-to-energy transform of (9.36) and (9.37) (Althorpe et al. 1997). Alternatively, one can start with the TIW equation (9.24), with $G^+(E)$ taken to be the Green's function which propagates $\chi(0)$ throughout the entire coordinate space of the $A + BC$ reaction (shown in Fig. 9.1). We give here a summary of the latter derivation, since it helps to explain the propagation method used to solve the equations. The full derivation is given by Althorpe, Kouri, and Hoffman (1997a).

The first step in deriving the TIW version of the RPD equations (the "TIW RPD equations") is to decide how best to include V_{pq}. We mentioned at the end of Section 9.2.3 that the most efficient way of including an absorbing potential into the TIW equation is to include one having the form of (9.32), which is similar to the work of Mandelshtam and Taylor (1995a,b). We thus replace V_{pq} by an energy dependent absorbing potential $\Gamma_{pq}(E)$, which takes the form of the imaginary part of equation (9.32):

$$\Gamma_{pq}(\phi) = -iV_{pq}\Delta H \sin \phi. \tag{9.40}$$

Unlike Mandelshtam and Taylor, we do not include the real part of equation (9.32) into $\Gamma_{pq}(E)$ because this part does not absorb the wave packet and makes the resulting TIW RPD equations more difficult to solve.

To include $\Gamma_{pq}(E)$ into equation (9.24), we write out the Lippmann-Schwinger equation

$$G^+(E) = G_r^+(E) - G^+(E)\Gamma_{pq}(E)G_r^+(E) \tag{9.41}$$

which relates $G^+(E)$, the Green's function corresponding to the Hamiltonian H, to $G_r^+(E)$, the Green's function corresponding to the Hamiltonian $H + \Gamma_{pq}(E)$. The TIW RPD equations are then obtained by substituting (9.41) into (9.24), to yield

$$\xi_r^+(E) = \frac{i}{2\pi}G_r^+(E)\chi(0), \tag{9.42}$$

$$\xi_p^+(E) = -G^+(E)\Gamma_{pq}(E)\xi_r^+(E), \tag{9.43}$$

where

$$\xi^+(E) = \xi_r^+(E) + \xi_p^+(E). \tag{9.44}$$

Equations (9.42) to (9.44) are the TIW equivalents to equations (9.36) to (9.39), as may be confirmed by considering the propagation of $\xi_r^+(E)$ and $\xi_p^+(E)$, the TIW equivalents to $\chi_r(t)$ and $\chi_p(t)$. The solution $\xi_r^+(E)$ is confined to region r of Figure 9.4 because it is propagated by the Green's function $G_r^+(E)$ which includes the absorbing potential $\Gamma_{pq}(E)$. Thus in the reactant-interaction region, $\xi_r^+(E)$ is equal to $\xi^+(E)$. The solution $\xi_p^+(E)$ is confined to region p of Figure 9.4 because, from equation (9.44), $\xi_p^+(E)$ is zero in the strong-interaction region (where $\xi^+(E) = \xi_r^+(E)$). Thus in the product region $\xi_p^+(E)$ is equal to $\xi^+(E)$ for $r > Q_p$. The energy dependent source $\Gamma_{pq}(E)\xi_r^+(E)$ of equation (9.43) thus propagates through region p, traveling in the outward direction only. Note that for clarity, we have omitted the absorbing potentials V_r and V_p from equations (9.42) and (9.43); it is always straightforward to include V_r and V_p using equations (9.30) and (9.31).

Having derived the TIW RPD equations, we now show how they can be solved efficiently by splitting up the Chebyshev propagator given in (9.25) and (9.26). As with the time dependent RPD equations, it is the equation for region p, equation (9.43), which is difficult to solve, since this equation contains the energy dependent source term $\Gamma_{pq}(E)\xi_r^+(E)$. We split up the Chebyshev propagator by expanding each of $\xi_r^+(E)$ and $\xi_p^+(E)$ as

$$\xi_\lambda^+(E) = \frac{1}{2\pi\Delta H \sin\phi}\sum_{n=0}^{N} e^{-in\phi}\,\eta_{\lambda n}, \qquad \lambda = r, p, \tag{9.45}$$

where ΔH, ϕ, and N have the same definitions as in equation (9.25). We can then make use of the "inverse Chebyshev transform" (Althorpe, Kouri, and Hoffman 1997a),

$$\Delta H \int_{-\pi}^{\pi} \xi^+(E) \, e^{in\phi} \sin \phi \, d\phi = \begin{cases} \eta_n & \text{for } n \geq 0, \\ 0 & \text{for } n < 0, \end{cases} \tag{9.46}$$

together with equation (9.41), to obtain recursion relations for η_{rn} and η_{pn} which are analogous to those for η_n in equation (9.26). Now there is a value n_{pq} of n such that for all $n < n_{pq}$ the overlap of η_{rn} with V_{pq} is zero (to our specified accuracy), because η_n has to travel through the reactant region and into the strong-interaction region before it reaches V_{pq}. Hence for $n < n_{pq}$ the recursion relations for η_{rn} are identical to the recursion relation for η_n given in equation (9.26) (where η_n is the n-dependent wave packet that would have been generated had one solved the original TIW equation (9.24)). Similarly each η_{pn} is zero for $n < n_{pq}$. For $n \geq n_{pq}$ the recursion relation for η_{rn} is

$$\eta_{rn} = \frac{1}{1 + V_{pq}} [2H_{\text{norm}} \eta_{rn-1} - (1 - V_{pq}) \eta_{rn-2}], \tag{9.47}$$

and the recursion relation for η_{pn} is

$$\eta_{pn} = 2H_{\text{norm}} \eta_{pn-1} - \eta_{pn-2} + \sigma_n, \tag{9.48}$$

where the n-dependent source term σ_n is given by

$$\sigma_n = V_{pq}(\eta_{rn} - \eta_{rn-2}). \tag{9.49}$$

It may be shown that

$$\eta_n = \eta_{rn} + \eta_{pn} \tag{9.50}$$

(to our specified accuracy). This last equation implies that the propagations of (9.47) and (9.48) can be terminated at $n = N$ because (as defined in Section 9.2.3) N is the value of n above which η_n, and hence each of η_{rn} and η_{pn} has exited the reaction coordinate space. The propagation of η_{pn} given in (9.48) is initiated at $n = n_{pq}$, since this is the lowest value of n at which σ_n is nonzero.

The recursion relations (9.47) and (9.48) ensure that η_{rn} propagates through region r only, and that η_{pn} propagates through region p only. Hence the factor $1/(1 + V_{pq})$ in equation (9.47) damps η_{rn} in the region of V_{pq} in such a way that η_{rn} is kept out of the product region, without being reflected back into the reactant-interaction region; in the reactant-interaction region η_{rn} is equal to η_n. From (9.50) we can see that η_{pn} is zero in the strong-interaction region (where $\eta_{rn} = \eta_n$) and that the source terms σ_n must therefore propagate through region p, traveling in the outward direction only (as shown by the vertical arrow in Fig. 9.4).

The method we have just described clearly takes full advantage of the form of equations (9.36) to (9.39). The propagation scheme in (9.47) and (9.48) requires only one propagation to be evaluated in region r, and one propagation in region p. When evaluating each propagation, H need only be represented by a basis set that spans the appropriate region of coordinate space. The two propagations are completely decoupled, since the only connection between equations (9.47) and (9.48) is the source term σ_n. The σ_n are of course generated in the propagation of η_{rn}, when they may be stored on disc to be retrieved later for propagating the η_{pn}.

We finish this section by mentioning that equation (9.47) can be evaluated on its own, without evaluating (9.48). This means that, if V_{pq} in (9.47) is replaced by $V_r + V_p$, then (9.47) may be used as an alternative to the damping method of Mandelshtam and Taylor—as a method of propagating $\xi^+(E)$ through the reaction coordinate space of Figure 9.1 and absorbing $\xi^+(E)$ with V_r and V_p. It makes no difference in such a calculation which of the two damping methods is used; the damping method of (9.47) must, however, be used when solving the TIW RPD equations (since the damping method of Mandelshtam and Taylor leads to a recursion relation for η_{pn} that is impractical to evaluate).

9.2.8 Further Partitioned Version of the RPD Method

We now describe the second version of the RPD method, which we mentioned at the end of Section 9.2.5. This version enables the coordinate space of the reaction to be partitioned in the most efficient way possible, so that the dynamics in the strong-interaction region can be treated completely separately from the dynamics in the reactant region and the dynamics in the product region. This version of the RPD method is based on a further partitioned generalization (Kouri et al. 1996; Althorpe, Kouri, and Hoffman 1997b) of Peng and Zhang's original RPD equations, which we give below. We will show how to solve these equations using the TIW method in the next section.

The further partitioned version of the RPD equations may be written (Althorpe, Kouri, and Hoffman 1997b)

$$i\hbar\frac{\partial\chi_s(t)}{\partial t} = H\chi_s(t) + W_{qs}\chi_s(t), \tag{9.51}$$

$$i\hbar\frac{\partial\chi_q(t)}{\partial t} = H\chi_q(t) - W_{qs}\chi_s(t) - iV_{rq}\chi_q(t) - iV_{pq}\chi_q(t), \tag{9.52}$$

$$i\hbar\frac{\partial\chi_r(t)}{\partial t} = H\chi_r(t) + iV_{rq}\chi_q(t), \tag{9.53}$$

$$i\hbar\frac{\partial\chi_p(t)}{\partial t} = H\chi_p(t) + iV_{pq}\chi_q(t), \tag{9.54}$$

where the absorbing potentials V_{rq} and V_{pq}, and the real potential W_{qs} (which we explain below), are placed as shown in Figure 9.5. In analogy to equations (9.36) and (9.37), equations (9.51) to (9.52) add up so that $\chi(t)$, given by

$$\chi(t) = \chi_s(t) + \chi_q(t) + \chi_r(t) + \chi_p(t), \qquad (9.55)$$

is the (exact) solution (within our specified accuracy) to the time dependent Schrödinger equation (9.18) for the $A + BC$ reaction. The labels p, q, r, and s used in equations (9.51) to (9.54) refer to the regions of coordinate space shown in Figure 9.5; note that the region r defined in Figure 9.5 is different from the region r defined previously in Figure 9.4. In the second version of the RPD method, one solves the RPD equations (9.51) to (9.54), in place of the original RPD equations (9.36) to (9.37). We now explain how each of the solutions $\chi_\lambda(t)$, $\lambda = p, q, r, s$, is confined to the corresponding region λ of Figure 9.5, starting with the solution $\chi_s(t)$ of equation (9.51).

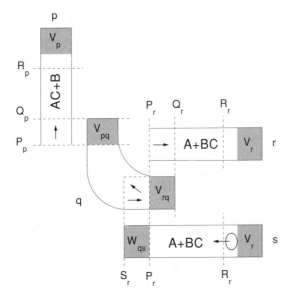

Figure 9.5 Coordinate space of the $A + BC$ reaction partitioned into regions p, q, r, and s in the second version of the RPD method. Region r of Figure 9.4 is here further partitioned into regions q, r, and s. Region p is the same as region p in Figure 9.4. Each of regions r and s encloses the reactant region and reactant asymptotic region of Figure 9.1. Region s also encloses part of the strong-interaction region at W_{qs}. Regions q and r overlap between P_r and Q_r, the overlap containing the absorbing potential V_{rq} in region q, and the source term $V_{rq}\chi_q(t)$ (the horizontal arrow) in region r. Regions q and s also overlap between P_r and Q_r, and between P_r and S_r, the latter overlap containing the reflecting potential W_{qs} in region s, and the source term $W_{qs}\chi_s(t)$ (represented by two arrows) in region q.

The solution $\chi_s(t)$ is confined to region s of Figure 9.1 because W_{qs} is defined to be a real potential which *reflects the entire wave packet back into the reactant region* (Althorpe, Kouri, and Hoffman 1998). The initial wave packet $\chi(0)$ thus travels through the reactant region, reflects off W_{qs}, travels back through the reactant region, and is finally absorbed by V_r. Thus $\chi_s(t)$ is confined to the reactant region but is not equal to $\chi(t)$ there, since, after reflecting off W_{qs}, $\chi_s(t)$ is mostly artificial. As we will explain later, the artificially reflected component of $\chi_s(t)$ is exactly canceled out by an equal and opposite component of $\chi_r(t)$. (Note that equations (9.51) to (9.54) hold regardless of whether W_{qs} is an absorbing potential, or a reflecting potential. We take W_{qs} to be a reflecting potential because it is always possible to choose a reflecting potential that fits inside the strong-interaction region; an absorbing potential must be of a certain length, which means that it does not always fit inside the strong-interaction region.)

The solution $\chi_q(t)$ is generated by propagating the time dependent source $W_{qs}\chi_s(t)$ (on the right of equation (9.51)) and is confined to region q by the absorbing potentials V_{rq} and V_{pq}, placed as shown in Figure 9.5. Since $W_{qs}\chi_s(t)$ is formed from the overlap of $\chi_s(t)$ with W_{qs}, it is located just inside the strong-interaction region (as indicated by dotted lines in Figure 9.5) so that $\chi_q(t)$ propagates throughout the strong-interaction region, eventually splitting into two pieces. One piece enters the product region, where it is absorbed by V_{pq}; the other piece enters the reactant region, where it is absorbed by V_{rq}.

Let us now suppose that instead of solving (9.51) to (9.54), we had solved the equations

$$i\hbar\frac{\partial\chi_s(t)}{\partial t} = H\chi_s(t) + W_{qs}\chi_s(t), \tag{9.56}$$

$$i\hbar\frac{\partial\chi_{qs}(t)}{\partial t} = H\chi_{qs}(t) - W_{qs}\chi_s(t), \tag{9.57}$$

where H and W_{qs} are defined exactly as in (9.51) to (9.54). Clearly, equation (9.56) is identical to equation (9.51), and hence $\chi_s(t)$ in equations (9.56) and (9.57) is identical to $\chi_s(t)$ in equations (9.51) to (9.54). After we had solved equation (9.56) for $\chi_s(t)$, we would have propagated the source term $W_{qs}\chi_s(t)$ (which is the same as in equation (9.51)) throughout the entire reaction coordinate space (as shown in Fig. 9.1) to yield the solution $\chi_{qs}(t)$. We know that

$$\chi(t) = \chi_s(t) + \chi_{qs}(t) \tag{9.58}$$

(to our specified accuracy), and hence that calculating $\chi_s(t)$ and $\chi_{qs}(t)$ is a valid (if rather inefficient) way of calculating $\chi(t)$.

The hypothetical calculation just described can now be used to explain how $\chi_r(t)$ is confined to region r and $\chi_p(t)$ to region p. We know that to our specified accuracy, the absorbing potentials V_{rq} and V_{pq} absorb $\chi_q(t)$ without affecting it at all in the strong-interaction region; hence we also know that $\chi_q(t)$ is equal to $\chi_{qs}(t)$ in the strong-interaction region, since $\chi_q(t)$ is related to $\chi_{qs}(t)$ in the same way that $\chi_r(t)$ of equations (9.36) to (9.37) is related to $\chi(t)$. It then follows from (9.58) that $\chi_s(t) + \chi_q(t)$ is equal to $\chi(t)$ in the strong-interaction region, and thus, from (9.55), that both $\chi_r(t)$ and $\chi_p(t)$ are zero in the strong-interaction region. This means that $\chi_r(t)$ is confined to region r, and $\chi_p(t)$ to region p (since it is impossible, say, for $\chi_r(t)$ to extend into region p from region r when it has to be zero in the intervening strong-interaction region). From (9.55) it follows that $\chi_p(t)$ is equal to $\chi(t)$ in the product region (and product asymptotic region), and that $\chi_s(t) + \chi_r(t)$ is equal to $\chi(t)$ in the reactant region (and reactant asymptotic region). Because of the latter equality, we see that the artificially reflected component of $\chi_s(t)$ (mentioned above) must be canceled out by an equal and opposite component in $\chi_r(t)$. We will illustrate this canceling out later in Section 9.3.4 with a numerical example.

It should be evident from Figure 9.5 that region q is identical to the region of coordinate space used in the method of Seideman and Miller (which is shown in Fig. 9.3). We have thus shown in this section how to split up the calculation of the reactive scattering wavefunction such that the most difficult part of the calculation (in the strong-interaction region) is made as small as is possible for a specified accuracy. In doing this, we are now taking full advantage of the partitioning of coordinate space shown in Figure 9.1. Provided that equations (9.51) to (9.54) can be solved efficiently, we can expect the version of the RPD method given in this section to be more efficient than any other wave packet method for calculating state-to-state reaction probabilities.

9.2.9 Further Partitioned Version of the TIW RPD Method

We now show how to solve the RPD equations (9.51) to (9.54) efficiently, using an adaptation (Althorpe, Kouri, and Hoffman 1997b, 1998) of the TIW RPD method that was given in Section 9.2.7 for solving (9.36) and (9.37). We first give the TIW version of (9.51) to (9.54) (which are related to equations (9.51) to (9.54) in the same way that the RPD equations (9.42) to (9.44) are related to the TIW RPD equations (9.36) to (9.39)), and then show how to solve these equations using a similar type of propagator to the one given in Section 9.2.7. The method we are about to describe has been tested on the Li + HF → LiF + H reaction, and it is this calculation that we discuss in Section 9.3.

The TIW version of equations (9.51) to (9.54) may be derived (Althorpe, Kouri, and Hoffman 1997b) in a similar manner to equations (9.42) to (9.44) by writing out the Lippmann-Schwinger equations,

$$G^+(E) = G_s^+(E) - G^+(E)W_{qs}G_s^+(E), \tag{9.59}$$

$$G^+(E) = G_q^+(E) - G^+(E)\Gamma_{rq}(E)G_q^+(E) - G^+(E)\Gamma_{pq}(E)G_q^+(E), \tag{9.60}$$

which relate $G^+(E)$, the Green's function corresponding to the Hamiltonian H, to $G_s^+(E)$, the Green's function corresponding to the Hamiltonian $H + W_{qs}$, and $G_q^+(E)$, the Green's function corresponding to the Hamiltonian $H + \Gamma_{rq}(E) + \Gamma_{pq}(E)$. The energy dependent absorbing potentials $\Gamma_{rq}(E)$ and $\Gamma_{pq}(E)$ are of the form of equation (9.40). Note that we have not multiplied the real potential W_{qs} by an energy dependent factor; it can be shown that for a real potential, simpler recursion relations are obtained without an energy factor.

To complete the derivation, one substitutes equation (9.60) into the right-hand side of equation (9.59), and then substitutes the resulting equation into (9.24), to obtain

$$\xi_s^+(E) = \frac{i}{2\pi} G_s^+(E)\chi(0), \tag{9.61}$$

$$\xi_q^+(E) = -G_q^+(E)W_{qs}\xi_s^+(E) \tag{9.62}$$

$$\xi_r^+(E) = -G^+(E)\Gamma_{rq}(E)\xi_q^+(E) \tag{9.63}$$

$$\xi_p^+(E) = -G^+(E)\Gamma_{pq}(E)\xi_q^+(E) \tag{9.64}$$

where

$$\xi^+(E) = \xi_s^+(E) + \xi_q^+(E) + \xi_r^+(E) + \xi_p^+(E). \tag{9.65}$$

Each of these equations is the TIW equivalent to the corresponding RPD equation in (9.51) to (9.55); each solution $\xi_\lambda^+(E)$, $\lambda = p, q, r, s$, is the TIW equivalent to the solution $\chi_\lambda(t)$. By using arguments similar to those given in Section 9.2.7, one can confirm that each $\xi_\lambda^+(E)$ is confined to region λ of Figure 9.5.

One can solve equations (9.61) to (9.64) by propagating the solutions $\xi_\lambda^+(E)$, $\lambda = p, q, r, s$, analogously to how one propagates the solutions $\xi_r^+(E)$ and $\xi_p^+(E)$ of (9.42) and (9.43). Thus $\xi_\lambda^+(E)$ is expanded as

$$\xi_\lambda^+(E) = \frac{1}{2\pi\Delta H \sin\phi} \sum_{n=0}^{N} e^{-in\phi}\eta_{\lambda n}, \qquad \lambda = p, q, r, s, \tag{9.66}$$

and the "inverse Chebyshev transform" equation (9.46) is used with equations (9.59) and (9.60) to derive the recursion relations for $\eta_{\lambda n}$. In analogy with the recursion relations of Section 9.2.7, one can show that there is a value n_{qs} of n such that when $n < n_{qs}$, the recursion relations for η_{sn} are identical to the recursion relations for η_n (of equation (9.26)); the functions $\eta_{\lambda n}$, $\lambda = p, q, r$, are thus all zero for $n < n_{qs}$. When $n \geq n_{qs}$, the recursion relations for $\eta_{\lambda n}$, $\lambda = p, q, r, s$, are

$$\eta_{sn} = 2H_{norm}\eta_{sn-1} - \eta_{sn-2} + 2W_{qs}\eta_{sn-1}, \tag{9.67}$$

$$\eta_{qn} = \frac{1}{1 + V_{tot}}[2H_{norm}\eta_{qn-1} - (1 - V_{tot})\eta_{qn-2} + \sigma_{qn}], \qquad (9.68)$$

$$\eta_{rn} = 2H_{norm}\eta_{rn-1} - \eta_{rn-2} + \sigma_{rn}, \qquad (9.69)$$

$$\eta_{pn} = 2H_{norm}\eta_{pn-1} - \eta_{pn-2} + \sigma_{pn}, \qquad (9.70)$$

where $V_{tot} = V_{rq} + V_{pq}$, $\sigma_{qn} = -2W_{qs}\eta_{sn-1}$, $\sigma_{rn} = V_{rq}(\eta_{qn} - \eta_{qn-2})$, and $\sigma_{pn} = V_{pq}(\eta_{qn} - \eta_{qn-2})$. It may be shown that

$$\eta_n = \eta_{pn} + \eta_{qn} + \eta_{rn} + \eta_{sn}, \qquad (9.71)$$

which confirms that solving (9.61) to (9.64) by the propagation scheme of (9.66) to (9.70) is equivalent to solving the TIW equation (9.24) by the Chebyshev propagator of (9.26). Equations (9.67) to (9.70) take full advantage of the form of equations (9.61) to (9.64), since it may be shown that for a given λ, the functions $\eta_{\lambda n}$ are confined to region λ of Figure 9.5. Thus when evaluating each of equations (9.67) to (9.70), the basis set used to represent H need only span region λ.

9.3   APPLICATION TO THE LI + HF → LIF + H REACTION

In the previous section we presented a simple theory of reactive scattering, whereby the coordinate space of a reaction is partitioned as shown in Figure 9.1 (for the two-dimensional $A + BC$ reaction). We then reviewed several wave packet methods that make use of this theory, ending in Section 9.2.9 with a description of the recently developed TIW RPD method. As we mentioned in Section 9.2.9, the TIW RPD method takes full advantage of the partitioning in Figure 9.1 by splitting the wave packet propagation into a set of separate propagations, each of which is evaluated in one region of Figure 9.1 as shown in Figure 9.5.

We now describe an application of the TIW RPD method to calculate the state-to-state reaction probabilities of the (fully three-dimensional) Li + HF → LiF + H reaction (for total angular momentum $J = 0$). The application is described in full in the paper by Althorpe, Kouri, and Hoffman (1998). We discuss here the basis sets and coordinate systems used in each region of the reaction, show "snapshots" of the wave packets at different stages of the calculation, and discuss how the calculated state-to-state reaction probabilities converge with respect to the surfaces-of-no-return P_r and P_p (of Figs. 9.1 and 9.5).

It is in order to examine the latter convergence that we chose to test the method on the Li + HF reaction. As we show in Figure 9.6, the Li + HF reaction has a very long-range potential energy surface (Parker, Pack, and Laganà 1993; Parker et al. 1995), with a well in the Li + HF entrance channel, and a double well with a barrier in the LiF + H exit channel. These features are known (Parker et al. 1995) to support

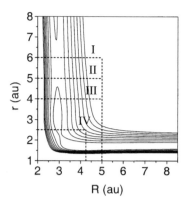

Figure 9.6 Cut through the Li + HF → LiF + H potential energy surface (Parker et al. 1995) plotted as a function of the reactant (Li + HF) Jacobi coordinates R and r, at $\theta = 74°$. The dashed rectangles enclose different choices of the strong-interaction region, each used in calculations I–IV, so the four horizontal dashed lines pass through four different values of P_p, and the two vertical dashed lines through two different values of P_r.

resonances in the strong-interaction region, which have long-range "tails" extending out into the reactant and product regions. One thus expects the results of a TIW RPD calculation on Li + HF to be very sensitive to P_r and P_p, especially at resonance energies.

9.3.1 Coordinates

For total angular momentum $J = 0$, the Li + HF → LiF + H reaction is a three-dimensional $A + BC → AC + B$ reaction, for which we can neglect the formation of $AB + C$ (i.e., LiH + F). The TIW RPD equations to be solved are thus the same equations (9.61) to (9.65) that were given in Section 9.2.9 (for the two-dimensional $A + BC$ reaction), with H now the exact three-dimensional Hamiltonian for Li + HF (with $J = 0$). The coordinate space of the reaction is partitioned into regions p,q,r, and s as is shown in Figure 9.5 (with Fig. 9.5 now serving as a two-dimensional schematic picture of the three-dimensional reaction).

For each region of Figure 9.5, we thus need a three-dimensional coordinate system with which to represent H of equations (9.67) to (9.70), noting that a different coordinate system can be used in each region. We choose to represent regions q, r, and s of Figure 9.5 in terms of the reactant (i.e., Li + HF) arrangement Jacobi coordinates (R, r, θ), of which r is the HF bond length, R the distance between Li and the HF center of mass, and θ the angle between R and r. The $(J = 0)$ Hamiltonian in these coordinates is (Althorpe, Kouri, and Hoffman 1997c)

$$H = -\frac{\hbar^2}{2\mu}\frac{1}{R}\frac{\partial^2}{\partial R^2}R - \frac{\hbar^2}{2m}\frac{1}{r}\frac{\partial^2}{\partial r^2}r + \left[\frac{\hbar^2}{2\mu R^2} + \frac{\hbar^2}{2mr^2}\right]\mathbf{j}^2 + V(R, r, \theta), \quad (9.72)$$

where μ and m are the appropriate reduced masses, and $V(R, r, \theta)$ is the potential. We choose to represent region p of Figure 9.5 in terms of the product (i.e., LiF + H) arrangement Jacobi coordinates (R', r', θ'), which are defined analogously to (R, r, θ).

It is very unlikely that the coordinates just given are the most efficient coordinates that could have been chosen for each region of the Li + HF reaction. In region q, for example, it would probably be better to use hyperspherical coordinates (Pack and Parker 1987), or coordinates based on the transition state geometry of Li + HF. We have chosen to use Jacobi coordinates because they give a simple expression for H (see equation (9.72)) and because they enable us to test an important advantage of the TIW RPD method—that different coordinate systems can be used in each region p, q, r, s. Without using the TIW RPD method, it would be very difficult to represent the entire coordinate space of the Li + HF reaction in terms of, say, reactant arrangement Jacobi coordinates. These coordinates are a good representation of H in the reactant and strong-interaction regions but a very inefficient, unstable representation of H in the product region (where they attempt to describe LiF + H states as highly distorted Li + HF states). Using the TIW RPD method, it is simple to get round this "coordinate problem," representing region p (and hence the product region) in terms of product arrangement Jacobi coordinates, as we have done above.

We finish this section with a reminder that in the Jacobi coordinate system, the vector R lies along the z-axis of a body-fixed (BF) frame of reference. Within the BF frame the quantum numbers describing the internal states of the isolated molecules Li + HF (or LiF + H) are v—the vibrational excitation of the HF (or LiF) stretch, and j—the rotational angular momentum of HF (or LiF).

9.3.2   Basis Sets

In each region p, q, r, s of Figure 9.5, we represent the Hamiltonian on a three-dimensional grid of discrete points, distributed in the coordinates (R, r, θ), or (R', r', θ'). We construct the grids in a simple way (see below) that takes into account some general properties of the dynamics. Each representation gives a Hamiltonian matrix with a sparse, banded structure, whose action can be evaluated without having to store the entire matrix in memory. We now give details of the grid representations, starting with the grid used to represent the r coordinate in region r.

Region r encloses the reactant region (and a small part of the reactant asymptotic region). Hence throughout region r the atoms remain in the Li + HF ($A + BC$) configuration. As R decreases (i.e., as Li approaches HF), the potential $V(R, r, \theta)$ mixes different rotational and vibrational levels of HF but does not (by definition) mix in any LiF + H states (with the Li + HF states). Thus a good basis set for the r coordinate in region r is likely to be the vibrational wavefunctions of the isolated HF molecule. We construct the grid representation of the r coordinate by taking the HF vibrational wavefunctions, and diagonalizing the coordinate operator with them; this yields a discrete variable representation (DVR) (Bacic and Light 1989) of r, which is a linear combination of HF vibrational wavefunctions. We construct a similar DVR of r' in region p, which is a linear combination of LiF vibrational wavefunctions.

Like region r, region s encloses the reactant region (and part of the reactant asymptotic region)—but it also encloses part of the strong-interaction region (at W_{qs}). Thus, at first sight, it appears that the DVR just given for region r cannot be used to represent the r coordinate in region s, since such a basis does not properly describe the dynamics in the strong-interaction region. We know from Section 9.2.8, however, that the effect of W_{qs} on the dynamics of region s is entirely artificial: Whatever effect W_{qs} has on $\xi_s^+(E)$ is canceled out by an equal and opposite effect that the source term $W_{qs}\xi_s^+(E)$ has on $\xi_q^+(E)$. We also know that W_{qs} can be chosen in any way we like, provided that it reflects the wave packet out of the strong-interaction region. We thus choose W_{qs} of the form

$$W_{qs}(R, r, \theta) = V(\overline{R}, r, \theta) - V(R, r, \theta) + V_{\text{ramp}}(R), \qquad (9.73)$$

where $V_{\text{ramp}}(R)$ is a reflecting ramp, positioned as shown in Figure 9.7, and $\overline{R}$ is a value of R located in the reactant region (to the right of R_r in Fig. 9.5). With this choice of $W_{qs}(R, r, \theta)$, the total potential $W_{qs}(R, r, \theta) + V(R, r, \theta)$ (which governs the dynamics of $\xi_s^+(E)$ at W_{qs}) is clearly equal to the potential $V(R, r, \theta)$ at the point $\overline{R}$ in the reactant region, plus $V_{\text{ramp}}(R)$. This potential keeps $\xi_s^+(E)$ in the Li + HF configuration throughout the whole of region s, with the result that the vibrational wavefunctions of HF are a good basis set for representing r in region s. We thus represent r in region s with the same DVR that is used to represent r in region r.

In region q, the dynamics is of course not restricted to one configuration of atoms, so a more general basis set is required. We choose to represent r in region q by an equally spaced grid of discrete points. This representation does not attempt to describe

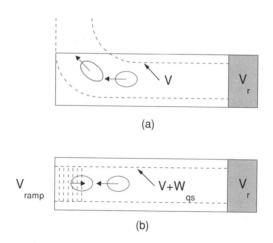

(a)

(b)

Figure 9.7   Schematic representation of (*a*) the potential energy surface $V(R, r, \theta)$ in region s of the Li + HF calculation, and (*b*) the total potential $V(R, r, \theta) + W_{qs}(R, r, \theta)$ in region s, with $W_{qs}(R, r, \theta)$ taking the form of equation (9.73). The ovals represent the propagating wave packet η_{ns} as it (*a*) tries to continue into the strong-interaction region and (*b*) is reflected back into the reactant region by $V_{\text{ramp}}(R)$.

well any feature of the dynamics but is reasonably efficient, since it enables the action of the kinetic energy operator to be evaluated via a fast Fourier transform (FFT) algorithm. We construct the basis, not as a DVR based on sine-functions (which is probably the best known way of constructing an equally spaced grid representation; see Colbert and Miller 1992), but as a discrete distributing approximating functional (or DDAF) representation (Hoffman et al. 1996, and references therein; Wei et al. 1998). In the DDAF representation, a function $f(r)$ is approximated by,

$$f(r) \simeq \Delta \sum_{l}^{N_l} \overline{\delta}(r - r_l) f(r_l),$$ (9.74)

where $\overline{\delta}(r - r_l)$ is the (continuous) DAF approximation to the Dirac delta function $\delta(r - r_l)$, $\{r_l\}$ is a set of equally spaced grid points along r, and Δ is the grid spacing. It can be shown that (9.74) is exact in the limit that the number of grid points $N_l \to \infty$. The DDAF representation of the kinetic energy operator is given by

$$K_{ll'} = -\frac{\overline{h}^2 \Delta}{2m} \overline{\delta}''(r_l - r_{l'}),$$ (9.75)

where $\overline{\delta}''(r_l - r_{l'})$ is the second derivative of $\overline{\delta}(r - r_l)$ with respect to r. The DDAF representation of the potential energy is simply $V_{ll'} = \delta_{ll'} V(r_l)$ (which is the same as in the DVR). Recent tests (Wei et al. 1998) have shown that, for a given number of grid points N_l, the DDAF representation is at least as accurate as the corresponding sine-DVR. We prefer the DDAF representation to the DVR, first, because the kinetic energy matrix $K_{ll'}$ is more banded in the DDAF representation than in the DVR, and, second, because the DDAF yields as good an approximation to $f(r)$ off the grid points as on the grid points. This latter advantage is particularly useful when transforming from one grid representation to another (see Section 9.3.3 below).

In each of regions p, q, r, s, we also represent R (or R') using the DDAF representation. We represent θ (in each region) in terms of a Gauss-Legendre DVR—which is the natural discrete basis to use in θ. As mentioned above, we evaluate the action of the DDAF kinetic energy operators using an FFT algorithm, details of which are given by Huang et al. (1994).

9.3.3 Transforming from Reactant to Product Coordinates

Since we represent regions p and q in terms of different coordinates, we must transform the terms σ_{pn} of equation (9.70) from the reactant arrangement Jacobi coordinates (R, r, θ) (in terms of which the σ_{pn} are calculated in region q) to the product arrangement Jacobi coordinates (R', r', θ') (in terms of which the σ_{pn} must be expressed in order to act as source terms in region p).

For any 3D grid representation, the transformation of σ_{pn} between the two coordinate systems takes the simple form

$$\sigma_{pn}(R'_i, r'_j, \theta'_k) = \sum_{lmn} \Delta(\overline{R}_i, R_l) \Delta(\overline{r}_j, r_m) \Delta(\overline{\theta}_k, \theta_n) \sigma_{pn}(R_l, r_m, \theta_n), \tag{9.76}$$

where (R_l, r_m, θ_n) is a grid point in (R, r, θ), (R'_i, r'_j, θ'_k) is a grid point in (R', r', θ'), and $(\overline{R}_i, \overline{r}_j, \overline{\theta}_k)$ is the grid point (R'_i, r'_j, θ'_k) expressed as function of (R, r, θ). The symbol Δ represents an approximation to the Dirac delta function which is consistent with the basis set used in region q. Thus in our calculations $\Delta(\overline{R}_i, R_l)$ and $\Delta(\overline{r}_j, r_m)$ are taken to be DAFs (as given by equation (9.74)), and $\Delta(\overline{\theta}_k, \theta_n)$ is taken to be an expansion of $\delta(\overline{\theta}_k - \theta_n)$ in terms of Legendre polynomials. Evidently the sum over grid points in (9.76) has only to include those points in region q for which η_{qn} overlaps V_{pq} (and for which σ_{pn} is therefore nonzero).

There is a property of the Chebyshev propagator (which is shared by the TIW RPD propagator of equations (9.66) to (9.70)), which enables the number of σ_{pn} terms, and hence the number of times that (9.76) must be evaluated, to be reduced by a factor of M (where M is an integer). This property is that when the spread of energies contained in the initial wave packet $\chi(0)$ is less than the spectral range of H, only a fraction of the basis functions η_n (or $\eta_{\lambda n}$, $\lambda = p, q, r, s$) are linearly independent (Althorpe, Kouri, and Hoffman 1997a). The expansion of $\xi_q^+(E)$ in equation (9.66) may then be approximated as

$$\xi_q^+(E) \simeq \frac{M}{2\pi\Delta H \sin\phi} \sum_j e^{-in_j\phi} \eta_{qn_j} \tag{9.77}$$

where the sum over j includes only every Mth value of n, n_j. This approximation does not change the recursion relation for η_{qn} (which remains as in equation (9.68)), but does change the recursion relation for η_{pn} to

$$\eta_{pn} = 2H_{\text{norm}}\eta_{pn-1} - \eta_{pn-2} + \sigma_{pn} \qquad \text{for } n = n_j$$

$$= 2H_{\text{norm}}\eta_{pn-1} - \eta_{pn-2} - \sigma_{pn-2} \qquad \text{for } n - 2 = n_j$$

$$= 2H_{\text{norm}}\eta_{pn-1} - \eta_{pn-2} \qquad \text{otherwise.} \tag{9.78}$$

Clearly, this (approximate) recursion relation for η_{pn} includes only one 1/Mth the number of terms σ_{pn} as does the (exact) recursion of (9.70). In the calculations on Li + HF reported below, accurate results were obtained by applying (9.77) and (9.78) with M set to 16.

9.3.4   Results and Discussion

We now describe four separate TIW RPD calculations on the Li + HF reaction (calculations I–IV), each of which was carried out by implementing the theory of Section 9.2.9 as just described in Sections 9.3.1 to 9.3.3. Calculations I to IV differ in respect to the locations of the surfaces-of-no-return P_p and P_r (and hence also in respect to Q_p and Q_r; see Fig. 9.5) but are in all other respects identical. The locations of P_p and P_r are shown by dashed lines in Figure 9.6, which shows a cut through the potential energy surface at a fixed value of θ. We emphasize that θ was *not* fixed in the calculations but was represented by a basis spanning $θ = 0 → π$ (as described in Section 9.3.2). We give a selection of the parameters used in calculation II in Table 9.2; the parameters used in calculations I, III, and IV may be inferred from Table 9.2 (by scaling R_{max}, R_{min}, and r_{max}, and hence N_R and N_r). A complete description of all the parameters used in the calculations is given by Althorpe, Kouri, and Hoffman (1998).

Figures 9.8 and 9.9 show "snapshots" of the wave packet at different stages of calculation II. Although the details of Figures 9.8 and 9.9 are specific to the Li + HF reaction, the overall dynamics of the wave packets is consistent with the (general) description, given in Sections 9.2.8 and 9.2.9, of how each wave packets propagates in region p, q, r, or s.

Beginning with Figures 9.8*a* and 9.8*b*, we see that $η_{sn}$ propagates through the reactant region in the direction of the strong-interaction region (Fig. 9.8*a*) before reflecting off W_{qs} and traveling back through the reactant region (Fig. 9.8*b*). We know from Section 9.2.8 that after reflecting off W_{qs}, $η_{sn}$ is entirely artificial and must be canceled out by an equal and opposite component in $η_{rn}$. Thus from the resemblance of $η_{sn}$ in Figure 9.8*b* to $η_{rn}$ in Figure 9.8*c* (which shows a snapshot taken at the same value of n as Fig. 9.8*b*), we see that the "artificial" component of $η_{rn}$ is evidently much larger than the "physical component." The physical component of $η_{rn}$, obtained by adding $η_{rn}$ to $η_{sn}$, is shown in Figure 9.9*a*. The reason the artificial component of $η_{rn}$ is much larger than the physical component is simply that at the value of n at which Figures 9.8*b* and 9.8*c* were taken, $η_n$ (the wave packet that would have been generated had we expanded $G^+(E)$ directly) is only just starting to emerge from the strong-interaction region into the reactant region. Later (at a larger value of n) more

TABLE 9.2   Selected parameters used in calculation II

	Grid Dimensions				Grid Points		
	R_{min}	R_{max}	r_{min}	r_{max}	N_R	N_r	$N_θ$
s	4.0	18.0	0.7	3.5	164	6	20
q	1.2	8.0	0.7	8.0	79	81	20
r	5.0	18.0	0.7	3.5	152	6	20
p	5.0	15.0	1.5	4.0	111	6	20

	Expansion Parameters		Damping Parameters	
N	$\bar{H}$	$ΔH$	R_W	R_V
6000	0.5	0.5	1.0	3.0

Note: Arbitrary units are used. The labels p, q, r, s refer to the regions illustrated schematically in Figure 9.5. R_W is the width of the reflecting potential W_{qs}; R_V is the width of each of the absorbing potentials V_r, V_p, V_{rq} and V_{pq}. All other parameters are either self-explanatory or are defined in the text.

of η_n will have entered the reactant region, by which time the artificial component of η_{rn} will have been almost completely absorbed by V_r.

The other two snapshots, Figures 9.9b and 9.9c, show the propagation through region p of *one* of the source terms σ_{pn} of equation (9.78), first in Figure 9.9b, just after it is introduced into region p, and then in Figure 9.9c, after it has propagated for some time. Thus σ_{pn} is initially located at the start of the product region (to the left in Fig. 9.9b), and then it propagates outward, spreading considerably until it extends throughout the whole product region (as shown in Fig. 9.9c). One reason why σ_{pn} spreads so much is that in the product region the wave packet contains a much wider range of velocities than it does in the reactant region (the reduced mass of Li + HF

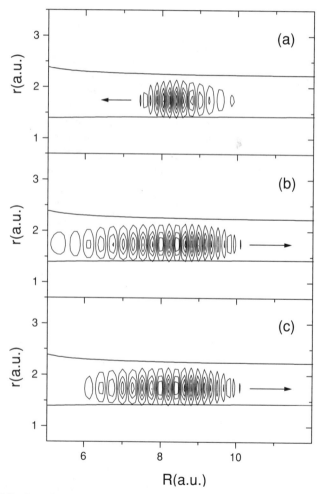

Figure 9.8 Snapshots of the wave packet at different stages of calculation II, showing (*a*) η_{sn} at $n = 450$, (*b*) η_{sn} at $n = 1050$, and (*c*) η_{rn} at $n = 1050$. The arrows indicate the directions in which the packets are traveling.

being about five times that of LiF + H). Of course, in calculation II, we did *not* propagate just the one source term shown in Figures 9.9*b* and 9.9*c*; we propagated *all* the source terms (simultaneously) according to the recursion relation (9.78).

In Figure 9.10 we present the state-to-state reaction probabilities, $P(j_0 = 0, v_0 = 0 \to j = 0, v = 0)$, obtained from calculations I to IV. The results are shown in pairs in order to facilitate comparison. We see that all four calculations are in good overall agreement, predicting a broad peak centered around 0.43 eV and a set of narrower peaks between 0.55 and 0.7 eV. This overall structure is consistent with the results of other calculations (Göğtas, Balint-Kurti, and Offer 1996) on the Li + HF reaction.

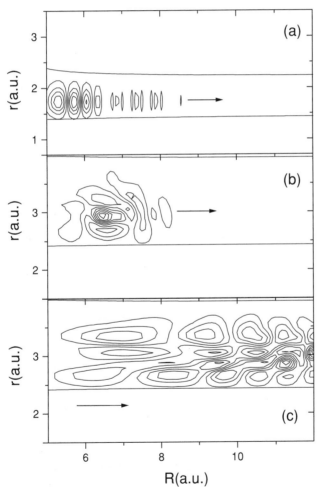

Figure 9.9   Same as Figure 9.8, showing (*a*) $\eta_{sn} + \eta_m$ at $n = 1050$, (*b*) σ_{pn_j}, $n_j = 1936$, just after being introduced into region *p*, at $n = 1950$, and (*c*) σ_{pn_j}, $n_j = 1936$, after having spread across region *p*, at $n = 2250$.

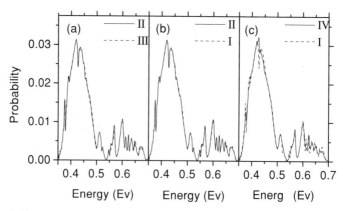

Figure 9.10   State-to-state reaction probabilities $P(j_0 = 0, v_0 = 0 \rightarrow j = 0, v = 0)$ for the reaction Li + HF$(j_0, v_0) \rightarrow$ LiF(j, v) + H (total angular momentum, $J = 0$), obtained from the TIW RPD calculations, calculations I–IV.

As we mentioned above, the Li + HF potential energy surface is known to support resonances that can extend down both the product and the reactant channel. When applying the TIW RPD method to Li + HF, one might therefore expect poor convergence with respect to the locations of P_r and P_p, since, at a given resonance energy, P_r and P_p would have to be moved outward until the strong-interaction region enclosed the tail of the resonance wavefunction. We see from Figures 9.10a and 9.10b, however, that the state-to-state reaction probabilities for Li + HF converge very well with respect to the location of P_p. Similar results (not shown here) were obtained for convergence with respect to P_r. When P_p is moved outward from 4 to 5 au (Fig. 9.10a), the fine structure of some of the resonances (e.g., the peak located around 0.56 eV) changes only slightly. When P_p is moved from 5 to 6 au (Fig. 9.10b), smaller changes occur at the same energies as when P_p is moved from 4 to 5 au. Calculations I to III have thus yielded numerically stable results of the state-to-state reaction probabilities, whose convergence can be estimated from Figures 9.10a and 9.10b.

We see from Figure 9.10c that even when P_r and P_p are moved in to 4.25 and 2.5 au (in calculation IV), the results are still in good qualitative agreement with the results obtained with P_r and P_p at 5 and 6 au (in calculation I). Thus, in the language of Sections 9.2.1 to 9.2.9, calculation IV yields the correct results but at a significantly lower "specified accuracy" than, say, calculation I. Since the accuracy of calculation IV is low enough to noticeably alter the structure of the peaks (especially above 0.55 eV), we can consider calculation IV to be an approximate application of the TIW RPD method. This type of application is particularly useful in calculations of more complicated reactions, since one can obtain an approximate overview of the state-to-state reaction probabilities before investing time in a more precise calculation.

The results of this section demonstrate that the TIW RPD method yields numerically stable results when applied to the Li + HF reaction. Accurate state-to-state reaction probabilities are obtained with the strong-interaction region enclosed in a box of dimensions 5 × 6 au; good qualitative state-to-state reaction probabilities are obtained with box of dimensions 4.25×2.5 au. These results imply that the TIW RPD method yields good approximations to the resonance wavefunctions, even though at some resonance energies the wavefunction is undoubtedly being truncated by V_{rq} and V_{pq}, which effectively impose outgoing boundary conditions on the wavefunction at P_r and P_p. Thus the TIW RPD method (or any other method based on the theory of Sections 9.2.8 and 9.2.9) will almost certainly yield accurate state-to-state reaction probabilities when applied to other chemical reactions, including reactions in which the dynamics is influenced by a long-range potential energy surface. We will be testing the method on other reactions in future work.

9.4 CONCLUSIONS AND FURTHER WORK

In this chapter we have given a general introduction to time dependent and time independent wave packet approaches to reactive scattering. We have reviewed in detail much of the recent work that has been done in this area, by presenting a simple theory of reactive scattering in which the coordinate space of any chemical reaction is partitioned into a strong-interaction region, reactant region, and product region(s). We have shown how to make full use of the theory by calculating state-to-state reaction probabilities using the recently developed time independent wave packet (TIW) reactant-product decoupling (RPD) method, which obtains the state-to-state reactive scattering wavefunction from a set of *completely separate* wave packet propagations, each of which is confined to just one region of the reaction. We have demonstrated the method by applying it to the (fully three-dimensional) Li + HF $\rightarrow$ LiF + H reaction (for J = 0).

The reactant-product decoupling (RPD) approach on which the TIW RPD method is based was only introduced very recently into reactive scattering theory. There is already enough evidence, however, to suggest that RPD methods are going to prove the most efficient wave packet methods for calculating state-to-state reaction probabilities. Numerical tests, such as the one reported here on the Li + HF reaction, have already shown that RPD methods are simple to apply and yield converged, numerically stable results. Because RPD methods separate the calculation in the product region from the calculation in the reactant region, they not only reduce the size of the Hamiltonian matrix but also remove the need to represent the entire coordinate space of a reaction in terms of one set of coordinates.

Future work on wave packet methods in state-to-state reactive scattering will probably concentrate on developing RPD methods and applying them to larger reactions. Future work on wave packet methods in cumulative reactive scattering (i.e., the calculation of rate constants) will probably concentrate on further developing wave packet versions of the method of Seideman and Miller (1992) (see Section 9.2.5). It is convenient that both the latter method and RPD methods like the TIW RPD method represent the strong-interaction region of a reaction in the same way (see Figs. 9.3 and 9.5), so any basis set, coordinate system, or dynamical approximation developed for use in the method of Seideman and Miller can also be used in the strong-interaction region part of the TIW RPD method.

Recent wave packet calculations (which did not use the RPD approach) have yielded the first (fully six-dimensional) state-to-state reaction probabilities for the four-atomic $H_2 + OH$ (Zhu et al. 1996) and $H_2O + H$ reactions (Zhang and Light 1996). Total reaction probabilities have also been calculated for the OH + CO reaction (an important reaction in atmospheric and combustion chemistry) by Zhang and Zhang (1995). The development of RPD-based wave packet methods will make it possible to calculate state-to-state reaction probabilities for many more four or even five atom reactions, including reactions involving several heavy atoms. For example, we know that it will be possible to calculate state-to-state reaction probabilities for the OH + CO reaction, since a TIW RPD calculation on this reaction will be cheaper than the calculation of the total reaction probabilities just mentioned. Other four-atom reactions which will almost certainly be tractable by RPD-based wave packet methods include $CN + H_2$, $Cl + H_2O$, Cl + HCN, and NH + NO. Approximate applications of RPD-based methods, in which different approximations are made in different regions of a reaction, will enable calculations to be carried out on larger reactions, such as Cl $+ CH_4$ and $Cl + C_2H_6$. These calculations will help theoreticians to interpret the results of recent molecular beam experiments, which are now using lasers to probe state-to-state reactive scattering in greater detail than ever before.

ACKNOWLEDGMENTS

The research for this chapter was partially supported under National Science Foundation grant CHE-9700297, R.A. Welch Foundation grant E-0608, the Petroleum Research Fund, administered by the American Chemical Society, and the Ames Laboratory, which is operated for the Department of Energy by Iowa State University, under Contract 2-7405-ENG82.

REFERENCES

Althorpe, S. C., Kouri, D. J., and Hoffman, D. K. (1997a). *J. Chem. Phys.* **106**, 7629–7636.

Althorpe, S. C., Kouri, D. J., and Hoffman, D. K. (1997b). *J. Chem. Phys.* **107**, 7816–7824.

Althorpe, S. C., Kouri, D. J., and Hoffman, D. K. (1997c). *Chem. Phys. Lett.* **275**, 173–180.

Althorpe, S. C., Kouri, D. J., and Hoffman, D. K. (1999). *J. Phys. Chem.* A**102**, 9494.

Althorpe, S. C., Kouri, D. J., Hoffman, D. K., and Zhang, J. Z. H. (1997). *J. Chem. Soc., Faraday Trans.* **93**, 703–708.

Bacic, Z., and Light, J. C. (1989). *An. Rev. Phys. Chem.* **40**, 469–498.

Bowen, H. F., Kouri, D. J., Mowrey, R. C., Yinnon, A. T., and Gerber, R. B. (1993). *J. Chem. Phys.* **99**, 704–720.

Brown, G. (1969). *Unified Theory of Nuclear Models*. North-Holland, Amsterdam.

Clary, D. C., ed. (1985). *The Theory of Chemical Reaction Dynamics*. Norwell: Kluwer Academic Publishers, Dordrecht, Netherlands.

Colbert, D. T., and Miller, W. H. (1992). *J. Chem. Phys.* **96**, 1982–1991.

Feit, M. D., and Fleck, J. A., Jr. (1984). *J. Chem. Phys.* **80**, 2578–2584.

Göğtas, F., Balint-Kurti, G. G., and Offer, A. R. (1996). *J. Chem. Phys.* **104**, 7927–7939.

Goldberger, M. L., and Watson, K. M. (1964). *Collision Theory*. Wiley, New York, Chs. 3–5.

Gray, J. C., Fraser, G. A., Truhlar, D. G., and Kulander, K. C. (1980). *J. Chem. Phys.* **73**, 5726–5733.

Hoffman, D. K., Huang, Y., Zhu, W., and Kouri, D. J. (1994). *J. Chem. Phys.* **101**, 1242–1250.

Hoffman, D. K., Marchioro II, T. L., Arnold, M., Huang, Y., Zhu, W., and Kouri, D. J. (1996). *J. Math. Chem.* **20**, 117–140.

Huang, Y., Iyengar, S. S., Kouri, D. J., and Hoffman, D. K. (1996). *J. Chem. Phys.* **105**, 927–939.

Huang, Y., Kouri, D. J., Arnold, M., Marchioro II, T. L., and Hoffman, D. K. (1994). *Comput. Phys. Comm.* **80**, 1–16.

Huang, Y., Kouri, D. J., and Hoffman, D. K. (1994). *J. Chem. Phys.* **101**, 10493–10506.

Huang, Y., Zhu, W., Kouri, D. J., and Hoffman, D. K. (1993). *Chem. Phys. Lett.* **206**, 96–102.

Kemble, E. C. (1958). *The Fundamental Principles of Quantum Mechanics*. Dover, New York.

Kosloff, R., and Cerjan, C. (1984). *J. Chem. Phys.* **81**, 3722–3729.

Kosloff, R., and Kosloff, D. (1986). *J. Comput. Phys.* **63**, 363–376.

Kouri, D. J., Arnold, M., and Hoffman, D. K. (1993). *Chem. Phys. Lett.* **203**, 166–174.

Kouri, D. J., and Hoffman, D. K. (1995). *Few Body Systems* **18**, 203–212.

Kouri, D. J., Hoffman, D. K., Peng, T., and Zhang, J. Z. H. (1996). *Chem. Phys. Lett.* **262**, 519–525.

Kouri, D. J., Huang, Y., Zhu, W., and Hoffman, D. K. (1994). *J. Chem. Phys.* **100**, 3662–3671.

Leforestier, C., and Wyatt, R. E. (1983). *J. Chem. Phys.* **78**, 2334–2344.

Lippmann, B., and Schwinger, J. (1950). *Phys. Rev.* **79**, 469–480.

Mandelshtam, V., and Taylor, H. S. (1995a). *J. Chem. Phys.* **102**, 7390–7399.

Mandelshtam, V., and Taylor, H. S. (1995b). *J. Chem. Phys.* **103**, 2903–2907.

McCullough, E. A., and Wyatt, R. E. (1971). *J. Chem. Phys.* **54**, 3578–3592.

Miller, W. H., Schwartz, S. D., and Tromp, J. W. (1983). *J. Chem. Phys.* **79**, 4889–4898.

Mowrey, R. C., Bowen, H. F., and Kouri, D. J. (1987). *J. Chem. Phys.* **86**, 2441–2442.

Mowrey, R. C., and Kouri, D. J. (1986). *J. Chem. Phys.* **84**, 6466–6473.

Neuhauser, D., and Baer, M. (1989). *J. Chem. Phys.* **91**, 4651–4657.

Neuhauser, D., Baer, M., Judson, R. S., and Kouri, D. J. (1989). *J. Chem. Phys.* **90**, 5882–5884.

Neuhauser, D., Baer, M., Judson, R. S., and Kouri, D. J. (1991). *Comput. Phys. Comm.* **63**, 460–481.

Neuhauser, D., Baer, M., and Kouri, D. J. (1990). *J. Chem. Phys.* **93**, 2499–2505.

Neuhauser, D., Judson, R. S., Kouri, D. J., Adelman, D. E., Shafer, N. E., Klein, D. A., and Zare, R. N. (1992). *Science* **257**, 519–522.

Pack, R. T., and Parker, G. A. (1987). *J. Chem. Phys.* **87**, 3888–3921.

Parker, G. A., Laganà, A., Crocchianti, S., and Pack, R. T. (1995). *J. Chem. Phys.* **102**, 1238–1250.

Parker, G. A., Pack, R. T., and Laganà, A. (1993). *Chem. Phys. Lett.* **202**, 75–81.

Peng, T., and Zhang, J. Z. H. (1996). *J. Chem. Phys.* **105**, 6072–6074.

Seideman, T., and Miller, W. H. (1992). *J. Chem. Phys.* **96**, 4412–4422.

Sun, Y., Judson, R. S., and Kouri, D. J. (1989). *J. Chem. Phys.* **90**, 241–250.

Sun, Y., Mowrey, R. C., and Kouri, D. J. (1987). *J. Chem. Phys.* **87**, 339–349.

Tal-Ezer, H., and Kosloff, R. (1984). *J. Chem. Phys.* **81**, 3967–3971.

Tannor, D. J., and Weeks, D. E. (1993). *J. Chem. Phys.* **98**, 3884–3893.

Wei, G. W., Althorpe, S. C., Zhang, D. S., Kouri, D. J., and Hoffman, D. K. (1998). *Phys. Rev.* **A57**, 3309–3316.

Zhang, D. H., and Light, J. C. (1996a). *J. Chem. Phys.* **104**, 6184–6191.

Zhang, D. H., and Light, J. C. (1996b). *J. Chem. Phys.* **105**, 1291–1294.

Zhang, D. H., and Zhang, J. Z. H. (1995). *J. Chem. Phys.* **103**, 6512–6519.

Zhu, W., Dai, J., Zhang, J. Z. H., and Zhang, D. H. (1996). *J. Chem. Phys.* **105**, 4881–4884.

Zhu, W., Peng, T., and Zhang, J. Z. H. (1997). *J. Chem. Phys.* **106**, 1742–1748.

CHAPTER 10

Laser-Excited Wave Packets in Semiconductor Heterostructures

KARL LEO and MARTIN KOCH

10.1 INTRODUCTION

In the recent years the dynamics of wave packets have been investigated in detail in various physical systems. The simplest way to set up an electronic wave packet is by excitation with a short laser pulse with sufficient band width to cover the various states the wave packet is built from. As described in the other chapters of this book, most of the experiments performed so far have been using atoms and molecules.

At a first glance semiconductors seem not to be very useful for performing wave packet experiments, mainly by two reasons: First, semiconductors share the most characteristic feature of the solid state: continuous energy bands. Thus wave packets in semiconductors, in general, consist of many different energy levels that usually lead to complicated dynamics, without any pronounced features in the temporal and spatial dynamics which make the investigation in molecules so interesting and fruitful.

A second reason why semiconductors are not well suited are the short dephasing times. A prerequisite for observing wave packet dynamics is that the coherence between the different constituents of the wave packet is kept as long as possible. This coherence will be destroyed by electronic dephasing processes. The corresponding dephasing times are in solids usually by orders of magnitude faster than in atoms or molecules due to the interaction with phonons and other scattering processes. Thus it can be expected that wave packet dynamics in solids or, in particular, semiconductors, will be limited to a very narrow time window. Typically the dephasing times of the semiconductor continuum states are well below a picosecond, which is several orders of magnitude faster than in atoms or molecules in the vapor phase.

Despite these seemingly unattractive features, semiconductors offer several unique features for wave packet experiments and have been investigated intensively in the last decade. The key feature which has spurred this interest is the fact that by using semiconductor heterostructures, *one has great freedom in designing virtually every potential shape*. Thus heterostructures allow one to perform (at least in principle) simple, textbook experiments that are not possible in other physical systems. For

The Physics and Chemistry of Wave Packets, Edited by John Yeazell and Turgay Uzer
ISBN 0-471-24684-0 © 2000 John Wiley & Sons, Inc.

TABLE 10.1   Overview of the properties of electronic wave packets in atoms and semiconductors

	Atoms	Semiconductors
Spatial displacement	Few Å	Tens to hundreds of nm
Energy scale	eV	1–100 meV
Dephasing time	> ns	fs to ps

instance, it is comparatively easy to realize a *quantum well* with rectangular potential shape as known from quantum mechanics textbooks. The properties of wave packets in atoms and semiconductors are comparatively summarized in Table 10.1.

Another effect that helps realize wave packet experiments in semiconductors is the formation of excitons. Even in a two- (*quantum well*) or one-dimensional (*quantum wire*) structure, the remaining degrees of freedom lead to a continuous density of states. Thus only a semiconductor heterostructure with confinement in all three directions (*quantum box*) allows one to perform wave packet experiments with discrete levels. However, the formation of excitons mediated by the Coulomb coupling between electrons and holes leads to the formation of discrete states (energetically below the continuum band edge); this allows one to perform wave packet experiments with discrete transitions even in one-, two-, or three-dimensional semiconductors.

We will show in various parts of this chapter that the presence of excitons leads to pronounced modifications of the simple potential structures that can be realized using semiconductor heterostructures. For instance, the anticrossing of the discrete states in coupled double quantum wells is markedly changed by this effect (see Section 10.2.4). On the other hand, wave packet experiments in excitonic resonances can be used to obtain new insight into the properties of these resonances (see Section 10.2.3).

Another feature that makes wave packets in semiconductors interesting are the broad possibility of applications. One example (which will be discussed in Section 10.2.4) are Bloch oscillations in semiconductor superlattices which emit widely tunable THz radiation. They might thus serve as the working principle of a tunable THz emitting device.

In the following discussion we will consider in more detail the basic properties of semiconductor heterostructures and their excitonic resonances.

10.1.1 Semiconductor Heterostructures

The band edge of a three-dimensional bulk semiconductor is characterized by continuous electron and hole bands with a continuous density of states (DOS). Because of the alternative growth of semiconductors with larger and smaller band gaps, structures with reduced dimensionality can be realized. A simple example is the quantum well (QW) already mentioned above (see Fig. 10.1): A semiconductor layer

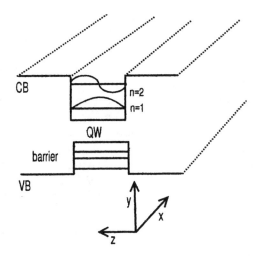

Figure 10.1 Schematic picture of a semiconductor quantum well.

with a smaller band gap is sandwiched between layers with larger band gaps.[1] We denote the growth direction, which is the direction perpendicular to the layers, the z-direction. The electrons and holes are confined in this direction, but they can still freely move in x- and y-directions. The total density of states of this systems now results from discrete states corresponding to the z-direction and a two-dimensional continuum from the motion in x- and y-directions. Thus every discrete state is connected to a 2D subband. The total density of states is constant for a 2D system; the subbands thus form a staircase. If one restricts the motion in y-direction as well, a quantum wire with 1D subbands will result; if the restriction is in all three directions, a quantum box is obtained.

A crucial point in the realization of quantum structures is the interface quality. The realization of these model structures is only possible if well-defined heterointerfaces without trap levels can be realized. A detailed discussion of this topic is beyond the scope of this chapter. However, at least for 2D structures two crucial properties can be achieved: (1) By using semiconductors with similar structural and electronic properties (e.g., GaAs for the well and $Al_xGa_{1-x}As$—with x as the Al molar ratio—as the barrier), one can achieve interfaces that are largely free of structural defects and electronic interface states. (2) By using sophisticated growth techniques such as molecular beam epitaxy (MBE) (e.g., see Parker 1985; Göbel and Ploog 1990), it is possible to control the layer thickness and the interface roughness down to one

---

[1]We address here the case where the band edges of the material with smaller band gaps are located in between the bands of the material with larger band gaps (called a *type I heterostructure*). It is also possible for one of the bands (electrons or holes) to form a barrier while the others form a potential well (*type II heterostructure*) where the band gaps do not overlap. We will restrict ourselves to type I heterostructures.

monolayer. The fulfillment of these two requirements allows the growth of nearly perfect quantum structures. The description of the electronic states in such structures can be made simply by using the so-called envelope approximation (Bastard 1988). In this approximation it is assumed that the wavefunction can be written as a product of the original eigenstates of the bulk semiconductors, the spatially extended Bloch function, and an *envelope function* that is brought about by the heterostructure. For virtually all calculations it is sufficient to solve the Schrödinger equation for the envelope function and ignore the influence of the heterostructure on the Bloch functions. If one treats, for simplicity, only the z-direction, this equation is equivalent to the one-dimensional Schrödinger equation used to solve the textbook quantum well problems. The motion in x- and y-direction is taken into account by multiplying the z-direction wave function with plane waves for the x- and y-direction. Recently also the confinement of the photonic wavefunction has been demonstrated in semiconductor microcavities. These structures will be discussed in more detail in Section 10.2.5.

It is not our intention to dwell on the possibilities of using heterostructures to design artificial new potential structures, so-called bandstructure engineering (Capasso et al. 1986). We want just to mention the most refined recently published example on intraband lasers (Faist et al. 1994; Scamarcio et al. 1996) which contains hundreds of layers that control the intricate transport, relaxation, and emission features needed to achieve laser action in these structures.

10.1.2 Excitons in Semiconductors

While in atoms and molecules the optical transitions between electronic or vibronic energy levels give rise to sharp absorption lines, semiconductors show distinct optical resonances only as a consequence of the Coulomb attraction between electron and hole. A theoretical model describing these resonances (the so-called excitons) was developed back in 1957 by Elliott (Elliott 1957). After several approximations (effective mass approximation, no short range exchange interaction, vanishing excitation density) his early theory leads to the so-called Wannier equation which is mathematically equivalent to the Schrödinger equation for the hydrogen atom. The Wannier equation successfully describes the hydrogen like series of sharp absorption lines which is observed below the semiconductor band gap (Ulbrich 1988). In GaAs, for example, the optically allowed s-like exciton transitions are found at energies $E_g - E_n$, where E_g is the energy of the band gap and $E_n = -m_r e^4 / 2\hbar^2 \varepsilon_0^2 n^2$ is the binding energy of the nth s-like exciton (m_r is the reduced electron-hole mass; the other constants have their usual meaning).

Excitonic effects are even more prominent in quantum well structures. This is caused by quantum confinement which increases the probability of finding an electron and a hole at the same spatial position. The confinement therefore considerably enhances the oscillator strength of the excitonic resonances and leads to higher binding energies. Consequently, in quantum wells, excitonic transitions are usually observable even at room temperatures and in the presence of disorder-induced inhomogeneous line broadening.

It is often forgotten over the success of this simple model that excitonic transitions can not be regarded as being similar to optical transitions of isolated atoms: Excitons represent a many-body problem and reflect the collective properties of the electronic system in a solid. Elliotts simplified approach becomes particularly questionable for the description of nonlinear optical experiments. With increasing excitation density several many-particle effects (Haug and Koch 1993), such as screening of the Coulomb interaction, exchange interaction, and carrier–carrier scattering, alter the exciton problem considerably and the Wannier equation becomes inadequate.

Yet, because of its dominance, the excitonic $1s$ transition is often modeled as a two-level absorber. Wave packet experiments that involve more than one excitonic transition are frequently described using a multi-level model. In this chapter we will encounter some systems where the interference dynamics is satisfactorily reproduced by this crude approach. There are, however, distinct cases for which the multi-level model completely fails; an example is set by the dynamics of an excitonic wave packet observed in a time-resolved four-wave mixing experiment (Section 10.2.3). In those cases a variety of many-particle effects have to be considered to obtain a satisfactory description of the experimental data.

10.2 WAVE PACKET EXPERIMENTS IN SEMICONDUCTORS

10.2.1 Systems Where Wave Packets Have Been Seen

Wave packet experiments in semiconductors have been performed only in the last decade. A prerequisite for these experiments was the development of tunable ultrafast laser sources, which allow to resonantly excite specific excitonic transitions. The first experiments were actually performed on excitons in the bulk semiconductor AgBr by Langer et al. (1990). This system offers comparatively long dephasing time at low temperatures so that the wave packet dynamics could be observed in luminescence.

The first wave packet experiments in heterostructures were performed soon after by Göbel et al. (1990) investigating excitons in quantum wells. These experiments have studied samples where the excitonic transitions are split in multiple peaks due to areas of the quantum well with different thickness. Initially there was a discussion whether these experiments were caused by wave packet interference (quantum beats) or simply by polarization interference of coherent polarization radiated from two different islands. Later experiments by Koch et al. (1993a) have, however, shown that these different island transitions are at least partially coupled and, hence, form a coherent quantum superposition. After the initial experiments a large number of different systems were investigated, ranging from light-hole heavy-hole interference (Leo et al. 1990a; Feuerbacher et al. 1990; Koch et al. 1993b; Dekorsy et al. 1996; Joschko et al. 1997), free exciton–bound exciton interference (Leo et al. 1991a), magnetoexciton inteference (Cundiff et al. 1996), excited state interference in heavy-hole excitons (Feldmann et al. 1993; Koch et al. 1996), to wave packet experiments where pronounced spatial dynamics were visible, such as coherent oscillations in coupled double quantum wells (Leo et al. 1991b) and Bloch oscillations

(Feldmann et al. 1992; Leo et al. 1992a). For both systems with spatial dynamics, it was shown that the oscillating macroscopic dipole moments lead to the emission of THz radiation (Roskos et al. 1992; Waschke et al. 1993). Recently a direct measurement of the spatial displacement of Bloch-oscillating wave packets was made (Lyssenko et al. 1997), proving that the original predictions of Zener about the spatial dynamics of these wave packets were correct.

All the examples mentioned above are wave packets that consist only of electronic states. Recently experiments where spin states are superimposed have been successfully performed (Oestreich and Rühle 1996). An interesting fact about these spin wave packets is that their dephasing time is about two orders of magnitude longer than that of their electronic counterparts, thus enabling to observe coherent dynamics over time ranges of several hundred picoseconds. Another example for wave packets going beyond electronic states are wave packets formed by the superposition of coupled electron-phonon modes in microcavities (Koch et al. 1998b).

In our review of wave packet experiments in semiconductor heterostructures, we will limit ourselves to a few examples. In particular, we will cover the following topics:

- Wave packets that give information about the electronic structure of the semiconductor band edge. We will discuss wave packets composed of the ground and higher excitonic states (Section 10.2.3).

- Wave packets with particular spatial dynamics. We will discuss the textbook wave packet experiment *per se*: coherent oscillations in double quantum wells, and one of the most disputed and oldest wave packet experiments in solid state physics: Bloch oscillations (Section 10.2.4).

- Mixed electronic-nonelectronic wave packets quantum interference of coupled electron-photon states in a strongly coupled semiconductor microcavity (Section 10.2.5).

10.2.2 Four-Wave Mixing

The experimental technique used throughout this chapter is degenerate four-wave mixing (FWM). The popularity of this technique for investigating wave packets in semiconductors results basically from the facts that (1) it detects only coherent contributions of the optical signal, thus suppressing the strong incoherent response of semiconductors, and (2) stray light which is difficult to avoid in semiconductor samples with often imperfect surfaces is suppressed, since the signals are detected in a background-free direction.

A diagram of a typical FWM setup is shown in Figure 10.2. It describes a particularly simple version of FWM, using the so-called self-diffracted geometry (Yajima and Taira 1979). Two laser pulses with wave vectors k_1 and k_2 are focused onto the sample under a small angle. The temporal delay t between the two pulses can be varied by sending either one over a variable delay line. The pulse that arrives first sets up a macroscopic polarization $P^{(1)}(t)$ in the material. The second pulse then acts twofold. First, its electric field interferes with $P^{(1)}(\tau)$ and creates a real population of

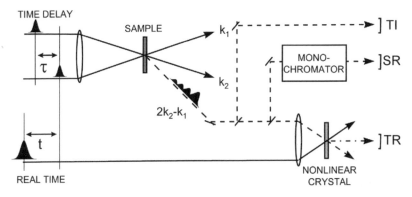

Figure 10.2 Experimental setup for a nonlinear four-wave mixing experiment with time-integrated (TI), spectrally resolved (SR), and time-resolved (TR) detection. (Reprinted with permission from Koch et al. 1996.)

carriers, and second, it generates a third-order polarization $P^{(3)}(t,\tau)$. This nonlinear polarization is the source of the so-called FWM signal emitted in the direction $2k_2 - k_1$. It arises from a variety of nonlinear optical effects such as phase space filling, screening, excitation-induced dephasing, and band gap renormalization.

The different schemes for detection of the FWM signal include the time-integrated (TI) detection using a slow detector (upper right) the spectrally resolved (SR) detection by means of a monochromator and the time-resolved (TR) detection via up-conversion. In the later scheme the signal is time resolved by mixing it in a nonlinear crystal with a third strong reference pulse (lower right). The time delay t of the third pulse corresponds to the real time of the emission process.

If more than one optical transition is excited, quantum beats can be observed in the time-integrated FWM intensity as a function of τ. In addition beats are observed in the time-resolved emission, namely as a function of t for a fixed time delay τ. The spectrally resolved FWM signal for each time delay τ then shows distinct resonances that correspond to the excited transitions.

Although the experiments described here have been performed in different laboratories, the laser pulses were always provided by a commercially available Kerr-lens mode-locked Ti : sapphire laser producing pulses of about 100 fs duration with a repetition frequency of 80 MHz.[2] All experiments are performed at sample temperatures of 10 K or below. For an extensive description of the techniques, we refer to Shen (1984), Kuhl et al. (1989), Koch et al. (1996), and Shah (1996).

10.2.3 Excitonic Wave Packets in Quantum Wells

In this section we investigate the coherent dynamics of a wave packet composed of all bound and part of the unbound excitonic eigenstates associated with a *single*

---

[2]For the coupled double quantum well experiment, a tandem-pumped dye laser system was actually used (Leo et al. 1992b).

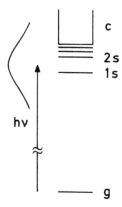

Figure 10.3 Schematic representation of the optical excitation of an excitonic wave packet (g = ground state, c = continuum states). (Reprinted with permission from Feldmann et al. 1993.)

one-particle transition. A schematic illustration of the excitation process is shown in Figure 10.3. Depicted are the crystal ground state g, the optically allowed s-like transitions and the unbound scattering states that form the ionization continuum. The laser spectrum with center frequency $h\nu$ that energetically exceeds the excitonic binding energy and hence covers all bound states and part of the continuum states is shown on the very right. We call a wave packet resulting from an excitation scheme like this an *excitonic wave packet* (EWP).

The results presented next are obtained on a strained InGaAs/GaAs multiple quantum well sample. The structure comprised 20 periods of 10 nm thick $In_{0.08}Ga_{0.92}As$ wells and 40 nm thick GaAs barriers. The photoluminescence

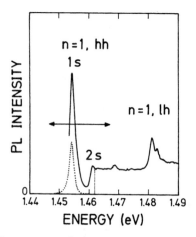

Figure 10.4 Photoluminescence excitation spectrum (solid line) together with the luminescence spectrum (dotted line) of the InGaAs/GaAs sample. The dotted vertical line indicates the onset of the continuum transitions. The horizontal arrow indicates the spectral width (FWHM) of the laser pulses. (Reprinted with permission from Feldmann et al. 1993.)

excitation (PLE) spectrum, which is plotted in Figure 10.4, showed the $n = 1$ hh $1s$ transition at 1.4544 eV with a line width of 1.7 meV. The $n = 1$ hh $2s$ exciton merges into the hh continuum above 1.4619 eV. The large blue shift of the lowest $n = 1$ lh transition which is found at 1.4813 eV is mainly due to inherent strain within the InGaAs layers. Due to this blue shift an EWP can be created from the heavy-hole states without complications due to light-hole states.

Time-Integrated Detection In Figure 10.5 we plot time integrated FWM transients obtained for different detunings Δ of $h\nu$ with respect to the $1s$ heavy-hole transition excitonic transition. For negative detunings the laser spectrum does not noticeably cover the $2s$ resonance and the higher lying continuum states. Hence only the $1s$ transition is excited, and we observe an unmodulated, slowly decaying FWM signal. It is interesting to note that under this excitation condition a retarded signal rise is observed that leads to a considerably delayed signal maximum. This feature arises from an interplay of inhomogeneous broadening and many-particle effects (Wegener et al. 1990).

If the laser spectrum is tuned to higher energies, oscillations can be observed in the signal amplitude. These oscillations are most pronounced for $\Delta = 4.4$ meV when the laser pulses excite not only the $1s$ exciton but also the higher lying bound states and part of the hh ionization continuum. For photon energies at the band edge, that is, for

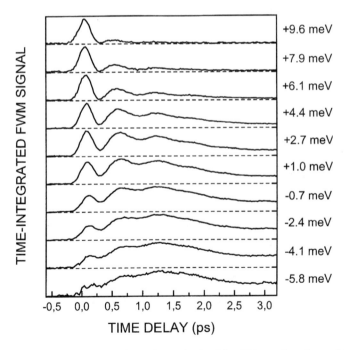

Figure 10.5 Time-integrated FWM transients obtained for different detunings of the central laser frequency with respect to the excitonic $1s$ transition. (Reprinted with permission from Koch et al. 1996.)

$\Delta = 7.9$ meV, the FWM signal consists only of a fast initial peak and a very weak, longer-living component that is only weakly modulated.

In the following discussion the characteristic beating behavior is considered in more detail. To this end we plot a FWM transient for $\Delta = 4.4$ meV on an expanded scale in Figure 10.6. After a fast initial increase the signal quickly collapses, which results in a pronounced minimum around 300 fs. Subsequently the signal recovers, which leads to a second maximum around a time delay of 600 fs. What follows is a slow decay superimposed by damped oscillations whose amplitude is considerably smaller than that of the first "dip."

In principle, these observations can be explained as follows: For time delays smaller than 150 fs, the contributions of the different excitonic transitions are in phase and combine constructively. This results in an enhanced signal amplitude as compared to what would have been expected from the excitation of the 1s transition alone (Feldmann et al. 1993). The various contributions, however, run quickly out of phase due to their different transition frequencies. This leads to a destructive interference between the 1s resonance, on the one hand, and all the higher-lying excited states, on the other hand. The destructive interference reflects itself in the pronounced signal minimum at 300 fs. The signal subsequently recovers because the contribution from the 1s transition will be in phase again with the contributions from the higher bound and unbound transitions after a time, which roughly corresponds to the inverse excitonic binding energy. Four-wave mixing experiments at the band edge of semiconductor quantum wells can therefore be used to estimate excitonic binding energies (Koch et al. 1996; Braun et al. 1997). This can be particularly helpful in the case of inhomogeneous broadening where it is difficult or impossible to extract the binding energy from the absorption spectrum of the sample.

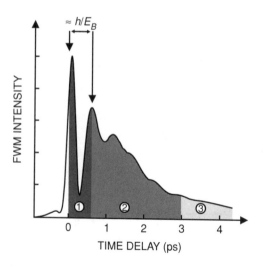

Figure 10.6   Time-integrated FWM signal obtained for a detuning of $\Delta = +4.4$ meV. The dynamics of the excitonic wave packet can be coarsely divided into three regimes. (Reprinted with permission from Koch et al. 1996.)

With increasing time delay, however, the contribution from the continuum states gradually fades for two reasons: First, continuum states dephase more quickly as compared to bound excitonic states. Second, the continuum states represent a continuous distribution and destructively interfere (apparent dephasing; see Feldmann et al. 1993; Cundiff et al. 1996). Thus the end of the temporal region in which all excited continuum states contribute to the signal (region 1) is marked by the second maximum.

The signal modulation observed in the following (region 2) arises from the interference between the contributions of the $1s$ and the $2s$ exciton. For time delays larger than 2.5 ps (region 3), the $2s$ exciton has lost its coherence, and the signal arises from the $1s$ transition only. This illustrative interpretation which divides the dynamics into three temporal regimes (the boundaries between which are of course smooth) is based on calculations using the phenomenological multi-level optical Bloch equations and assuming different dephasing times for the individual transitions (Feldmann et al. 1993).

The characteristic interference pattern displays itself also in the spatial dynamics of the excitonic noneigenstate. Figure 10.7 shows the absolute square of the wave function $|\psi(r, t)|^2$ as a function of time t and electron-hole separation r (Haas n.d.). Since the optical excitation creates the electron and hole at the same spatial position, only the dynamics at $r = 0$ is of relevance in an FWM experiment. Qualitatively $|\psi(r = 0, t)|^2$ shows the same beating behavior as the experimental FWM transient. Observed is a pronounced maximum around $t = 0$ resulting from the excitation of continuum states followed by damped oscillations with a period approximately corresponding to the inverse $1s$-$2s$ splitting. The fact that FWM intensity is sensitive to the overlap of the electron and hole wavefunctions has the interesting side effect that it allows to monitor the ultrafast electron-hole separation induced by an external electric field. Recent wave packet experiments on a semiconductor superlattice in which the field-induced exciton ionization was

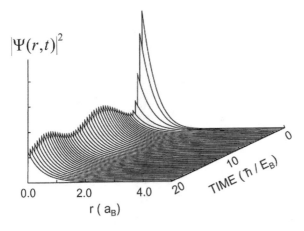

Figure 10.7 Spatial dynamics of an excitonic wave packet.

monitored in the time domain clearly show that continuum states react more sensitively to the presence of an electric field than the excitonic $1s$ state (Plessen et al. 1996).

We can conclude that the above multi-level approach describes the dynamics of the time-integrated four-wave mixing signal quite well. The multi-level model has the advantage that by its simplicity it gives physical insight into the origin of the experimentally observed beating behavior. Yet, as we will see next, it fails in the description of the excitonic wave packet dynamics observed in a time-resolved FWM experiment. For the time-resolved FWM signal, that is, for $|P^{(3)}(t, \tau)|^2$, the multi-level optical Bloch equations predict qualitatively the same beating behavior as observed in the above time-integrated measurements.

Time-Resolved Detection The time-resolved FWM transients experimentally obtained for different time delays τ are displayed in the three-dimensional plot in Figure 10.8. We observe a photon echolike (Shen 1984) nonlinear response peaked along the $t = 2\tau$ line. The amplitude of the "photon-echo" oscillates as a function of time delay τ which agrees with the signal modulation observed in Figure 10.6. The time-resolved response for a fixed time delay differs in two ways from the expectations for the multi-level model. First, it is unmodulated, and second, the full width at the half maximum Δt of the echo does not correspond to the inhomogeneous line width Γ_{inhom} of the $1s$ exciton as extracted from the linear spectrum; it is somehow shorter as can be seen by

$$\Delta t = \frac{2\sqrt{2}\,h\ln 2}{\pi \Gamma_{inhom}} \tag{10.1}$$

(Cundiff et al. 1992). These features can nevertheless be reproduced when solving the full semiconductor Bloch equations (SBE) (Lindberg et al. 1992; Schäfer et al. 1993) including many-particle effects as well as inhomogeneous broadening. In agreement with the experimental transients the SBE calculations show an unmodulated echo with a width that is shorter than expected from the above relation (see Jahnke et al. 1994).

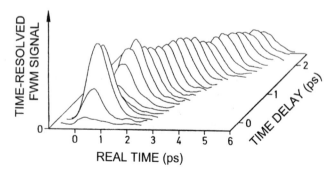

Figure 10.8 Time-resolved FWM signal for different time delays τ. (Reprinted with permission from Koch et al. 1996.)

This clearly demonstrates the many-particle nature of excitonic resonances and hence the complexity of semiconductors compared to atomic systems.

Although the multi-level approach fails here partly for the description of the interference between the excitonic transitions associated with a *single* one-particle transition, it was shown to generally give a satisfactory description of the interference observed between 1*s* resonances associated with *different* one-particle transitions, even if the time-resolved FWM dynamics is considered (Koch et al. 1992; Dekorsy et al. 1994). We will therefore use the optical Bloch equations again in the following sections when discussing the dynamics of electronic and mixed electron-photon wave packets.

10.2.4 Wave Packets with Spatial Dynamics

Coupled Double Quantum Wells: Coherent Oscillations The coupled double-well system is probably the most simple system for observing wave packet dynamics. Figure 10.9 displays a schematic picture of a double-well system (Leggett et al. 1987). Such double-well systems are known from many very different areas of physics. One particularly well-known example is the NH_3 molecule, where the two states are separated by a potential well in real space. Many other situations, like the motion of defects in solids or some types of chemical reactions, can be described as such a system with a geometrical degree of freedom. In a more general picture, one can also treat problems with a nongeometrical degree of freedom (e.g., spin) in a similar manner. The properties of coupled double wells have been addressed in a large number of theoretical works (for a review, see Leggett et al. 1987).

The coupling of the two states in the double-well potential leads to new eigenstates. For full resonance of the localized states (corresponding to $\varepsilon = 0$ in Fig. 10.9), the new eigenstates of the system are a symmetric (bonding) lower state and an asymmetric (antibonding) upper state. The wavefunctions of these states are then given by (Leggett et al. 1987)

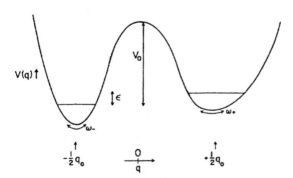

Figure 10.9 Schematic picture of a generalized double-well system. (After Leggett et al. 1987.)

$$\Psi_{symm} = \frac{1}{\sqrt{2}}(\Psi_1 + \Psi_2) \quad \text{and} \quad \Psi_{antisymm} = \frac{1}{\sqrt{2}}(\Psi_1 - \Psi_2) \tag{10.2}$$

with an energy-splitting ΔE caused by the repulsion of the two states. A wave packet located in the left well (i.e., the probability $P_L(t=O)=1$) can be prepared by a superposition of the two wavefunctions:

$$\Psi = \Psi_{symm} e^{-eE_1 t/\hbar} + \Psi_{antisymm} e^{-iE_2 t/\hbar}. \tag{10.3}$$

The probability of being in the left well is then oscillating in time

$$P(t) = |<\Psi|\Psi_1>|^2 = \cos^2\left(\frac{2\pi}{\tau_{DQW}} t\right), \tag{10.4}$$

with a period

$$\tau_{DQW} = \frac{h}{\Delta E}. \tag{10.5}$$

This oscillation manifests the phase coherence between the two states and has no classical analogue (Leggett et al. 1987).

The semiconductor double quantum well (DQW) system as discussed here is an example where coherent oscillations can be directly observed. The upper part of Figure 10.10 gives a schematic picture of an asymmetric double quantum well structure (a-DQW) realized, for example, in the GaAs/AlGaAs material system. The wide well (WW) and narrow well (NW) then consist of GaAs, the barrier between the wells, and the cladding layers of *AlGaAs* with a wider band gap. Shown in this schematic picture is the case where the electric field is tuned so that the two electronic levels are in resonance. It is important to note that the hole levels are then out of resonance, in contrast, to the case of symmetric double wells. Figure 10.10 shows the

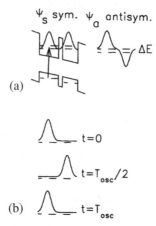

Figure 10.10   (*a*) Double well at resonance with symmetric (*left*) and asymmetric (*right*) eigenstates. (*b*) Dynamics of a photoexcited wave packet.

wavefunction of the lower delocalized level (symmetric) and of the upper delocalized level (antisymmetric), as expressed in equation (10.2).

Resonant photo-excitation of, for example, the WW with a short laser pulse (vertical arrow) with a bandwidth that covers the two delocalized levels will create a wave packet localized in the WW consisting of a linear superposition of the two wavefunctions. This wave packet then oscillates between the wells (as schematically shown in the lower part of Fig. 10.10) with the time period given by (10.5). The idealized picture given in Figure 10.10 is modified in a real system due to collisions and relaxation, which will lead to a damping of the oscillation. For an experimental observation of these coherent oscillations, the period has to be in the picosecond regime: Otherwise, the QW exciton dephasing time constants of a few picoseconds will prevent the observation. It is thus obvious that the direct observation of coherent oscillations requires an experimental technique that allows resonant excitation and subpicosecond time resolution (Leo et al. 1991a). In the following, time-resolved four-wave mixing (FWM) and pump-probe (PP) experiments will be employed for the observation of coherent oscillations.

Sample Design and Properties We have investigated two different sample structures (Leo et al. 1992b). Both structures contain 10 periods of an a-DQW system and were grown by molecular beam epitaxy. The first sample consists of a 170 Å GaAs wide well, followed by a 17 Å $Al_{0.35}Ga_{0.65}As$ barrier and a 120 Å GaAs narrow well; the second sample of a 150 Å GaAs wide well, followed by a 25 Å $Al_{0.20}Ga_{0.70}As$ barrier and a 100 Å GaAs narrow well. The samples are grown by molecular beam epitaxy on n-doped oriented GaAs substrate. The samples were carefully characterized by continuous wave (cw) photoluminescence (PL and PLE spectroscopy at a lattice temperature of $T_L = 2K$. Both samples display very sharp excitonic transitions (FWHM less than 1 meV for the WW, indicating that the roughness of the interfaces is well below one monolayer). A high-resolution transmission electron microscopy (TEM) study of the 170/17/120 sample gave well widths of 541 monolayers (ML), namely 153 Å for the wide well (WW) and 39 ± 1 ML for the narrow well (NW), which is in excellent agreement with the PL studies. The barrier thickness determined by TEM was 6 ML/17 Å, that is, identical to the design value. It should be noted that the experiments described would not have been possible if the design value had been missed by only one monolayer: For five ML, the splitting would have had to be larger to excite both transitions with the laser source used; for seven ML, the oscillations would have had to be overdamped by scattering and relaxation.

One attractive feature of wave packet experiments in semiconductor heterostructures is the fact that they can be easily investigated under the influence of electric fields. For this purpose the active structure is located in the intrinsic region of a *pin*-junction.[3] After diode processing, bias voltages of only a few volts across the

---

[3]The results presented here have been obtained with a Schottky junction instead of a *pin*-junction. For this purpose a semitransparent metal film was evaporated on the samples.

pin-junction can create fields that easily reach 100 kV/cm. The fields can be changed during the optical investigation.

Using these tunable fields, one can directly trace the anticrossing of the transitions by absorption spectroscopy. Figure 10.11 displays the absorption spectra of the 170/17/120 sample in the vicinity of the first heavy-hole (hh) and the light-hole (lh) transition of the NW. For a low electric field (+0.8 V), the spectrum shows the first hh and lh exciton transitions of the NW. As the field is increased, a new transition appears below the hh exciton peak (+0.6 V). This peak gains strength and the original hh exciton peak loses strength until both have about equal intensity at resonance (+0.3 V). The splitting between the two peaks is about 3 meV. With further increases in field, the upper energy peak rapidly moves to higher energy and loses strength, while the lower energy peak moves slowly and acquires the strength of the original hh exciton peak. The spectra reflect the eigenstate probability distribution in the wells. At electronic resonance, both states are delocalized and have nearly equal probability in both wells; out of resonance, the delocalization gradually disappears, and only the transitions within a well are important. The observation of the anticrossing in both hh and lh transition clearly shows that the anticrossing is taking place between electron states, not between hole states.

Figure 10.12 shows corresponding spectra for the WW transitions. The anticrossing of the electronic levels is similarly observed as in the NW transition shown in Figure 10.11. One surprising observation is that the resonance of the electronic levels is obtained *at higher fields* in the WW transition than in the NW transition: This effect is a direct manifestation of the fact that we are dealing with

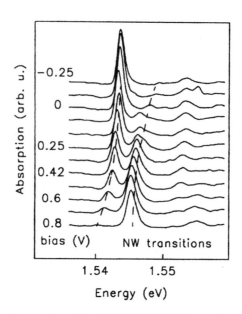

Figure 10.11 Absorption spectra of the 170/17/120 DQW for various bias voltages. Shown is the region of the transitions to the hole states of the narrow well.

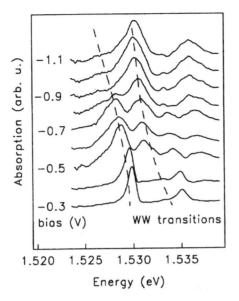

Figure 10.12   Absorption spectra of the 170/17/120 DQW for various bias voltages. Shown is the region of the transitions to the hole states of the wide well.

excitonic transitions and not with single-particle transitions. For single-particle transitions there is no reason why the resonance as probed from the WW should be at another field than the resonance probed from the NW. The effect of the different resonance fields is caused by the different binding energies of direct and indirect excitons (Fox et al. 1990). Figure 10.13 shows a schematic explanation of the

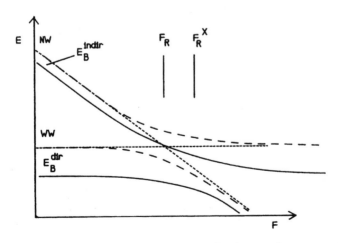

Figure 10.13   Schematic explanation of the influence of excitons on the resonance in coupled double wells. Energies without excitonic effects (long dashed line); energies with excitonic effects (solid lines); energies if the anticrossing is neglected (short dashed line).

influence of the excitonic interaction on the electronic resonance in a semiconductor quantum well system. The WW electronic state is chosen as the energy reference. The dotted lines show the energy of the states without coupling between the wells. In this case the states of the WW and NW cross at a given field F_R. The two states will anticross if the coupling through the barrier is included (dashed lines). The resonance will be approximately at the field F_R. The resonance field changes when the excitonic binding energies are included (solid lines). Figure 10.13 shows the case where one probes the transitions from the wide well. The exciton of the WW will then have a larger (by about a factor of two) binding energy than the NW exciton. Above the anticrossing the situation is exchanged, leading to a shift of the resonance to higher field (F'_R). The situation for transitions to the NW hole state is analogous with the binding energy difference interchange, leading to a resonance shift to lower energy. It is obvious from these results that the coherent oscillations, as observed below using photo excitation, do not probe single-particle wave packets but excitonic wave packets.

Time-Resolved Experiments The data in Figure 10.14 are the results of a pump-probe experiment in the 150/25/100 sample with larger splitting ($\Delta E = 5$ meV). In such an experiment the transmission is probed with a weak probe pulse as a function of time delay after a strong pump pulse. The excitation is chosen resonantly to the WW transition. Temperature (about 5 K) and density (about 5×10^8 cm^{-2}) are kept low to avoid fast dephasing of the transition due to carrier–phonon or carrier–carrier scattering. At delay $t = 0$, a step of the transmission is visible, which is due to the bleaching of the absorptive transitions by the photo-excited carriers. At electric fields close (10.4 kV/cm) or slightly above the resonance (14.2 kV/cm), the transmitted

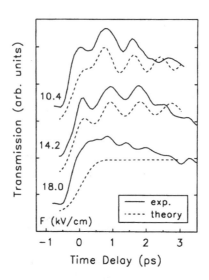

Figure 10.14 Pump-probe data for the 150/25/100 sample for various electric fields.

signal for delay time $t > 0$ shows a strong modulation that reflects the oscillation of the wave packet. The oscillation period is about 800 fs, as expected by equation (10.5) for the observed splitting of about 5 meV at full resonance. Moving further away from resonance (18.0 kV/cm), the modulation becomes weaker and the oscillation period shorter (about 500 fs). These effects show clearly the increasing localization of the wavefunctions in the wells and the increase in splitting between the levels at greater distances from the electronic resonance. Pump-probe experiments resonant to the NW exciton transitions display similar oscillations within approximately the same oscillation period. However, the oscillations are more strongly damped for NW excitation, possibly for two reasons: (1) the creation of free carriers in the WW, which lead to a more rapid dephasing of the wave packet, and (2) the higher influence of interface roughness on the hole state in the NW. The dashed lines in Figure 10.14 show the theoretical results obtained with the model described below.

The pump-probe results are confirmed by transient self-diffracted FWM experiments. The solid lines in Figure 10.15 show results for the 150/25/100 sample with larger splitting. Close to the resonance of the levels (at about 11 kV/cm), the FWM signal clearly shows an oscillatory modulation, which becomes weaker moving away from resonance.

Comparison with Theory A comparison with theory yields information about the scattering processes that are relevant for the dephasing observed in the PP and FWM experiments. In the model discussed here (Leo et al. 1991b), the coupled quantum well system, is described as a three-level system with the hole state defined as level $|0>$

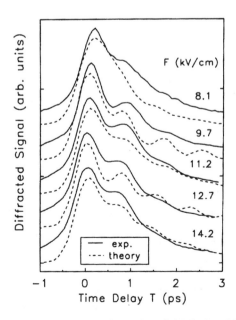

Figure 10.15 Four-wave mixing data for various fields in the 150/25/100 sample.

and the two delocalized states as $|1>$ and $|2>$. The relevant relaxation rates for this problem are γ_1 and γ_2, which describe the dephasing of the two upper levels to the ground state (i.e., the decay of the density matrix elements n_{01} and n_{02}), and a relaxation rate γ_3, which describes the dephasing due to relaxation between the two delocalized excited states (i.e., the decay of n_{12}). The signals for the PP and the FWM experiments are calculated using a density matrix theory in a perturbation ansatz. The pump-probe and four-wave mixing signals I_{PP} and I_{FWM} are then proportional to the third-order polarization in direction k_1 and $2k_2 - k_1$, respectively. For δ-shaped pulses and purely homogeneous broadening, we obtain the following relations:

$$I_{PP} = \text{const} \times w_1^2 + w_2^2 + w_1 w_2[1 + \cos(\Delta E t)\exp(-\gamma_3)] \qquad (10.6)$$

and

$$I_{FWM} = \text{const} \times (w_1^2 e^{-2\gamma_1 t} + w_2 e^{-2\gamma_2 t} + 2w_1 w_2 \cos(\Delta E)\exp[-(\gamma_1 + \gamma_2)t]), \quad (10.7)$$

where w_1 and w_2 are the spectral weights of the transitions. The most important result is that the phase relaxation between the delocalized levels γ_3 is responsible for the damping of the oscillation in the PP experiment, in contrast to the FWM experiment where the relaxation to the ground states γ_1 and γ_2 lead to the damping.

The dashed lines in Figures 10.14 and 10.15 are theoretical calculations based on a numerical model of the nonlinear response similar to equations (10.6) and (10.7) but with finite pulse width (Leo et al. 1992a). For the PP data shown in Figure 10.14, a comparison of this theory and experiment yields $\gamma_3 = 4ps^{-1}$. This value is in reasonable agreement with theoretical calculations of the relaxation between delocalized levels in a DQW structure (Liu et al. 1989).

The dashed lines in Figure 10.15 show theoretical (dashed lines) four-wave mixing traces for the 150/25/100 sample for various electric fields. The theoretical calculations agree reasonably well with the experiment. However, some differences remain. For instance, the theory predicts that the modulation is not maximal for resonance but for the traces at fields slightly off the resonance. This behavior is not observed in the experiment. Note that the parameters in the theory ($\gamma_3 = 4$ ps^{-1}, $\gamma_1° = 3.9$ ps^{-1}, $\gamma = 2.1$ ps^{-1}) are identical to the parameters for the fit of the time-resolved transmission data.

Figure 10.16 shows the splitting between the electronic levels ΔE as a function of the electric field. The experimental values are obtained by inserting the oscillation periods from the fits of the time-resolved data in (10.6). The theoretical values are from a calculation of the structure including excitonic interactions (Leo et al. 1992b). The experimental data clearly show the predicted nonmonotonic dependence on the field with a minimum at about 10.5 keV/cm. The absolute minimum, however, is about 5 meV, whereas the theory predicts 4 meV. This deviation is probably caused by a barrier thickness that is somewhat lower than the design value. Also the dependence on the electric field shows some deviation. Possible reasons are that the theory does not include all transitions relevant for the problem; for example, the light-hole states and higher states are omitted.

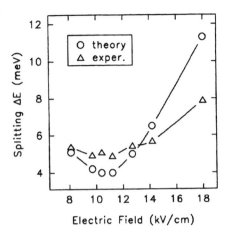

Figure 10.16 Splitting between the electronic levels plotted in the electric field. Theory ($\circ$) and data ($\triangle$) obtained by fits to the pump-probe and four-wave mixing data.

Superlattices: Bloch Oscillations

Bloch oscillations (BO) in periodic potentials are another example of wave packets with pronounced spatial dynamics. In solid-state physics textbooks, BO are usually discussed in a semiclassical picture based on the eigenstates of a periodic potential without field, the delocalized Bloch states. It was shown by Bloch (1928) that the k-space velocity of an electron in a periodic potential subject to a static field F is linearly proportional to the field:

$$\hbar\frac{dk}{dt} = eF. \tag{10.8}$$

Since the energy dispersion of a periodic solid is also periodic in k, it is immediately obvious (as was first pointed out by Zener 1934) that an electron will perform a periodic motion in energy. The electron will gain energy until it is at the upper edge of the band; thereafter it will lose energy until it reaches the center of the second Brillouin zone.

The real-space velocity can be easily calculated if one assumes, for example, a harmonic dispersion relation for the band:

$$E(k) = \frac{\Delta}{2} - \frac{\Delta}{2}(\cos(kd)) \tag{10.9}$$

(where Δ is the bandwidth). This results in a real-space velocity which is given by

$$v_R = \frac{1}{\hbar}\frac{\partial E}{\partial k} = \frac{d\Delta}{2\hbar}\sin(kd), \tag{10.10}$$

that is, the electron moves back and forth (see Fig. 10.17, after Ashcroft and Mermin 1981). It can then easily be shown that the spatial position of the electron is harmonically changing with time:

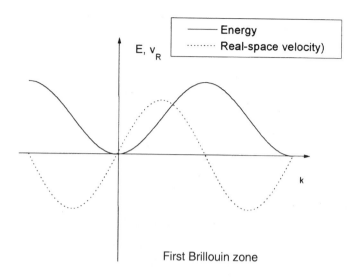

Figure 10.17 Bloch oscillations in the semiclassical picture: An electron at $k = 0$ starts to move with constant velocity in k-space once the field is turned on. It gains energy until the edge of the first Brillouin zone. When it leaves the first Brillouin zone, the energy starts to decrease and the velocity (dashed line) becomes negative. When it reaches the center of the second Brillouin zone, it has returned to its original spatial position.

$$z(t) = z_0 - \frac{\Delta}{2eF} \cos(wt), \qquad (10.11)$$

where

$$\omega = \frac{eFd}{\hbar} \qquad (10.12)$$

is the period of the oscillation (Zener 1934). The total amplitude of the oscillating electron is the field-induced localization length

$$L = \frac{\Delta}{eF}. \qquad (10.13)$$

The Gedankenexperiment just described has not successfully been performed in bulk solids. For metals it can be estimated that the BO frequencies are many orders of magnitude lower than the scattering rates (Ashcroft and Mermin 1981). BO have also not been observed in bulk semiconductors, although the situation is much more favorable for BO compared to metals.

A key development for experimental studies of BO was the invention of the semiconductor superlattice (SL) (Esaki and Tsu 1970). By the alternating growth of semiconductor layers with small and large band gaps, a lattice with much larger lattice constant (and therefore much smaller band width) can be designed (see Fig. 10.18). In superlattices the frequency-space equivalent of BO has been observed: the so-called Wannier-Stark ladder (see Fig. 10.18 lower part). The eigenstates with applied electric field form a ladder (Wannier 1969) with energies

$$E = E_0 + neFd, \tag{10.14}$$

where n is an integer number and E_0 is the energy of a particular reference state. The Wannier-Stark ladder has been directly observed in optical experiments in semiconductor superlattices (Mendez et al. 1988; Voisin et al. 1988). The heavy-holes are rather strongly localized because of their great mass. Thus the optical transitions between the heavy-hole states, and the electrons trace the ladder spectrum as given by (10.14). If the field is swept, a fan chart of the transitions is obtained (see Fig. 10.19).

It is straightforward to conduct an optically excited wave packet experiment by exciting the WSL with a short laser pulse. It is obvious from equations (10.12) and

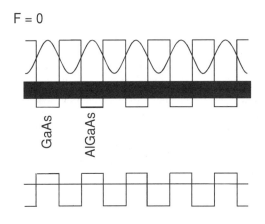

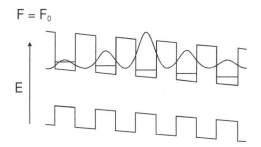

Figure 10.18 Schematic picture of a type I semiconductor superlattice without applied field (upper part) and with applied field (lower part). The probabilities amplitudes of the electronic states are also sketched.

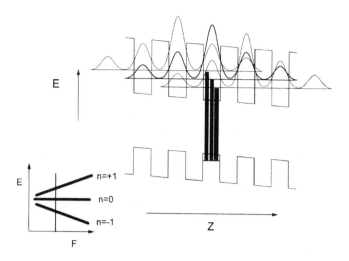

Figure 10.19 Scheme of the optical transitions in a biased semiconductor superlattice characterized by spatially direct and indirect transitions, leading to a fan chart when the field is swept (inset).

(10.14) that the temporal periodicity of photo-excited wave packets will be identical to the one in the transport Gedanken experiment discussed above. However, it is not a priori obvious that the spatial dynamics are the same. It has been theoretically shown (Bastard and Ferreira 1989) that the optically generated wave packets can undergo breathing mode motions that have spatial dynamics very different from transport BO.

Two limiting cases are shown in Figures 10.20 and 10.21. Displayed are single-particle calculations of wave packets dynamics for excitation with the laser centered well below the $n = 0$ transition of the WSL (Fig. 10.20), which leads to a well-localized wave packet with a well-pronounced center-of-mass dynamics, and for excitation symmetric to the $n = 0$ transition (Fig. 10.21), associated with a breathing-mode motion without any center-of-mass motion. We will show below how both types of motion can be experimentally observed in the optical experiments discussed here. We will first discuss four-wave mixing results that have been used to observe the temporal dynamics of the Bloch wave packets (Feldmann et al. 1992; Leo et al. 1992a; Leisching et al. 1994). As a second example, we will discuss results that have directly shown the spatial dynamics (Lyssenko et al. 1997, 1998). For a review of Bloch oscillation experiments, see Leo (1998).

Bloch oscillations have recently performed in a very different system (Ben Dahan et al. 1996). Atoms were placed in a lattice formed by the standing wave of two counterpropagating laser beams. By tuning one of the beams, a constant force can be generated on these atoms. It was shown that these atoms beautifully perform Bloch oscillations with periods about 10 orders of magnitude slower than in the semiconductor experiments discussed here. These experiments clearly show the huge differences in dephasing which limit coherent experiment of the electronic states in solids to the time range of a few picoseconds.

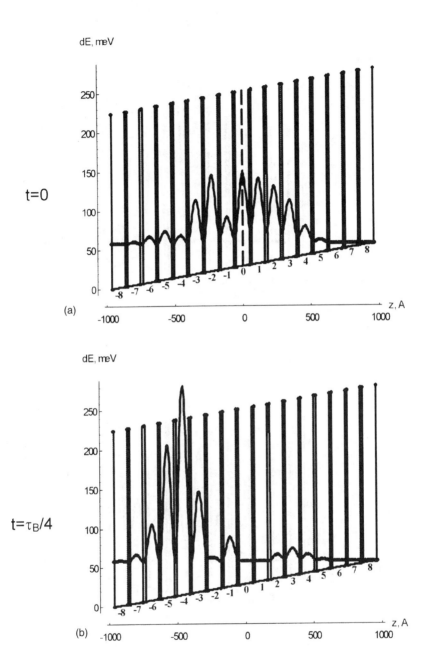

Figure 10.20 Single-particle calculation of the wave packet dynamics for excitation below the center of the Wannier-Stark ladder. (*a*) The wave packet at the time of excitation; (*b*) the wave packet after one-quarter of a Bloch oscillation. Excited are the $n = -2$, -1, and 0 transitions. The vertical dashed line marks the center of mass of the wavefunction. (Reprinted with permission from Leo 1992a.)

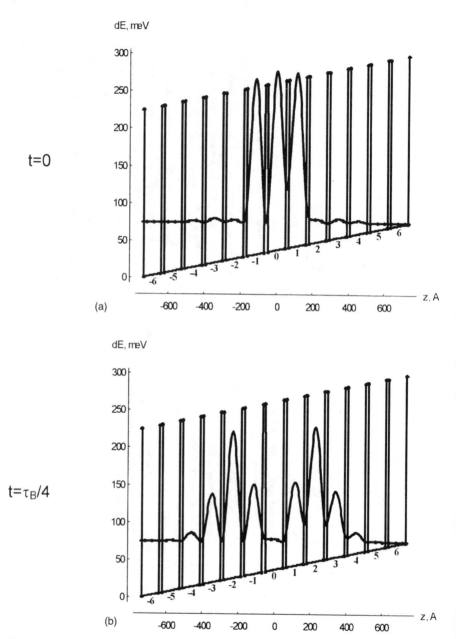

Figure 10.21 Single-particle calculation of the wave packet dynamics for excitation symmetric to the center of the Wannier-Stark ladder. (*a*) The wave packet at the time of excitation; (*b*) the wave packet after one-quarter of a Bloch oscillation. Excited are the $n = -1$, 0, and +1 transitions.

Experimental Technique and Samples The experiments are performed in a series of $GaAs/Al_{0.3}Ga_{0.7}As$ superlattices which have been described in detail elsewhere (Leisching et al. 1994). We discuss here results taken on SL with 67 and 97 Å well widths and a 17 Å barrier width. A Kronig-Penney calculation yields miniband (MB) widths of 38 and 20 mev, respectively. The samples are labeled in the following discussion according to their well/barrier widths, that is, 97/17.

Figure 10.22 shows peak positions in the low-temperature photocurrent spectrum of the 97/17 sample as a function of the bias voltage (Leo et al. 1992a). For higher fields, the spectra clearly show a Wannier-Stark ladder with spectral shifts as expected for the design parameters of the sample. For fields higher than about 10 kV/cm, the transitions shift linearly with applied field. For lower fields, deviations from the linear field dependence are observed. Wannier-Stark ladders of both hh and lh transitions are visible.

Figure 10.23 displays transient FWM signals as a function of the electric field (Leo et al. 1992a). The excitation energy is at 1.537 eV slightly below the center of the superlattice transitions. The signals show a periodic modulation with a period that decreases with increasing electric field, as expected for BO. Up to four oscillations of the wave packet in the Brillouin zone are visible for higher fields. For lower fields the oscillation is damped out quickly due to the long oscillation period compared to the dephasing time of the excitation. The overall decay time of the FWM signals, which can be related to the dephasing of the excitation, is about 1.5 ps. It is independent of the field in the range studied here.

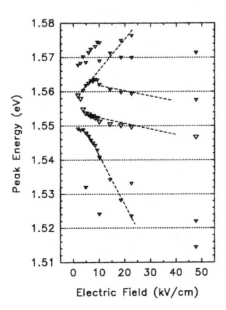

Figure 10.22 Wannier-Stark ladder fan chart of the 97/17 sample. (Reprinted with permission from Leo et al. 1992a.)

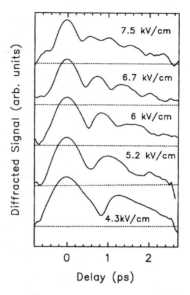

Figure 10.23   Transient FWM signal for the 97/17 superlattice for various electric fields. (Reprinted with permission from Leo et al. 1992a.)

Figure 10.24 displays the energies calculated from the oscillation periods observed in the FWM signals using equation (10.12) (triangles) in comparison to the splitting expected for the sample parameters (dashed line). The energy splitting as derived from the oscillation period shows the expected linear dependence on the electric field. The proportionality constant is in good agreement with that expected from the parameters

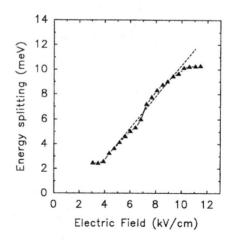

Figure 10.24   Splitting of the WSL as determined from the experimentally observed BO frequency (symbols) and theoretically predicted for the sample parameters and field. (Reprinted with permission from Leo et al. 1992a.)

of the superlattice. One encouraging feature of the observations is the large tuning range of the oscillation frequency. A tuning range of more than 400% is attainable, which is much larger than the tuning range in the double quantum well experiments discussed above. This is caused by the much slower localization of the wavefunction in a superlattice as compared to the case of the double quantum well. In the latter the wavefunctions are already nearly completely localized if the oscillation frequency is tuned about 50% off the resonance frequency.

Direct Determination of the Spatial Dynamics   The experiments just discussed demonstrate the temporal beating of the wave packet with BO frequency. However, they do not demonstrate the harmonic spatial motion as expected from the Gedankenexperiment idea of Zener outlined above. It has been shown that it is possible to directly solve the spatial motion of the Bloch wave packets (Lyssenko et al. 1997). The basic idea is quite simple. The spatial motion of the electrons will lead to a macroscopic dipole moment due to the center-of-mass displacement of the electron wavefunctions relative to the localized holes. This dipole moment is superimposed onto the static field in the SL (see Fig. 10.25). The modulation of the total field by the oscillating dipole moment can be detected by using the WSL itself as a sensitive field sensor: The oscillation of the electric field leads to an oscillation of the transition energies. The most important feature of these oscillations of the WSL transitions is the fact that the energy shift can be related to the relative displacement $z(t)$ of the center-of-mass of the electron and hole wavefunction. It can be shown that

$$z(t) = \frac{\varepsilon_0 \varepsilon_r}{n e^2 n_{\text{well}}} \Delta E(t), \qquad (10.15)$$

where n_{well} is the carrier density, ε_r is the relative dielectric constant, and ΔE is the energy shift of a Wannier-Stark ladder transition with WSL index n as given in (10.14).

The only parameter needed in determining $z(t)$ is the carrier sheet density per well n_{well}, which can be determined with high precision. A dynamic treatment of the modulation of the WSL transitions by the THz field of the oscillating electrons predicts similar motions of the WSL peaks as discussed for the static case.

The oscillation of the transient spectra is visible in the data of Figure 10.26 which show transient FWM spectra for the 67/17 sample. Visible are the center transition of the WSL (hh_0), the first transition below the center (hh_{-1}) and two transitions above the center. The hh_{-1} transition shows a shift to lower energy during the first part of the BO cycle and shifts back during the second part of the Bloch cycle. The hh_0 center transition stays at constant energy. The hh_{+2} transition shifts first to higher energy and then to lower energies, with an amplitude which is larger than for the hh_{-1} transition.

Figure 10.27 shows the displacement of the wave packet as derived from the peaks shifts. The motion of the wave packet is well described by an oscillation with a total amplitude of approximately 160 Å. In the derivation of the amplitude, the loss of coherent carriers due to dephasing has to be taken into account. The solid line is a

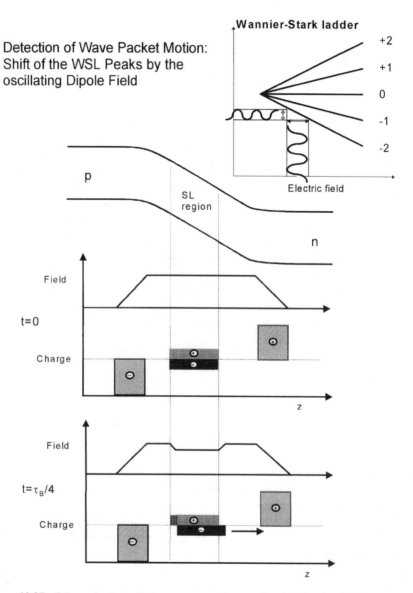

Figure 10.25 Schematic of the displacement experiment using the Wannier-Stark ladder as a field sensor: When both electrons and holes are at the same place, the field is given by the static field created by the space charges of the pn junction. If the electron wave packets move to the right, the field is reduced (bottom); for motion to the other side, the field would be enhanced. The oscillating field associated with the electron oscillation causes a motion up and down the WSL fan chart (upper right corner).

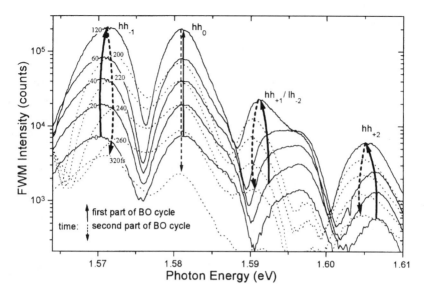

Figure 10.26   Shift of different transitions of the Wannier-Stark ladder for various delay times. Solid arrows indicate the behavior during the first part of a Bloch oscillation cycle, dashed arrows during the second. The relative delay times with respect to the start of the Bloch cycle (0 fs) are indicated.

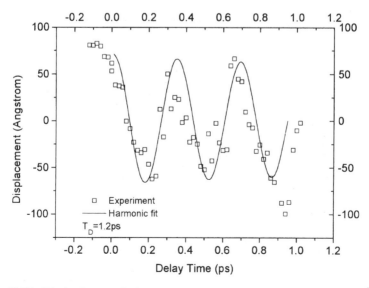

Figure 10.27   Bloch electron displacement as a function of delay time for a 67 Å/17 Å superlattice. The solid line is a result of the model described in the text.

harmonic function assuming a dephasing time of 1.2 ps. This damping constant is obtained for the given experimental conditions from independent measurements (Valusis et al. 1996). Within error, the displacement as a function of time follows a harmonic function, as expected for a miniband with harmonic dispersion (10.11).

Figure 10.28 displays the dependence of the total amplitude on the electric field. The experimental data are depicted as the circles. The solid line is the semiclassical result given by equation (10.13). The amplitude is decreasing with electric field as predicted by Zener. The triangles in Figure 10.28 show a calculation using a model that takes into account excitonic effects (Dignam et al. 1994) with the parameters of our experiment. The amplitudes are somewhat lower than the semiclassical limit and in good agreement with the experimental data. The main reason for the lower amplitudes predicted by the excitonic theory (in comparison to the semiclassical results) are not the electron-hole coupling but the fact that the laser excitation chosen here was close to the center of the WSL.

Control of the Amplitude by Changing Laser Excitation: Tuning between Breathing Modes and Spatial Oscillations As discussed above, the displacement dynamics and the absolute amplitude of the optically excited BO strongly depend on the composition of the wave packet by the laser excitation. Examples for a single-particle picture have been discussed above. The general picture remains the same if excitonic interactions are taken into account (Dignam et al. 1994): The upper limit of the amplitude as given by the semiclassical theory (10.13) can nearly be reached if the center of the excitation is chosen well below or above the center of the WSL. For excitation close to the center of the WSL, the amplitude of the wave packet reaches a minimum.

In recent experiments it was shown (Sudzius et al. 1998) that the spatial amplitude of the oscillating wave packets can be continuously tuned between symmetric

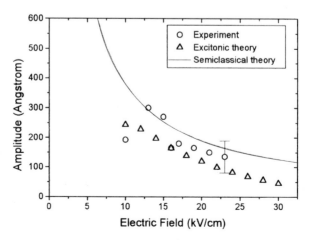

Figure 10.28 Total oscillation amplitude as a function of the electric field. Experiment (○), semiclassical theory (—), and theory including excitonic interactions (△).

breathing-mode oscillations and a harmonic spatial motion with the amplitude as predicted by Zener.

Figure 10.29 displays the spectral shift of the hh_{-1} peak with regard to delay time for various excitation conditions. The relative position of the laser center wavelength is defined in units of the WSL splitting, with ω_c as the distance relative to the (experimentally observed) hh_0 transition.

Figure 10.30 shows the amplitude (as derived from the FWM peak shift) as a function of the excitation energy. The total amplitude of the center-of-mass oscillation of the wave packet is expressed in units of the WSL period (in this case 84 Å).

The data clearly show a minimum of the amplitude for excitation near the center of the WSL. For excitation well below and above the center of the WSL, the amplitudes reaches several periods of the SL. The semiclassical amplitude as given by (10.13) would be about 3.4 times the SL period. For excitation well below the center of the WSL, the experimental amplitude comes close to the semiclassical limit.

The experimental data are compared to theoretical results (lines in Fig. 10.30) based on a theory by Dignam et al. (Dignam 1994) described above. The theoretical curves also show a minimum for excitation close to the center of the WSL. However, this minimum is only pronounced if dephasing is taken into account (dashed and solid lines in Fig. 10.30). The minimum amplitude reaches then values that agree well with

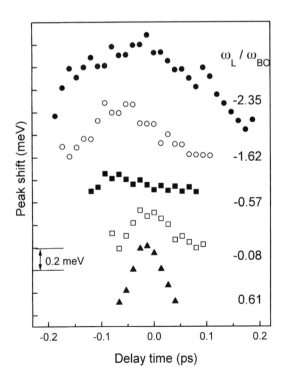

Figure 10.29 Shift of the hh_{-1} peak plotted in delay time for various excitation conditions.

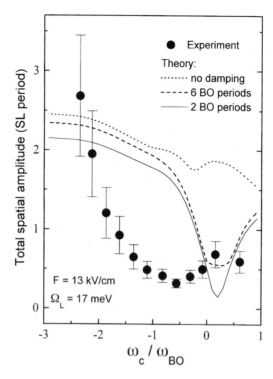

Figure 10.30 Experimentally determined amplitude of the Bloch oscillating wave packet as a function of the excitation energy at a field of 13 kV/cm and an excitation density (in the first well) of 1.2×10^9 cm^{-2} (dotted line). Theoretical calculation of the amplitude for various field, with the thick line corresponding to the field of the experiment (solid line). The energy is given in units of the WSL splitting relative to the experimentally observed energy of the hh$_0$ transition.

the experimental value. The inclusion of damping also improves the agreement of the energy of the amplitude minimum. The fact that the oscillation does not reach zero amplitude for excitation close to the center (as predicted from the single-particle calculations) is caused by the fact that the excitonic interaction influences the oscillator strength of the WSL ladder transitions and thus destroys the balance between transitions with positive and negative n required for the creation of a symmetric breathing-mode wave packet.

The influence of damping on the total amplitude predicted by theory is caused by a subtle effect of wave packet physics. The energy spacing of the excitonic WSL levels is not equal due to excitonic effects. The exciton binding energies of different WSL transitions differs due to the dependence of the Coulomb coupling on overlap. The minimum of the amplitude for excitation near the $n = 0$ transition is caused by the equal excitation of levels with negative and positive n leading to two components of the wave packet moving in opposite direction, and with the dipole moments canceling each other. The wave packet thus initially performs a breathing-mode-type motion.

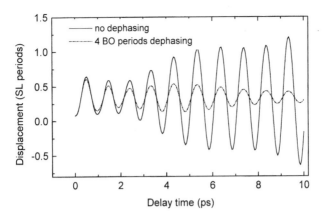

Figure 10.31 Effect of the dephasing in the time evolution of displacement for symmetric excitation of the wave packet when excitonic effects are included. No damping (solid line); damping time of 4 BO periods (dashed line).

Due to the different frequencies caused by the exciton-induced nonequidistance of the energy ladder, the motion of the two constituents does not cancel anymore after some delay time. The motion continuously evolves from a breathing-mode motion into an asymmetric motion.[4] Figure 10.31 shows the effect of the dephasing in the time evolution of dipole for the wave packet excited with the laser central frequency at the $n = 0$ WSL state. The development of the oscillating dipole amplitude with increasing delay time when dephasing is neglected is clearly visible. In a real system, dephasing reduces the number of coherent carriers, so the amplitude can only be derived for small delay times. The large amplitudes to the out-of-phase motion of the constituents of the wave packet has not developed for these small delay time, and thus the low amplitudes of the initial breathing-mode motion is detected in the experiment. This is a point where the detailed investigation of wave packet dynamics is hindered by the very fast dephasing times in semiconductors.

Another major difference between experiment and theory is that theory predicts that the minimum of the amplitude is reached for excitation above the center of the WSL, whereas the experiment observes a minimum amplitude below the center. A possible explanation is, for example, that the theory does not take light-hole excitation into account. However, the strong delocalization of light holes due to their much lower mass leads to spatial oscillations similar to those of the electrons. These oscillations will contribute to the total dipole moment which is measured here.

10.2.5. Electron-Photon Wave Packets in Cavities

Introductory Remarks In the previous sections purely electronic wave packets have been investigated. These studies were performed on heterostructures in which the

---

[4]For a simulation of the wave packet motion, refer to the WWW page http://ppprs1.phy.tudresden.de

electronic wavefunction is confined and the electronic density of states is tailored. In this section we will investigate "mixed electron-photon wave packets" in semiconductor microcavities (SMC). The idea of SMC is to additionally confine the photonic wavefunction. By growing two semiconductor Bragg mirrors consisting of many quarter wave stacks separated by a spacer layer with an extension on the order of a wavelength, it is possible to tailor also of the photonic density of states (Burstein et al. 1995).

Initially the research on SMC was spurred by the desire for very low threshold lasers and has very soon led to the so-called vertical cavity surface emitting lasers (VCSELs) (Jewell et al. 1989; Yokohama 1992). Yet the work on microcavities is not only exciting for the large number of technical applications but also from the viewpoint of fundamental research. This includes the possibility to alter the spontaneous emission of an absorber placed inside the microcavity and even to make radiative decay a reversible process. The latter is achieved in the so-called strong coupling limit when a photon emitted by the absorber remains in the cavity long enough to get reabsorbed, subsequently reemitted, and so on. This regime of oscillatory energy transfer which was first studied in atomic systems (Haroche 1992) is realized if the coupling frequency g between absorber and cavity exceeds dissipative channels, namely the dephasing rate γ of the electronic excitation and the escape rate κ of photons from the cavity (inverse cavity lifetime). The energy transfer between the intracavity field α (photonic subsystem) and the polarization $P^{(1)}$ (electronic subsystem) then leads to oscillations in the light intensity coherently emitted by the system after it has been excited by a short laser pulse. Figure 10.32 shows the squared magnitudes of the intracavity field $|\alpha(t)|^2$ (solid line) and the macroscopic first-order polarization $|P^{(1)}(t)|^2$ (dashed line) after excitation with a short laser pulse (gray area). The data in Figure 10.32 are the outcome of a numerical calculation modeling a two-level atom in a cavity (see Koch 1998a for details). The frequency domain counterpart of the intensity oscillations are two Lorentzian shaped lines that show up in the transmission spectrum (Weisbuch et al. 1992). These lines, which are called *normal modes*, represent mixed electron-photon states. If the cavity mode is in resonance with the absorber, the mixing is maximum and the two normal modes have equal strength.

Although the SMC system conceptually differs from electronic systems some analogies to the electronic a-DQW system discussed in Section 10.2.4 can be drawn. While in the a-DQW system an optically created wave packet is initially localized in the WW in the microcavity system, all the energy is initially deposited in the intracavity field. In the a-DQW the electronic wave packet then oscillates back and forth between the two wells, whereas in a microcavity the energy is periodically exchanged between α and $P^{(1)}$.

Yet the special mixed electron-photon nature of the microcavity system cannot be concluded from its linear dynamics shown in Figure 10.32. A detector outside the cavity will monitor the dynamics of $|\alpha(t)|^2$. The oscillations in $|\alpha(t)|^2$, however, bear no special feature that possibly could identify them as originating from an oscillatory energy transfer between the photonic and the electronic subsystem. In fact the light

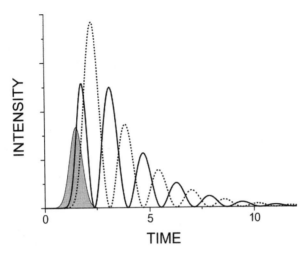

Figure 10.32 Squared magnitudes of the intracavity field (solid line) and the macroscopic first-order polarization (dashed line) after excitation with a short laser pulse (gray area). The calculation is performed for a slight electronic dephasing, infinite cavity life time, and no detuning. (See Koch 1998a for details.)

intensity coherently emitted by many other systems that consist of two optical transitions looks basically identical (Koch et al. 1992). Two simple examples are an electronic three-level system consisting of one ground state and two excited states (and vice versa) and a 2 × 2-level system composed of two uncoupled two-level absorbers. Examples for a three-level system in semiconductors include the heavy-hole light-hole system in a quantum well and the a-DQW system. A 2 × 2-level system is formed, for example, by excitons bound to two different impurities (Lyssenko et al. 1993). After they have been excited by a short laser pulse, all of the systems above coherently emit light whose intensity exhibits oscillations with a period given by the inverse energy splitting between the two transitions. Although the oscillations are essentially indistinguishable from each other, they all have a different physical meaning. In the case of a three-level system, the oscillations are real quantum beats and reflect the temporal evolution of an electronic wave packet. In the case of a 2 × 2-level system, in contrast, no wave packet is created, and the oscillations are due to a polarization interference (the electromagnetic fields emitted independently by the two transitions interfere at the position of the detector). The above-mentioned systems are likewise indistinguishable in the frequency domain, since all show a linear transmission spectrum that consists of two peaks.

A distinction can be reached on the basis of the nonlinear dynamics as observed in a FWM experiment with time-resolved detection (Koch et al. 1992, 1996). As mentioned in Section 10.2.2, excitation of more than one optical transition results in beats not only in the total FWM intensity as a function of τ but also in the time-resolved signal as a function of t. The signal extrema, however, are found for different times for the three- and the 2 × 2-level system, respectively. The different patterns become apparent if the positions of the maxima (or minima) are plotted versus t and τ. While the maxima follow the condition $t = \tau + nT_b$ for a three-level system (where n is an integer and T_b is the beat period), they obey $t = 2\tau + nT_b$ in the case of a 2×2-level system. Using this method, it is possible to confirm the quantum beat nature of Bloch oscillations (Dekorsy et al. 1994) and the "islands beats" (Koch et al. 1993). Moreover we can conclude that the modulation of the echo amplitude observed in Figure 10.8, which is characteristic for an excitonic wave packet, reflects a real quantum interference.

Next we will study the dynamics of electron-photon wave packets in a SMC. In particular, we want to address the question if the special nature of an electron-photon wave packet can be concluded from its nonlinear dynamics. We will see that the oscillation pattern obtained in a time-resolved FWM experiment is complex, differs from that of purely electronic multi-level systems, and reflects the mixed electron-photon character of the normal modes. To enlighten the underlying physics, the experimental data will be compared to calculations on the basis of the optical Bloch equations.

Sample The sample investigated is a high-quality semiconductor microcavity; a schematic drawing is shown in Figure 10.33a. It contains two 15 nm GaAs quantum wells positioned at the antinode of a λ Al$_{0.3}$Ga$_{0.7}$As spacer. The Bragg mirrors consist of 15 and 20.5 pairs of Al$_{0.11}$Ga$_{0.89}$As quarter-wave stacks. Due to a slightly wedge-shaped spacer layer, the spectral position of the cavity resonance varies across the sample. This allows the cavity mode to be brought in and out of resonance with the excitonic transition simply by varying the position on the sample. Figure 10.33b shows the spectral positions of the two

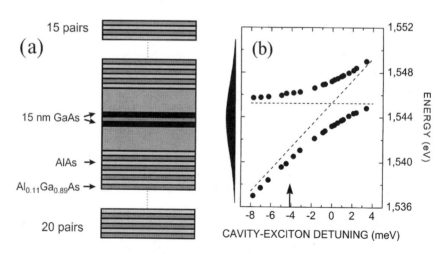

Figure 10.33 (a) Structure of the semiconductor microcavity. (b) Position of the normal modes in the reflectivity spectra as a function of the detuning (after Wang et al. 1995). The vertical arrow marks the position at which the experiment is performed. The laser spectrum is indicated on the left side. (Reprinted with permission from Koch 1998a.)

normal modes versus the cavity-exciton detuning. The experimental data derived from the reflection spectra are shown as dots. A pronounced anticrossing is observed because the cavity is tuned across the exciton resonance in analogy to what was observed for the purely electronic states in the case of the a-DQW (Fig. 10.11). For the experiments presented here, the cavity is tuned slightly below the hh resonance (the position is marked by the small vertical arrow) to avoid complications due to light hole excitons. Under this condition the normal mode splitting amounts to 6.1 meV. The spectrum of the laser is indicated on the left side of Figure 10.33*b*.

Experimental Data Figure 10.34 shows the time-integrated signal which is dominated by pronounced normal mode oscillations. The oscillation period agrees well with the normal mode splitting which can be determined form the FWM spectrum (inset of Fig. 10.34). The spectrum shows two peaks of different strength. Since the cavity mode is detuned with respect to the exciton resonance, the mixing is not complete. Hence the weaker mode at 1.5431 eV is more like that of an exciton, while the stronger mode at 1.537 eV is more like that of a cavity. In the following discussion the FWM signal is time-resolved for various time delays between $\tau = -0.1$ and $+1.45$ ps. Figure 10.35 shows two of the obtained transients. The signal envelope—which is determined by a variety of effects including dephasing, disorder, many-particle, and cavity effects—is superimposed by normal mode oscillations. Their mean frequency corresponds well to frequency of the oscillations observed in the time-integrated measurement and likewise to the splitting observed in the FWM spectrum. To obtain the oscillation pattern, many of these transients are taken. The temporal positions of the minima extracted from these curves are plotted as dots in Figure 10.36. It is apparent from the solid lines the signal minima follow a $t = 2\tau$ slope. At a first glance the (t, τ) pattern observed here might look identical to that of a 2×2-level system. Yet

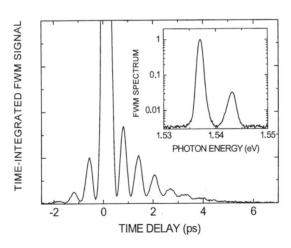

Figure 10.34 Normal mode oscillations in the time-integrated FWM signal. FWM spectrum for $\tau = -0.56$ ps; the two peaks correspond to the normal modes. (Reprinted with permission from Koch et al. 1998b.)

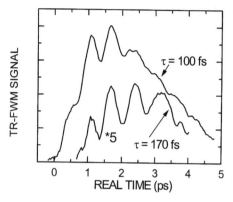

Figure 10.35 Time-resolved FWM signal for time delays τ of 100 and 170 fs. (Reprinted with permission from Koch et al. 1998b.)

a close inspection reveals distinct differences. The signal minima do not follow the $t = 2\tau$ slope exactly but undulate around this slope. In addition a crossover behavior, namely a mixing between two adjacent $t = 2\tau$ lines, can be observed around $\tau = 0.5$ ps and $t = 2.8$ ps. Hence the (t,τ)-behavior observed here differs from the strict $t = 2\tau$ behavior of the 2×2-level system and likewise from the $t = \tau$ behavior known from three-level systems.

Comparison with Theory We now compare the experimental results to analytical and numerical model calculations on the basis of the optical Bloch equations which are extended to allow for microcavity effects (see Koch et al. 1998b for details). In the previous sections the FWM intensity detected in the far field was proportional to the

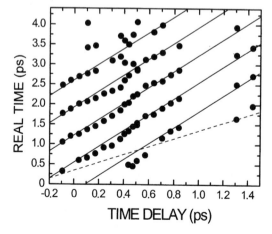

Figure 10.36 Positions of the minima in the time-resolved FWM transients plotted versus t and τ. (Reprinted with permission from Koch et al. 1998b.)

squared magnitude of the third-order polarization. In the presence of a cavity, the conversion of a polarization into a photon is more subtle. The third-order polarization is the source for an intracavity field $\alpha^{(2k_2-k_1)}$ associated with the direction $2k_2 - k_1$. One therefore has to go one step further and calculate $|\alpha^{(2k_2-k_1)}(t, \tau)|^2$, part of which leaves the cavity and is detected as FWM signal. With the assumption of a δ-pulse excitation, an analytical expression for $\alpha^{(2k_2-k_1)}$ can be derived (Yajima and Taira 1979). If one, for simplicity, neglects dissipation and assumes that the cavity mode is in resonance with the absorber and also with the central photon energy, one obtains

$$\alpha^{(2k_2-k_1)}(t, \tau) = \Theta(t)\Theta(t - \tau)E_1^* E_2^2 \tilde{g}^3 e^{-i\omega(t-2r)}$$

$$\times \frac{g}{4}[(t - \tau)\sin(g(t - 2\tau)) - \frac{1}{g}\cos(gt)$$

$$+ \frac{1}{4g}\cos(g(3t - 2\tau)) + \frac{3}{4g}\cos(g(t - 2\tau))]e^{i(2k_2 - k_1)r}. \quad (10.16)$$

The oscillation pattern is determined by the four terms in the rectangular brackets. Each of the four terms describes a different (t, τ) behavior. The first term, however, which describes a $t = 2\tau$ behavior, differs from the others, since it linearly increases with time. After one oscillation period at the latest, it exceeds all the others and imprints its $t = 2\tau$ behavior onto the oscillation pattern. Its increase is caused by the fact that in the presence of a cavity the polarization is not driven by the external laser pulse as in free space but by the intracavity field which persists for much longer times and due to its periodical energy exchange with the excitonic polarization. The effect of α on the nonlinear response is similar to that of Coulomb exchange terms known from semiconductors (Kim et al. 1992) and polymers (Meier et al. 1996). If dissipation is not considered, these terms also cause an infinite signal rise (Lindberg et al. 1992).

The origin of the $t = 2\tau$ behavior observed here differs from that of "$t = 2\tau$ cases" discussed in the literature. Without a cavity the third-order polarization of a two-level system is proportional to $e^{-i\omega(t-2\tau)}$. The phase factor in this exponential results from the phase conjugation (Shen 1984). In the case of an inhomogeneous line broadening, this phase factor leads to the photon echo, in the case of a 2×2-level it leads to the strict $t = 2\tau$ pattern (note that a 2×2-level system is the very first step toward inhomogeneous broadening). Following the calculation that leads to equation (10.16) reveals that the $t = 2\tau$ behavior of the first term results in a similar way from the phase conjugation as the above exponential. In some sense the $t = 2\tau$ behavior observed here reflects the fact that the only nonlinearity in the cavity is that of a two-level system. The oscillation pattern is further complicated by the combined action of the other terms. The fact that at times at which the first term is zero the other terms must not vanish can lead to deviations around the $t = 2\tau$ slope.

To visualize these analytical results, the optical Bloch equations have been solved numerically assuming 100 fs pulses. For the case of no detuning and no dissipation, the minima form a series of straight lines with a $t = 2\tau$ slope. If a slight detuning is introduced to account for the experimental conditions, the (t, τ) pattern gets more complex. The theoretical oscillation pattern obtained for a detuning of -1.5 meV (Fig. 10.37) reproduces the main features observed in the experimental (t, τ) pattern quite well. The minima perform slight undulations around the $t = 2\tau$ slope, and there is a suggestion of a crossover behavior in the vicinity of $\tau = 0.9$ ps. The crossover repeats itself in τ with the oscillation frequency. If dephasing is introduced, the undulations around the $t = 2\tau$ slope tend to get more pronounced; in this case they are observable for even larger detunings. The introduction of inhomogeneous line broadening or a moderate local field does not change the (t, τ) structure significantly.

One advantage of numerical calculations is that it is possible to study regimes that are difficult or impossible to investigate experimentally. With increasing cavity detuning the nonlinear emission at the excitonlike mode becomes extremely small and the FWM spectrum is dominated by the cavitylike mode (Wang et al. 1995). Consequently oscillations in TR-FWM are weak and difficult to measure; a (t, τ) pattern would be extremely hard to obtain. The numerical results, however, show an interesting and unusual (t, τ) behavior if the cavity detuning exceeds the coupling frequency (2.1 meV in our case; Wang et al. 1995). With increasing detuning the oscillation frequency increases, which simply reflects the larger normal mode splitting. Correspondingly the crossover appears more frequently as a function of time delay τ. While the sections between two subsequent crossover regions get shorter and shorter, their slope decreases until it approaches zero, and adjacent sections merge into a horizontal line. The latter situation is shown in Figure 10.38 where the (t, τ) pattern for a cavity detuning of -8 meV is plotted. Since detuning is clearly important in this case, the analytical expression for $\alpha^{(2k_2 - k_1)}$ given above does not give an intuitive explanation. Yet, Figure 10.38 shows that there are still interesting aspects of electron-photon wave packets that remain to be explored.

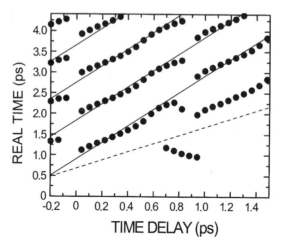

Figure 10.37 Theoretical (t, τ) pattern obtained for a slight cavity-exciton detuning of -1.5 meV. (Reprinted with permission from Koch et al. 1998b.)

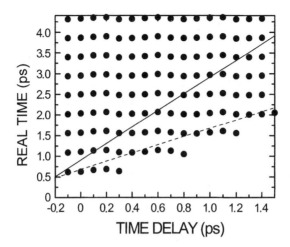

Figure 10.38 Theoretical (t, τ) pattern obtained for a cavity-exciton detuning of -8.0 meV. (Reprinted with permission from Koch et al. 1998b.)

10.3 CONCLUSION AND OUTLOOK

This chapter has reviewed some examples of wave packet experiments in semiconductor heterostructures. These experiments are interesting from both general physics and semiconductor physics viewpoints. The appeal for general physics stems from the fact that semiconductor heterostructures have made it possible to demonstrate some very basic textbook experiments that cannot be realized in other systems. One illustrative example is the coherent oscillations in a double quantum well, and this is the simplest example of wave packet experiments in general. However, the seemingly simple electronic states derived by the envelope approximation hide the intricate details of a dense many-body systems with strong interactions. Thus all the simple demonstration experiments show quite complicated features when inspected in detail. Nevertheless, in the future there will be many new and exciting experiments that make use of semiconductor heterostructures and their nearly unlimited freedom to realize virtually any potential shape. One has to keep in mind that pure demonstration experiments, such as the double quantum well experiment, are of limited interest if the only novelty is the choice of a more complicated structure. On the other hand, wave packet experiments can be used to obtain detailed information about electronic states in semiconductors.

Here we have the example of excitonic wave packets composed of the excitonic ground state and higher exciton states. The dynamics of these wave packets gives insight into the complex many-particle physics underlying excitonic transitions. There are many other opportunities for using wave packet effects to investigate optical excitations in semiconductors.

A future challenge will be to use wave packet dynamics for devices. Terahertz spectroscopy experiments have shown that both coherent oscillations in coupled double quantum well (Roskos et al. 1992) and Bloch oscillations (Waschke et al. 1993) emit tunable THz radiation. A premier example is the realization of an electrically driven Bloch oscillator, using principles that were demonstrated by an optical excitation of wave packets. Creating wave packets in semiconductors using transient electrical techniques is a challenge that has not yet been addressed. Also it is worth checking whether other devices can be realized using wave packet effects. For instance, one might be able to realize modulators with very high operating speeds that recover on a very short time scale by dephasing instead of recombination.

ACKNOWLEDGMENTS

We gratefully acknowledge T. C. Damen, E. O. Göbel, V. G. Lyssenko, F. Löser, J. Shah, M. Sudzius, and G. Valusis for collaborative work, T. Meier and S. Haas for performing accompanying calculations, and D. Dunlap, J. Feldmann, F. Jahnke, S. W. Koch, H. Kurz, P. Leisching, H. Roskos, P. Thomas, and G. von Plessen for fruitful discussions. The results presented here would not have been possible without the growth of excellent samples by J. E. Cunningham, P. Ganser, K. Köhler, and H. Nickel. Support from the Deutsche Forschungsgemeinschaft and the Volkswagen foundation is gratefully acknowledged.

REFERENCES

Ashcroft, N. W., and Mermin, N. D. (1981). *Solid State Physics.* Holt-Saunders, p. 225.

Bastard, G. (1988). *Wave Mechanics Applied to Semiconductor Heterostructure.* Les Editions de physique, Les Ulis, p. 63.

Bastard, G., and Ferreira, R. (1989). In *Spectroscopy of Semiconductor Microstructures*, vol. 206, eds. G. Fasol and A. Fasolino. NATO Advance Study Institute, Series B: Physics, Plenum Press, New York, p. 333.

Ben Dahan, M., Peik, E., Reichel, J., Castin, Y., and Salomon, C. (1996). *Phys. Rev. Lett.* **76**, 4508.

Bloch, F. (1928). *Z. Phys.* **52**, 555.

Braun, W., Bayer, M., Forchel, A., Zull, H., Reithmaier, J. P., Filin, A. I., Walck, S. N., and Reinecke, T. L. (1997). *Phys. Rev.* B**55**, 9290.

Burstein, E., and Weisbuch, C., eds. (1995). *Confined Electrons and Photons: New Physics.* Plenum Press, New York.

Capasso, F., Mohammed, K., and Cho, A. Y. (1986). *IEEE J. Quant. Electr.* **22**, 1853.

Cundiff, S. T., and Steel, D. G. (1992). *IEEE J. Quant. Electr.* **28**, 2423.

Cundiff, S. T., Koch, M., Knox, W. H., Shah, J., and Stolz, W. (1996). *Phys. Rev. Lett.* **77**, 1107.

Dekorsy, T., Leisching, P., Waschke, C., Köhler, K., Leo, K., Roskos, H. G., and Kurz, H. (1994). *Semicond. Sci. Technol.* **9**, 1959.

Dekorsy, T., Kim, A. M. T., Cho, G. C., Hunsche, S., Bakker, H. J., Kurz, H., Chuang, S. L., and Köhler, K. (1996). *Phys. Rev. Lett.* **77**, 3045.

Dignam, M. M., Sipe, J. E., and Shah, J. (1994). *Phys. Rev.* B**49**, 10502.

Elliott, R. J. (1957). *Phys. Rev.* **108**, 1384.

Esaki, L., and Tsu, R. (1970). *IBM J. Res. Dev.* **61**, 61.

Faist, J., Capasso, F., Sivco, D. L., Sirtori, C., Hutchison, A. L., and Cho, A. Y. (1994). *Science* **264**, 553.

Feldmann, J., Leo, K., Shah, J., Miller, D. A. B., Cunningham, J. E., Schmitt-Rink, S., Meier, T., von Plessen, G., Schulze, A., and Thomas, P. (1992). *Phys. Rev.* B**46**, 7252.

Feldmann, J., Meier, T., von Plessen, G., Koch, M., Göbel, E. O., Thomas, P., Bacher, G., Hartmann, C., Schweizer, H., Schäfer, W., and Nickel, N. (1993). *Phys. Rev. Lett.* **70**, 3027.

Feuerbacher, B. F., Kuhl, J., Eccleston, R., and Ploog, K. (1990). *Solid State Comm.* **74**, 1279.

Fox, A. M., Miller, D. A. B., Livescu, G., Cunningham, J. E., Henry, J. E., and Yan, W. Y. (1990). *Phys. Rev.* B**42**, 1841.

Göbel, E. O., Leo, K., Damen, T. C., Shah, J., Schmitt-Rink, S., Schäfer, W., Müller, J. F., and Köhler, K. (1990). *Phys. Rev. Lett.* **64**, 1801.

Göbel, E. O., and Ploog, K. (1990). *Prog. Quant. Electr.* **14**, 289.

Haas, S. (n.d.). Unpublished calculation for bulk GaAs.

Haroche, S. (1992). In *Fundamental Systems in Quantum Optics*, eds. J. Dalibard et al. Elsevier, Amsterdam.

Haug, H., and Koch, S. W. (1993). In *Quantum Theory of the Optical and Electronic Properties of Semiconductors*. World Scientific, Singapore.

Jahnke, F., Koch, M., Meier, T., Feldmann, J., Schäfer, W., Thomas, P., Koch, S. W., Göbel, E. O., and Nickel, H. (1994). *Phys. Rev.* B**50**, 8114.

Jewell, J. L., Huang, K. F., Tai, K., Lee, Y. H., Fischer, R. J., McCall, S. L., and Cho, A. Y. (1989). *Appl. Phys. Lett.* **55**, 424.

James, H. M. (1949). *Phys. Rev.* **76**, 1611.

Joschko, M., Woerner, M., Elsaesser, T., Binder, E., Kuhn, T., Hey, R., Kostial, H., and Ploog, K. (1997). *Phys. Rev. Lett.* **78**, 737.

Kim, D. S., Shah, J., Damen, T. C., Schäfer, W., Jahnke, F., Schmitt-Rink, S., and Köhler, K. (1992). *Phys. Rev. Lett.* **69**, 2725.

Koch, M., Feldmann, J., von Plessen, G., Göbel, E. O., Thomas, P., and Köhler, K. (1992). *Phys. Rev. Lett.* **69**, 3631.

Koch, M., Feldmann, J., Göbel, E. O., Thomas, P., Shah, J., and Köhler, K. (1993a). *Phys. Rev.* B**48**, 11480.

Koch, M., Hellmann, R., Cundiff, S. T., Feldmann, J., Göbel, E. O., Yakovlev, D. R., Waag, A., and Landwehr, G. (1993b). *Solid State Comm.* **88**, 515.

Koch, M., von Plessen, G., Feldmann, J., Göbel, E. O. (1996). *Chem. Phys.* **210**, 367.

Koch, M. (1998a). In *Festkörperprobleme/Advances in Solid State Physics*, vol. 37, ed. R. Helbig. Verlag Vieweg, Braunschweig/Wiesbaden, p. 169.

Koch, M., Shah, J., and Meier, T. (1998b). *Phys. Rev.* B**57**, R2049.

Kuhl, J., Honold, A., Schultheis, L., and Tu, C. W. (1989). In *Festkörperprobleme/Advances in Solid State Physics*, vol. 29, ed. U. Rössler. Vieweg-Verlag Braunschweig, p. 157.

Langer, W., Stolz, H., and von der Osten, W. (1990). *Phys. Rev. Lett.* **64**, 854.

Leggett, A. J., Chakravarty, S., Dorsey, A. T., Fisher, P. A., Garg, A., Zwerger, W. (1987). *Rev. Mod. Phys.* **59**, 1.

Leisching, P., Haring Bolivar, P., Beck, W., Dhaibi, Y., Brüggemann, F., Schwedler, R., Kurz, H., Leo, K., and Köhler, K. (1994). *Phys. Rev.* B**50**, 14389.

Leo, K., Damen, T. C., Shah, J., Göbel, E. O., and Köhler, K. (1990a). *Appl. Phys. Lett.* **57**, 19.

Leo, K., Shah, J., Göbel, E. O., Damen, T. C., Köhler, K., and Ganser, P. (1990b). *Appl. Phys. Lett.* **56**, 2031.

Leo, K., Damen, T. C., Shah, J., and Köhler, K. (1991a). *Phys. Rev.* B**42**, 11359.

Leo, K., Shah, J., Göbel, E. O., Damen, T. C., Schmitt-Rink, S., Schäfer, W., and Köhler, K. (1991b). *Phys. Rev. Lett.* **66**, 201.

Leo, K., Haring Bolivar, P., Brüggemann, F., Schwedler, R., and Köhler, K. (1992a). *Solid State Comm.* **84**, 943.

Leo, K., Shah, J., Damen, T. C., Schulze, A., Meier, T., Schmitt-Rink, S., Thomas, P., Göbel, E. O., Chuang, S. L., Luo, M. S. C., Schäfer, W., Köhler, K., and Ganser, P. (1992b). *IEEE J. Quant. Electr.* **28**, 2498.

Leo, K. (1998). *Semicond. Sci. Technol. Topical Rev.* **13**, 249.

Lindberg, M., Binder, R., and Koch, S. W. (1992). *Phys. Rev.* A**45**, 1865.

Liu, H. W., Ferreira, R., Bastard, G., Delalande, C., Palmier, J. F., and Etienne, B. (1989). *Appl. Phys. Lett.* **54**, 2082.

Lyssenko, V. G., Erland, J., Balslev, I., Pantke, K.-H., Razbirin, B. S., and Hvam, J. M. (1993). *Phys. Rev.* B**48**, 5720.

Lyssenko, V. G., Valusis, G., Löser, F., Hasche, T., Leo, K., Dignam, M. M., and Köhler, K. (1997). *Phys. Rev. Lett.* **79**, 301.

Lyssenko, V. G., Sudzius, M., Löser, F., Valusis, G., Hasche, T., Leo, K., Dignam, M. M., and Köhler, K. (1998). *Festkörperprobleme/Advances in Solid State Physics*, vol. 38, in press.

Meier, T., and Mukamel, S. (1996). *Phys. Rev. Lett.* **77**, 3471.

Mendez, E. E., Agullo-Rueda, F., and Hong, J. M. (1988). *Phys. Rev. Lett.* **60**, 2426.

Oestreich, M., and Rühle, W. W. (1996). *IEEE J. Selected Topics Quant. Electr.* **2**, 747.

Parker, E. H. C., ed. (1985). *The Physics and Technology of MBE.* Plenum, New York.

von Plessen, G., Meier, T., Koch, M., Feldmann, J., Thomas, P., Koch, S. W., Göbel, E. O., Goossen, K. W., Kuo, J. M., and Kopf, R. F. (1996). *Phys. Rev.* B**53**, 13688.

Roskos, H. G., Nuss, M. C., Shah, J., Leo, K., Miller, D. A. B., Fox, A. M., Schmitt-Rink, S., and Köhler, K. (1992). *Phys. Rev. Lett.* **68**, 2216.

Scamarcio, G., Capasso, F., Sirtori, C., Faist, J., Hutchison, A. L., Sivco, D. L., and Cho, A. Y. (1997). *Science* **276**, 773.

Schäfer, W., Jahnke, F., and Schmitt-Rink, S. (1993). *Phys. Rev.* B**47**, 1217.

Shah, J. (1996). *Ultrafast Spectroscopy of Semiconductors and Semiconductor Nanostructures.* Springer, New York.

Shen, Y. R. (1984). *The Principles of Nonlinear Optics.* Wiley, New York.

Sudzius, M., Lyssenko, V. G., Löser, F., Leo, K., Dignam, M. M., and Köhler, K. *Phys. Rev.* B**52** (thematic issue).

Ulbrich, R. G. (1988). In *Optical Nonlinearities and Instabilities in Semiconductors*, ed. H. Haug. Academic Press, San Diego.

Valusis, G., Lyssenko, V. G., Klatt, D., Pantke, K.-H., Löser, F., Leo, K., and Köhler, K. (1996). *Proc. 23rd Int. Conf. Phys. Semic.*, Berlin, eds. M. Scheffler and R. Zimmermann. World Scientific, Singapore, p. 1783.

Voisin, P., Bleuse, J., Bouche, C., Gaillard, S., Alibert, C., and Regreny, A. (1988). *Phys. Rev. Lett.* **61**, 1639.

Wang, H., Shah, J., Damen, T. C., Jan, W. Y., Cunningham, J. E., Hong, M., and Mannaerts, J. P. (1995). *Phys. Rev.* B**51**, 14713.

Wannier, G. H. (1969). *Phys. Rev.* **117**, 432.

Waschke, C., Roskos, H. G., Schwedler, R., Leo, K., Kurz, H., and Köhler, K. (1993). *Phys. Rev. Lett.* **70**, 3319.

Wegener, M., Chemla, D. S., Schmitt-Rink, S., and Schäfer, W. (1990). *Phys. Rev.* A**42**, 5675.

Weisbuch, C., Nishioka, M., Ishikawa, A., and Arakawa, Y. (1992). *Phys. Rev. Lett.* **69**, 3314.

Yajima, T., and Taira, Y. (1979). *J. Phys. Soc. Jpn.* **47**, 1620.

Yokoyama, H. (1992). *Science* **256**, 66.

Zener, C. (1934). *Proc. R. Soc. London* A**145**, 523.

INDEX